PART ONE
THEORETICAL IMMUNOLOGY

Santa Fe Institute Series List

Lecture Notes Volumes in the Santa Fe Institute Studies in the Sciences of Complexity

Volume I:
John Hertz, Anders Krogh, Richard G. Palmer, editors: Introduction to the Theory of Neural Networks

Volume II:
Gerard Weisbuch: Complex Systems Dynamics

Volume III:
Wilfred Stein and Francisco J. Varela, editors: Thinking About Biology

Volume IV:
Joshua M. Epstein: Nonlinear Dynamics, Mathematical Biology, and Social Science

Volume V:
H.F. Nijhout, Lynn Nadel, and Daniel L. Stein, editors: Pattern Formation in the Physical and Biological Sciences

Proceedings Volumes in the Santa Fe Institute Studies in the Sciences of Complexity

Volume I:
David Pines, editor: Emerging Synthesis in Science

Volume II:
Alan S. Perelson, editor: Theoretical Immunology, Part One

Volume III:
Alan S. Perelson, editor: Theoretical Immunology, Part Two

Volume IV:
Gary D. Doolen, senior editor: Lattice Gas Methods for Partial Differential Equations

Volume V:
Philip W. Anderson, Kenneth J. Arrow, and David Pines, editors: The Economy as an Evolving Complex System

Volume VI:
Christopher G. Langton, editor: Artificial Life

Volume VII:
George I. Bell and Thomas G. Marr, editors: Computers and DNA

Volume VIII:
Wojciech H. Zurek, editor: Complexity, Entropy, and the Physics of Information

Volume IX:
Alan S. Perelson and Stuart A. Kauffman, editors: Molecular Evolution on Rugged Landscapes: Proteins, RNA, and the Immune System

Volume X:
Christopher G. Langton, Charles Taylor, J. Doyne Farmer and Steen Rasmussen, editors: Artificial Life II

Volume XI:
John A. Hawkins and Murray Gell-Mann, editors: The Evolution of Human Languages

Volume XII:
Martin Casdagli and Stephen Eubank, editors: Nonlinear Modeling and Forecasting

Volume XIII:
Jay E. Mittenthal and Arthur B. Baskin, editors: The Principles of Organizations in Organisms

Volume XIV:
Danicl Friedman and John Rust, editors: The Double Auction Market: Institutions, Theories, and Evidence

Volume XV:
Andreas S. Weigend and Neil A. Gershenfeld, editors: Time Series Prediction: Forecasting the Future and Understanding the Past

Volume XVI:
George J. Gummerman and Murray Gell-Mann, editors: Understanding Complexity in the Prehistoric Southwest

Volume XVII:
Christopher G. Langton, editor: Artificial Life III

Volume XVIII:
Gregory Kramer, editor: Auditory Display: Sonification, Audification, and Auditory Interfaces

Volume XIX: George A. Cowan, David Pines, and David Meltzer, editors: Complexity: Metaphors, Models, and Reality

Volume XX:
David Wolpert, editor: The Mathematics of Generalization

Volume XXI:
P.E. Cladis and P. Palffy-Muhoray, editors: Spatio-Temporal Patterns in Nonequilibrium Complex Systems

Volume XXII:
Harold J. Morowitz and Jerome L. Singer, editors: The Mind, the Brain, and Complex Adaptive Systems

Volume XXIII: Bela Julesz and Ilona Kovacs, editors: Maturational Windows and Adult Cortical Plasticity

Volume XXIV:
Joseph A. Tainter and Bonnie Bagley Tainter, editors: Evolving Complexity and Environmental Risk in the Prehistoric Southwest

Volume XXV:
John B. Rundle, Donald L. Turcotte, and William Klein, editors: Reduction and Predictability of Natural Disasters

Volume XXVI:
Richard K. Belew and Melanie Mitchell, editors: Adaptive Individuals in Evolving Populations

Volume XXVII:
W. Brian Arthur, Steven N. Durlauf, and David A Lane, editors: The Economy as an Evolving Complex System II

Volume XXVIII:
Gcrald Myers, editor: Viral Regulatory Structures and Their Degeneracy

PART ONE
THEORETICAL IMMUNOLOGY

THE PROCEEDINGS OF THE THEORETICAL IMMUNOLOGY WORKSHOP, HELD JUNE, 1987 IN SANTA FE, NEW MEXICO

Alan S. Perelson, *Editor*
Theoretical Division
Los Alamos National Laboratory

Volume II

SANTA FE INSTITUTE
STUDIES IN THE SCIENCES OF COMPLEXITY

Advanced Book Program

CRC Press
Taylor & Francis Group
Boca Raton London New York

CRC Press is an imprint of the
Taylor & Francis Group, an **informa** business

Director of Publications, Santa Fe Institute: *Ronda K. Butler-Villa*

First published 1988 by Westview Press

Published 2019 by CRC Press
Taylor & Francis Group
6000 Broken Sound Parkway NW, Suite 300
Boca Raton, FL 33487-2742

First issued in hardback 2019

CRC Press is an imprint of the Taylor & Francis Group, an informa business

Visit the Taylor & Francis Web site at
http://www.taylorandfrancis.com

and the CRC Press Web site at
http://www.crcpress.com

Library of Congress Cataloging-in-Publication Data

Theoretical Immunology Workshop (1987 : Santa Fe, N.M.)
Theoretical immunology.

(Santa Fe Institute studies in the sciences of complexity ; v. 2-3)
Includes index.
1. Immunology--Mathematical models--Congresses.
2. Immunology--Philosophy--Congresses. 3. Immunologic diseases--
Mathematical models--Congresses. I. Perelson, Alan S., 1947-
II. Title. III. Series. [DNLM: 1. Allergy and Immunology--congresses.
2. Models, Theoretical--congresses.
QW 504 T3955t 1987]
QR182.2.M36T485 1987 616.07'9 88-6169
ISBN 0-201-15682-2 (pt. 1)
ISBN 0-201-15683-0 (pbk. : pt. 1)
ISBN 0-201-15687-3 (pt. 2)
ISBN 0-201-15688-1 (pbk. : pt. 2)

ISBN 13: 978-0-367-09111-8 (hbk)
ISBN 13: 978-0-201-15683-6 (pbk)

This volume was typeset using TEXtures on a Macintosh computer.

About the Santa Fe Institute

The *Santa Fe Institute* (SFI) is a multidisciplinary graduate research and teaching institution formed to nuture research on complex systems and their simpler elements. A private, independent institution, SFI was founded in 1984. Its primary concern is to focus the tools of traditional scientific disciplines and emerging new computer resources on the problems and opportunities that are involved in the multidisciplinary study of complex systems -- those fundamental processes that shape almost every aspect of human life. Understanding complex systems is critical to realizing the full potential of science, and may be expected to yield enormous intellectual and practical benefits.

All titles from the *Santa Fe Institute Studies in the Sciences of Complexity* series will carry this imprint which is based on a Mimbres pottery design (circa A.D. 950-1150), drawn by Betsy Jones.

ALAN S. PERELSON
Los Alamos National Laboratory, Los Alamos, NM, December 1987

Foreword

Immunology is one of the most exciting branches of biology. Advances are being made almost daily in a broad range of experimental areas. The current notariety of AIDS has made us all keenly aware, even in this age of modern medicine and miracle drugs, of the absolute necessity of a functioning immune system. Our inability to cure HIV infection or even to design therapies that eliminate its symptoms points out our lack of detailed understanding about the operation of the immune system.

The immune system is complex system of cells and molecules distributed throughout our bodies. Analogies have been drawn between the immune system and the nervous sytstem. Like the nervous system, the immune system performs pattern recognition tasks, learns and retains a memory of the antigens that it has fought. Many of the "players" in the immune system, the specific cells and molecules whose coordinate activity produce the phenomena of immunology, have been identified. The interactions between these cells and molecules are slowly being elucidated. The mechanisms that actually regulate the immune system are by and large still unknown.

Unlike many areas of biology, theoretical ideas have played a major role in the development of the field. Controversies such as instructive vs. selective theories of antibody formation, germ-line vs. somatic mutation models for the generation of antibody diversity, and regulatory circuits vs. idiotypic networks have dominated both the intellectual development of the field and determined the direction of much experimental effort. Quantitative theories, while playing a role in the design and interpretation of various assay systems (e.g., complement fixation, the precipitin reaction, the hemolytic plaque assay) have to date not been significant in the intellectual development of the field. This may be changing as the field addresses more quantitative issues such as the role of somatic hypermutation in the generation of antibody diversity, the role of receptor clusters in cell stimulation and desensitization, the effects of competition between solution phase antibodies and cell surface receptors of the same specificity for binding various ligands including anti-idiotypic antibodies, the effects of different concentrations of growth factors and of changes in receptor affinity and receptor number on cell stimulation, cell proliferation and cell differentiation, etc. Quantitative theories will also be essential for the development of global models of the immune system in which both the spatial and temporal aspects of humoral and cell-mediated responses are integrated into a single model, and may well be needed in order to understand the role of idiotypic networks in immune regulation.

In order to foster further development of the field of theoretical immunology the Santa Fe Institute and the Theoretical Biology and Biophysics Group, Los Alamos National Laboratory, sponsored a three-day workshop in Santa Fe, June 10-12, 1987. The workshop was made possible by a generous grant from the Office of Health and Environmental Research, U. S. Department of Energy. These two volumes represent the proceedings of that workshop.

Both the workshop itself and the production of these volumes could not have been done without the help of many people. Geoffrey Hoffmann, Ronald Mohler and George Bell were involved in the initial conception of the meeting, helped choose participants, and provided assistance in writing a grant proposal. The staff of the Santa Institute, and in particular Ginger Richardson, and Ronald A. Zee assumed the heavy burden of making local arrangements for the participants and handling the financial affairs of the meeting. Ronda K. Butler-Villa with great efficiency and proficiency changed our rather dull-looking manuscripts into the beautiful camera-ready form that appears here. Finally, I wish to thank both George A. Cowan, President, L. M. Simmons, Vice-President for Academic Affairs, and the Science Board for leading the Santa Fe Institute in many intellectually exciting areas including Theoretical Immunology.

One of the participants at the meeting, Petr Klein, whose work appears in this volume met a tragic death, along with his wife, in a head-on automobile collision some months after the meeting. His colleagues and I helped in the proofreading and final production of his contribution. Charles DeLisi, Chairman, Biomathematics Sciences, Mt. Sinai School of Medicine, New York, whom Petr worked with for two years, has provided the following remarks:

> One of the interesting facts which emerges on scanning the references in George Bell's pioneering papers on immune response dynamics, is early activity in mathematical immunology by Czechoslovokian scientists. A small group including Sterzl, Jílek and Hraba were breaking new ground theoretically and experimentally, but their research was relatively unknown in

the West. It was in this isolated but unusually innovative atmosphere that Petr Klein obtained his doctoral degrees in immunology and mathematics.

I first learned of Petr during one of Tomáš Hraba's visits to the National Institutes of Health (N.I.H.). Petr was soon thereafter awarded a Fogarty International Fellowship. When the Czechoslovokian government refused to allow him to accept it, he and his wife moved to Rome, and from there to Canada. His wife obtained a position with Jack Dainty at the University of Toronto and Petr joined my laboratory at the N.I.H.

Petr broke new scientific ground during his two years in Bethesda. He developed a new approach to predicting protein function from sequence properties, and also published the most reliable methods available for classifying membrane proteins and predicting secondary structure. by early 1985, his accomplishments were sufficiently impressive to enable him to secure a position in Canada near his wife.

I was hoping Petr would rejoin us in the United States during the next year or two, and I know many of our colleagues here had similar hopes. His gentle disposition and his scientific accomplishments made him a valued friend colleague. He will be greatly missed.

Contents

PART TWO

PART ONE
THEORETICAL IMMUNOLOGY

Cell Stimulation and Receptor Crosslinking

BYRON GOLDSTEIN
Theoretical Division, University of California, Los Alamos National Laboratory, Los Alamos, New Mexico 87545 USA

Desensitization, Histamine Release and the Aggregation of IgE on Human Basophils

INTRODUCTION

Since antibodies and antigens have valences of two or greater, they often form aggregates when they come together. Antibody-antigen aggregates play an important role in immune responses. When antibodies act as cell surface receptors, their aggregation is a requirement for signal generation. Some of the best studied examples of cellular responses being triggered by the aggregation of cell surface antibody arise in allergic reactions of the immediate hypersensitive type. Individuals who are allergic to a particular antigen produce immunoglobulin E (IgE) that is specific for that antigen. Some of the IgE binds to monovalent, high affinity Fc_ϵ receptors on the surface of mast cells and basophils. When surface-bound IgE is aggregated by the antigen, it triggers the cells to both release prestored granules that contain histamine, serotonin and other mediators of anaphylaxis, and to synthesize products that are rapidly secreted. Recently it has been shown that aggregation of IgE initiates many physical changes on rat basophilic leukemia (RBL) cells. It triggers internalization of IgE aggregates (Isersky et al., 1983), increased fluid-phase pinocytosis, transformation of the cell surface from finely microvillous to highly folded, cell spreading, an increase in the F-actin content of detergent extracted cell matrices (Pfeiffer et al., 1985), an increase in the immobile fraction of cell surface IgE

(Menon et al., 1986a; 1986b), and direct interactions between IgE-Fc_ϵ receptors and the cytoskeleton (Robertson et al., 1986). In this chapter, we shall review the relationship between IgE aggregation and some of the biological responses of basophils and mast cells. We shall focus on human basophils. In the chapters by Baird et al. and Oliver et al., and the recent reviews by Metzger et al. (1986), Pecht and Corcia (1987) and Oliver et al. (1987), the responses of RBL cells to IgE aggregation are discussed in detail.

THE BINDING OF IgE TO BASOPHIL SURFACES: SENSITIZATION

The Fc_ϵ receptor on human basophils binds IgE with high affinity. The equilibrium constant for the reaction, K_{Fc}, is of the order of $10^{10}M^{-1}$ (Malveaux et al., 1978). Similar values for K_{Fc} have been found for RBL cells (Conrad et al., 1975; Mendoza and Metzger, 1976; Sterk and Ishizaka, 1982) and mouse mast cells (Mendoza and Metzger, 1976; Sterk and Ishizaka, 1982). The large value for K_{Fc} reflects the slow dissociation of IgE from its Fc_ϵ receptor. For intact RBL cells, Kulczycki and Metzger (1974) found the IgE-Fc_ϵ receptor dissociation constant was $\leq 1.6 \times 10^{-5}s^{-1}$ at 37°C which corresponds to a half-life $\geq$ 19 hrs. More recently, Wank et al. (1983) determined the dissociation constant to be $4.3 \times 10^{-6}s^{-1}$ at 37°C which corresponds to a half-life of 45 hrs. By determining the forward rate constant and knowing the equilibrium constant, a similar value was estimated for the half-life of dissociation of IgE from Fc_ϵ receptors on human basophils (Goldstein et al., 1979). Isersky et al. (1979) followed the dissociation of IgE from Fc_ϵ receptors on RBL cells for over 100 hr and found that the dissociation could not be described by a single exponential. When they resolved their dissociation curves into two exponentials, they found that the fast component had a half-life of 7–12 hrs. and the slow component a half-life of 131–252 hrs. Whether the different rates of dissociation reflect Fc_ϵ receptor heterogeneity, a time-dependent modification of the IgE-Fc_ϵ receptor complex, or other events, is unknown.

The number of Fc_ϵ receptors per basophil varies widely in the human population, from $\sim 10^3$ to 10^6, with basophils from allergic individuals tending to have more Fc_ϵ receptors on their surfaces than basophils from nonallergic individuals (Conroy et al., 1977; Malveaux et al., 1978; Garcia et al., 1978). Allergic individuals also tend to have elevated levels of serum IgE (Conroy et al., 1977). Because of this, basophils from most allergic donors have their Fc_ϵ receptors saturated with IgE. The correlation between Fc_ϵ receptor number and serum IgE levels, raises the possibility that the solution IgE concentration can regulate the concentration of Fc_ϵ receptors on basophil surfaces. This has been directly demonstrated with RBL cells, where the incubation of these cells in high concentrations of IgE leads to an increase (up regulation) in the number of cell surface Fc_ϵ receptors (Furuichi et al., 1985; Quarto et al., 1985).

Basophils that have IgE on their surface that is specific for a known antigen are said to be sensitized to the antigen. A standard way to obtain sensitized basophils for *in vitro* studies is to incubate basophils that initially have unoccupied Fc_ϵ receptors in a solution containing specific IgE. This procedure is know as passive sensitization (Levy and Osler, 1966). A small number of allergic individuals and approximately 20% of nonallergic individuals have basophils with sufficient numbers of free Fc_ϵ receptors that they can be passively sensitized without having to first elute off the bound IgE. Dissociation of IgE from its Fc_ϵ receptors can be achieved through acid treatment that maintains the functional integrity of the basophils (Pruzansky et al., 1983).

IgE AGGREGATION AND MEDIATOR RELEASE

Monovalent ligands that bind to, but cannot bridge, IgE molecules do not trigger basophil degranulation (Becker et al., 1973; Siraganian et al., 1975). Ligands with valence of two or greater that can crosslink IgE, whether by binding to Fab or Fc portions of the IgE, can induce basophils to degranulate (Osler et al., 1968; Ishizaka et al., 1969; Siraganian et al., 1975; Dembo et al., 1978). Although aggregation is required, the formation of large aggregates of IgE is not necessary for triggering basophil degranulation. Segal et al. (1977) showed that preformed IgE-dimers (two IgE molecules covalently linked) triggered the degranulation of mast cells that had free Fc_ϵ receptors on their surfaces. Trimers and higher oligomers of IgEs also triggered degranulation. Similarly, when human basophils where passively sensitized with either IgE-dimers or -trimers they released histamine (Kagey-Sobotka et al., 1981).

Even though the smallest aggregate, the IgE-dimer, is capable of triggering histamine release, there is now considerable evidence that larger aggregates of IgE generate intrinsically different signals than the dimer. This was first proposed by Fewtrell and Metzger (1980), based on their observation that IgE-trimers and higher oligomers of IgE were effective stimulators of degranulation, while IgE-dimers were ineffective in stimulating RBL cells, even in the presence of D_2O, a potent enhancer of histamine release. Another striking difference in the response of RBL cells to dimers and higher oligomers is that clusters of two IgE molecules are predominantly mobile on the RBL cell surface, while clusters of more than two IgE molecules rapidly become immobile after formation (Menon et al., 1986a). These experiments, as well as experiments using a monoclonal anti-IgE to crosslink surface IgE (Menon et al., 1986b), indicate that on RBL cells, larger aggregates of IgE interact with the cytoskeleton while aggregates of two IgEs do not (see the chapter by Baird et al.).

Intrinsic differences in the signals generated by different size oligomers of IgE were first reported for human basophils by MacGlashan et al. (1983a) who observed that the pharmacologic agent indomethacin enhanced histamine release that was triggered by IgE-trimers, but not release that was triggered by IgE-dimers.

Recently, in a study of basophils from 14 nonallergic donors, MacGlashan et al. (1986) found marked differences in the responses of these cells to IgE-dimers. All the cells in the study had $> 10^4$ free Fc_ϵ receptors per basophil. When these cells were exposed to IgE-dimers and -trimers, they found that the basophils from 4 donors were approximately equally responsive to the dimers and trimers. In contrast, the basophils from the remaining 10 donors responded poorly to IgE-dimers, but were highly responsive to IgE-trimers. Based on their ability to respond to IgE-dimers, the basophils from the latter group were classified as low responders and the basophils from the former group as high responders.

MacGlashan et al. (1986) also studied the ability of basophils from both groups to release leukotrienes in response to dimers and trimers of IgE. When human basophils are exposed to agents that crosslink IgE, they synthesize and secrete leukotrienes (Grant and Lichtenstein, 1974). The kinetics of leukotriene and histamine release are similar, but leukotriene release consistently shows a small lag compared with histamine release. The lag reflects the different secretory paths these mediators of anaphylaxis take, since histamine is stored in preformed granules while leukotrienes must be synthesized before release. When high-responder basophils were exposed to either IgE-dimers or IgE-trimers, they released significant quantities of leukotrienes, with the trimers being more effective than the dimers in stimulating release. When low-responder basophils were stimulated with dimers of IgE, they were unable to release leukotriene, but on exposure to IgE-trimers, they released quantities of leukotriene similar to those found for antigen-induced release. When IgE-trimers are bound to low-responder basophils, they can generate a signal that dimers cannot.

Although oligomers of IgE are very useful, they have their limitations. They cannot be used to study signals that are generated only by large aggregates. Even when the oligomer is homogeneous, it will produce a distribution of crosslinked Fc_ϵ receptors on the basophil surface, e.g., exposing basophils to preformed trimers of IgE will produce trimers bound to one, two and three Fc_ϵ receptors. Since for a large range of IgE concentrations the binding of IgE to its Fc_ϵ receptor is slow, in many experiments it is difficult to separate the kinetics of binding from the kinetics of the biological response that the oligomer induced. Finally, it must be kept in mind that such oligomers produce infinitely long-lived crosslinks, and, therefore, their effects on basophils may be atypical. Another simple crosslinking reagent that avoids some of these problems, but of course has others, is the bivalent hapten.

BIVALENT HAPTENS AS TRIGGERS OF HISTAMINE RELEASE

We can divide ligands that bind to IgE into three broad classes: monovalent, bivalent and those with valence greater than two. Monovalent ligands, although they bind to IgE, trigger no basophil responses because they cannot aggregate IgE. Bivalent and higher valence haptens trigger a variety of basophil responses. Bivalent haptens can aggregate IgE, but only into linear chains or rings (see Figure 1). This makes them relatively simple objects to study theoretically. Haptens with valence greater than two, in addition to being able to form linear chains and rings, can form highly branched aggregates. Indeed, when higher valence ligands are exposed to bivalent cell surface receptors, such as IgE, the possibility arises that "superaggregates," i.e., very large ligand-receptor aggregates that span major portions of the cell surface, can form (Goldstein and Perelson, 1984). When the IgE concentration is high enough, theory predicts that if the ligand concentration is increased, a ligand concentration will be reached at which superaggregates form. This formation of superaggregates corresponds to a sol-gel phase transition. Whether haptens with valence three or greater are capable of generating biological signals that bivalent haptens cannot is unknown.

Bivalent haptens consisting of two haptenic groups separated by a spacer are useful tools for probing the response of basophils to IgE aggregation. They have been shown to degranulate rabbit (Siraganian et al., 1975) and human (Dembo et al., 1978) basophils sensitized with polyclonal IgE, as well as human basophils (MacGlashan et al., 1985), rat peritoneal mast cells (Kane et al., 1986) and RBL cells (Kane et al., 1986) sensitized with monoclonal IgE. To determine what the requirements are for forming a functional crosslink (an IgE aggregate capable of triggering histamine release or some other biological response) haptens with different length spacers and different flexibilities have been synthesized and their ability to induce degranulation has been tested (Siraganian et al., 1975; Kane et al., 1986). One question, for example, which may be answered with these haptens is whether there is an optimal separation distance at which two IgE molecules should be held to generate a maximal signal. Kane et al. (1985) showed that several bivalent haptens composed of short spacers (< 30 Å) connecting two dinitrophenyl (DNP) groups did not trigger RBL cell degranulation, even though in solution the haptens bound

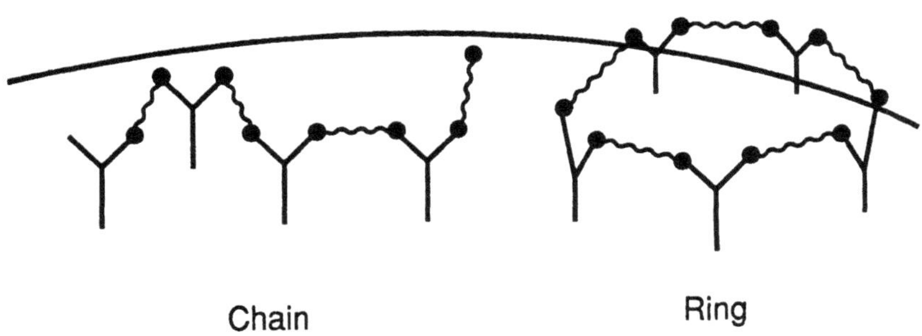

FIGURE 1 The only types of aggregates a bivalent hapten can form with IgE are linear chains and rings.

and crosslinked the sensitizing monoclonal anti-DNP IgE. One of these nonreleasing haptens was *N,N'*-di-2,4-DNP-L-lysine (α,ϵ-DNP-lysine). We used this hapten to trigger human basophils passively sensitized with the same monoclonal anti-DNP IgE that Kane et al. used to passively sensitize their RBL cells. Unlike the RBL cells, the human basophils did degranulate (MacGlashan et al., 1985). This suggests that the requirements for forming functional IgE aggregates are different on human basophils and RBL cells. Since RBL cells respond poorly to preformed IgE-dimers while high-responder human basophils release normally to them, a possible explanation for the difference seen with α,ϵ-DNP-lysine is that dimers are the dominant IgE aggregate that forms, but the dimers that form on RBL cells are not functional. Indeed, in solution binding studies this bivalent hapten appears to mainly aggregate anti-DNP IgE into closed dimers, i.e., rings consisting of two IgE and two haptens (Schweitzer-Stenner et al., 1987). Whether such a short hapten (its maximal length is $\sim$ 14 Å) can also form rings with surface bound IgE is unknown.

Not all dimers are nonfunctional on RBL cells. Basciano et al. (1986) showed that a monoclonal anti-Fc_ϵ receptor that only aggregated Fc_ϵ receptors into dimers, triggered RBL cells to degranulate. It is possible that to form functional IgE aggregates on RBL cells, the separation distance between IgE molecules must be in some appropriate range. If this is so, then although α,ϵ-DNP-lysine crosslinks IgE on RBL cells, they generate no signal because the separation distances between aggregated IgEs are too small (for a further discussion, see the chapter by Baird et al.).

THE BINDING CONSTANTS THAT CHARACTERIZE BIVALENT-HAPTEN IgE AGGREGATION

One attraction of working with a bivalent hapten and cells sensitized with a monoclonal IgE, is the possibility that one can determine through a series of physical measurements a relatively small set of binding constants that fully characterize the aggregation of IgE. For the binding in solution of bivalent DNP haptens to monoclonal anti-DNP IgE, considerable progress has been made using this approach (Erickson et al., 1986; Schweitzer-Stenner et al., 1987).

To illustrate the method, I will review the simplest case, the binding of a symmetric bivalent hapten to monoclonal IgE anchored to cell surface Fc_ϵ receptors. A model for the binding, which was first presented by Dembo and Goldstein (1978), makes the following assumption: there is no cooperativity between binding sites on an IgE molecule, i.e., the binding of a ligand to one site on an IgE does not change the binding properties of the second site on the IgE, as might occur if the hapten was large and partially blocked the second site. Because of this assumption, there is only one forward rate constant, k_+, and one reverse rate constant, k_-, for the reaction of a single binding site on a hapten in solution with a single IgE binding site, as illustrated in Figure 2. The single-site equilibrium binding constant $K = k_+/k_-$.

In Figure 3, we illustrate one of a set of addition reactions that build up linear chains. In general, the rate constants for forming a chain containing n IgEs can be a function of n. Indeed, the dissociation constants can be a function of the position within the aggregate (Hill, 1983). It is unknown whether these effects are important, but it is usual to ignore them and assume that all the crosslinking rate constants are identical (Dembo and Goldstein, 1978; Perelson and DeLisi, 1980). This "equivalent site" approximation reduces the infinite set of rate constants for crosslinking to two, k_{x+} and k_{x-}, and the infinite set of equilibrium constants to one, $K_x = k_{x+}/k_{x-}$.

In Figure 4, we illustrate the reactions that lead to ring formation. Again an approximation must be made to reduce the infinite set of rate constants to a small

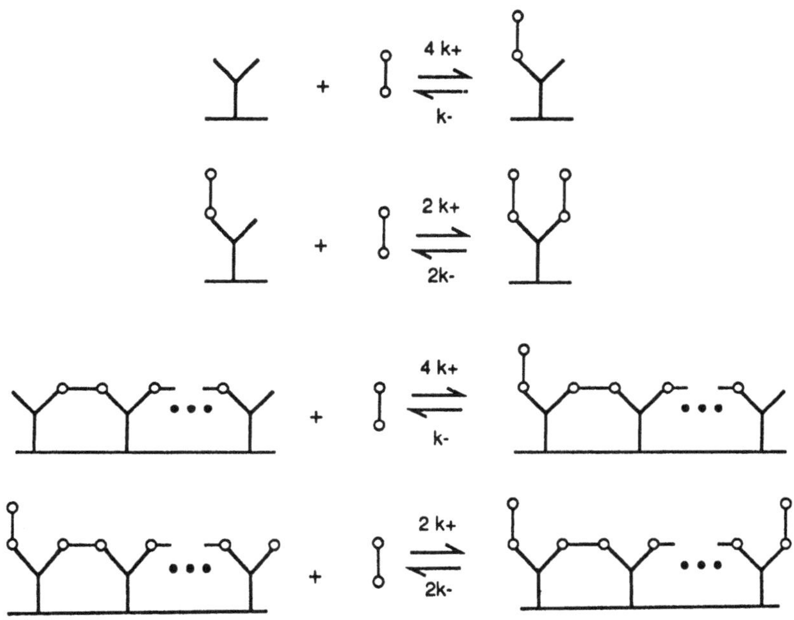

FIGURE 2 The binding of a bivalent hapten in solution to cell surface IgE. k_+ and k_- are the forward and reverse rate constants, respectively, for the binding of a single hapten site to, and the dissociation from, a single site on an IgE. The equilibrium binding constant $K = k_+/k_-$.

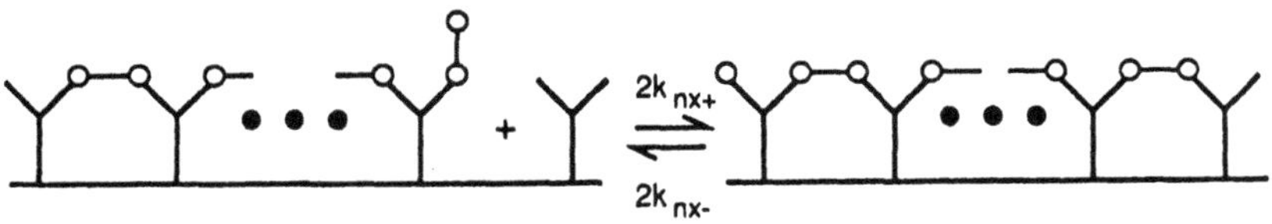

FIGURE 3 The crosslinking of a bivalent hapten-IgE complex to an IgE aggregate. k_{nx+} and k_{nx-} are the forward and reverse rate constants for the reactions. The equilibrium crosslinking constants $K_{nx} = k_{nx+}/k_{nx-}$. Usually it is assumed that $k_{nx+} = k_{x+}$, $k_{nx-} = k_{x-}$ and $K_{nx} = K_x$ for $n \geq 2$, where n is the number of IgEs in the final aggregate.

finite set. It is usual to assume that for all rings above a certain size ($n \geq m$) the dissociation constants are identical, i.e., $j_{n-} = j_{m-}$, and the forward rate constants $j_{n+} = (m/n)j_{m+}$ (Dembo and Goldstein, 1978).

In this approximation, we assumed that a chain containing m or more IgEs can be treated as a random walk of many steps. For a random walk of n steps in d dimensions starting at the origin, the probability of returning to a small neighborhood of the origin when n is large is inversely proportional to $n^{d/2}$. Therefore, for the random walk in two dimensions, the rate constant for ring closure is inversely proportional to n. In this approximation different parts of the chain are allowed to occupy the same position. A slightly more realistic model treats the chain as a self-avoiding random walk. Then $j_{n+} = (m/n)^z j_{m+}$, where in two dimensions z is somewhat greater than one (McKenzie, 1976). A third approach is simply to take the exponent z to be an unknown parameter that must be determined from experiments, just as we do the rate constants. We must certainly choose $m \geq 2$, since the process of forming a ring for $n = 1$ is quite different than the process for forming all other rings. For $n \geq 2$, rings are formed by crosslinking binding sites on two different IgEs (intramolecular bridging) while for $n = 1$, the formation of a ring requires the formation of an intermolecular bridge. Choosing $m = 2$, we obtain the minimum set of rate constants for describing ring formation, j_{1+}, j_{1-}, j_{2+} and j_{2-}. In addition, we need to assume a value for z, or treat it as an unknown parameter.

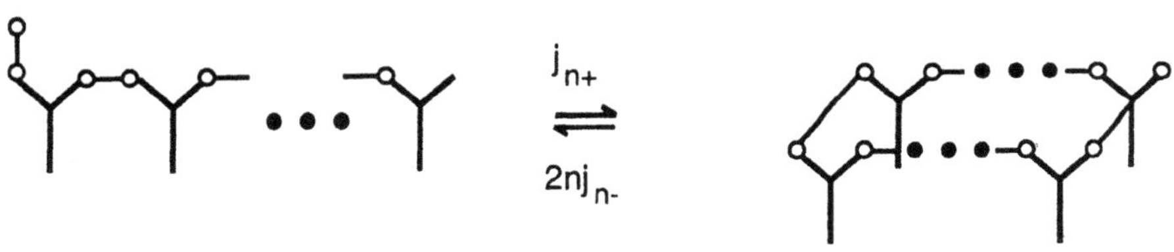

FIGURE 4 The formation of bivalent hapten-IgE ring complexes. j_{n+} and j_{n-} are the forward and reverse rate constants for the formation of a ring containing n bivalent haptens and n IgEs. The equilibrium constant for the reaction is $J_n = j_{n+}/j_{n-}$.

CALCULATING THE EQUILIBRIUM BIVALENT HAPTEN-IgE AGGREGATE SIZE DISTRIBUTION

For many haptens, the binding and aggregation reactions are fast compared to the kinetics of degranulation. This suggests that for these haptens, histamine release is controlled by the equilibrium distribution of crosslinks. Theories have been developed to calculate such equilibrium distributions for the noncooperative binding of symmetric (Dembo and Goldstein, 1978; Perelson and DeLisi, 1980), asymmetric (Wofsy et al., 1978), and mixed (Wofsy et al., 1978; Wofsy 1980; Perelson 1980) bivalent haptens to monoclonal antibody. Treatments have also been presented for symmetric bivalent haptens binding noncooperatively to polyclonal antibody (Wofsy and Goldstein, 1980) and binding cooperatively to monoclonal antibody (Wofsy and Goldstein, 1987). All the theories consider a dilute suspension of cells with bivalent antibody on their surfaces. The bivalent hapten can bind to the antibody, but the cells are taken to be so dilute that haptens cannot simultaneously bind to antibodies on different cells. Also it is assumed that the total concentration of receptors per cell, X_T, remains constant, e.g., internalization or shedding of antibodies, insertion of recycled antibodies, or interaction of antibodies with the cytoskeleton, is negligible for the times considered.

As I have discussed, for a symmetric bivalent hapten that binds to IgE noncooperatively, the minimum number of binding constants needed to describe the binding and aggregation is four, the equilibrium binding constant K, the equilibrium crosslinking constant K_x, and the equilibrium ring closure constants J_1 and J_2. This description assumes the equivalent site and random walk approximations can be applied. For short bivalent haptens < 100Å, $J_1 = 0$ because the haptens are too small to bridge both arms of the same IgE. For these haptens the minimum number of equilibrium parameters is three. As yet, there are no published determinations of a complete set of equilibrium binding constants for a bivalent hapten binding to cell surface IgE. For a series of bivalent DNP haptens binding to monoclonal IgE in solution, complete sets of equilibrium constants have been determined (Schweitzer-Stenner et al., 1987).

Erickson et al. (1986) have developed a fluorescence method to measure the fraction of cell surface IgE binding sites occupied by hapten (see also the chapter by Baird et al.). We (Erickson, Goldstein, Holowka, and Baird, unpublished results) are attempting to use this method to determine the equilibrium binding constants. We use the equilibrium binding model of Dembo and Goldstein (1978) to predict the fraction of sites bound in terms of the equilibrium binding constants, and then use nonlinear least squares data fitting methods to obtain estimates of the parameters from the data. Below, for symmetric bivalent haptens, I review the binding model of Dembo and Goldstein and show how it can be used to calculate quantities of interest, including the fraction of sites bound.

We start by expressing the concentrations of all aggregates in terms of the equilibrium binding constants and the free bivalent hapten and IgE concentrations. First we consider linear chains. In Figure 5, all linear chains that begin and end

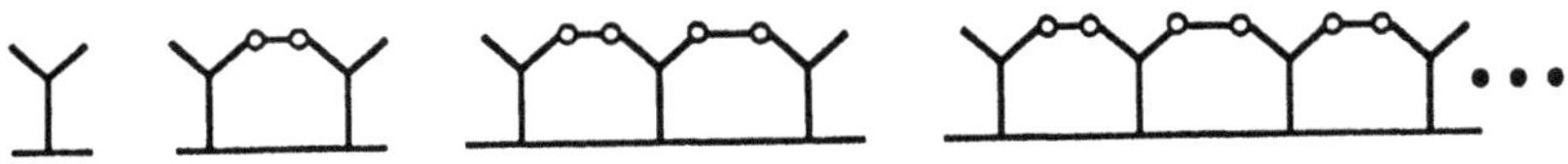

FIGURE 5 Bivalent hapten-IgE linear chains that both begin and end with a free receptor site.

with a free IgE binding site are depicted. If we let X be the equilibrium concentration of free cell surface IgEs (IgEs with both binding sites free) and C be the equilibrium concentration of free bivalent hapten in solution (haptens with both binding sites free), then from Figures 2 and 3, the equilibrium concentration of a complex of two IgEs crosslinked by a single bivalent hapten is $4KK_xCX^2$ and the equilibrium concentration of a complex of n IgEs crosslinked by $n-1$ haptens is $(4KK_xCX)^{n-1}X$.

In general, each end of a linear chain can be in one of two states, free or bound. Taking this into account, it follows that W_n, the equilibrium concentration of a linear chain containing n IgEs is

$$W_n = E^2(4KCK_xX)^{n-1}X \tag{1}$$

where from Figure 2 it follows that

$$E = 1+2KC \tag{2}$$

In some experiments it is useful to expose basophils to both monovalent and bivalent haptens. The monovalent haptens act to terminate the linear chains and prevent ring formation. To extend the theory to this case simply requires changing E in Eq. (2), since monovalent haptens can only bind to the ends of linear chains and can never be part of a ring. If in solution the free monovalent hapten concentration is C_M, and if the hapten binds to a single IgE site with an equilibrium constant K_M, as in Figure 6, then

$$E = 1+2KC+K_MC_M \tag{3}$$

It is useful to rewrite Eq. (1) in terms of W_1, the concentration of IgE that is not in aggregates or rings, and the following nondimensional parameter

$$\delta = \frac{4K_xX_TKC}{E^2} \tag{4}$$

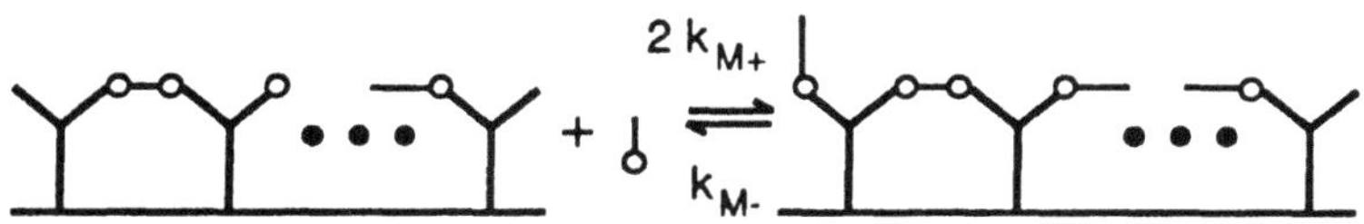

FIGURE 6 The binding of a monovalent hapten in solution to cell surface IgE. k_{M+} and k_{M-} are the forward and reverse rate constants, respectively, for the binding of a single hapten site in solution to, and the dissociation from, a single site on an IgE. The equilibrium binding constant $K_M = k_{M+}/k_{M-}$.

If we let $w = W_1/X_T$ and $w_n = W_n/X_T$, then

$$w_n = (\delta w)^{n-1} w \tag{5}$$

As indicated in Figure 4, a ring of n IgEs and n haptens is formed from a linear chain of n IgEs and n haptens. The equilibrium concentration of such a chain is $4KC(4KCK_x X)^{n-1}X$, and, therefore, R_n, the equilibrium concentration of the ring formed from such a chain is

$$R_n = 4KC(4KCK_x X)^{n-1} X \frac{J_n}{2n} \tag{6}$$

Introducing the nondimensional concentration $r_n = R_n/X_T$, we have that

$$r_n = \frac{J_n(\delta w)^n}{2nK_x X_T} \tag{7}$$

It can be seen from Eqs. (5) and (7) that for a given set of binding parameters, if we know δ and w, we can determine the concentration of all the aggregates. To determine δ, we must know all the equilibrium constants (K, K_M, K_x, J_n) and the concentrations C and C_M. To obtain equations for the three unknown concentrations, w, C, C_M, we use three conservation laws, i.e., that the total concentration of IgE, X_T, the total concentration of bivalent hapten, C_T, and the total concentration of monovalent hapten, C_{MT}, are constant.

$$X_T = \sum_{n=1}^{\infty} nW_n + \sum_{n=1}^{\infty} nR_n \tag{8a}$$

$$C_T = C + \frac{1}{E^2}\Bigg[(1+K_MC_M)^2\sum_{n=1}^{\infty}(n-1)W_n + 4KC(1+K_MC_M)\sum_{n=1}^{\infty} nW_n + (2KC)^2\sum_{n=1}^{\infty}(n+1)W_n\Bigg] + \sum_{n=1}^{\infty} nR_n \tag{8b}$$

$$C_{MT} = C_M + \frac{1}{E^2}\Big(2K_MC_M + 4KCK_MC_M + 2(K_MC_M)^2\Big)\sum_{n=1}^{\infty} W_n \tag{8c}$$

Eq. (8a) follows because there are n IgEs in a linear chain of concentration W_n and n IgEs in a ring of concentration R_n. Eq. (8b) follows because bivalent hapten is either free, in linear chains, or in rings. The term in brackets represents bivalent hapten in linear chains. The first sum is the concentration of bivalent hapten in chains with both ends free, or one end free and one end bound by monovalent ligand, or both ends bound by monovalent ligand; the second sum is the concentration of bivalent hapten in chains with one end bound to bivalent hapten and one end either free or bound to a monovalent hapten; and the last term is the concentration of bivalent hapten in chains with both ends bound to bivalent hapten. Eq. (8c) follows because monovalent hapten is either free or bound to the ends of linear chains. The three terms multiplying the sum in Eq. (8c) represent respectively chains with one end free and one end bound to monovalent hapten, one end bound to bivalent hapten and one end bound to monovalent hapten, and both ends bound to monovalent hapten.

To go further we must evaluate the sums in Eq. (8). From Eq. (5)

$$\sum_{n=1}^{\infty} W_n = X_T\sum_{n=1}^{\infty} w_n = X_T\frac{w}{1-\delta w} \tag{9a}$$

$$\sum_{n=1}^{\infty} nW_n = X_T\sum_{n=1}^{\infty} w_n = X_T\frac{w}{(1-\delta w)^2} \tag{9b}$$

We can evaluate these sums because we have made an equivalent site approximation and taken all the equilibrium crosslinking constants to be equal. For the sum over the ring concentrations, we have that

$$\sum_{n=1}^{\infty} nR_n = X_T\sum_{n=1}^{\infty} r_n = \sum_{n=1}^{\infty}\frac{J_n}{2K_x}(\delta w)^n \tag{10}$$

To evaluate the sum in Eq. (10), we must specify how J_n depends on the length of the ring. As we discussed, in a general random walk approximation, $J_n = (m/n)^z J_m$ for $n \geq m$. If we take $m = 2$, then we can write Eq. (10) in the following form:

$$\sum_{n=1}^{\infty} nR_n = \frac{\delta w}{2K_x}(J_1 - 2^z J_2) + \frac{2^z J_2}{2K_x}\phi(\delta w, z) \tag{11}$$

where $\phi(x, s)$ is a Truesdell function defined as follows:

$$\phi(x, s) = \sum_{n=1}^{\infty} \frac{x^n}{n^s} \tag{12a}$$

For $s = 1$,

$$\phi(x, 1) = -\log(1 - x) \tag{12b}$$

but in general the sum cannot be written in closed form. The series in Eq. (12a) is rapidly convergent near $x = 0$, but slowly convergent near $x = 1$. However, the sum can be converted to an integral which is rapidly convergent near $x = 1$ (Truesdell, 1945), so there are no difficulties in evaluating $\phi(x, s)$ numerically. In particular

$$\phi(x, s) = \frac{x}{\Gamma(s)} \int_0^{\infty} \frac{t^{s-1}}{\exp(t) - x} dt \tag{12c}$$

We use Eq. (8a) to eliminate the sum over ring concentrations in Eq. (8b). Then from Eqs. (9a), (9b) and (11), the conservation equations, Eqs. (8a), (8b) and (8c), become

$$1 = \frac{w}{(1 - \delta w)^2} + \frac{\delta w}{2K_x X_T}(J_1 - 2^z J_2) + \frac{2^z J_2}{2K_x X_T}\phi(\delta w, z) \tag{13a}$$

$$1 = \frac{C}{C_T} + \frac{X_T}{C_T}\left[1 + \frac{2KC - 1 - K_M C_M}{E} \, \frac{w}{1 - \delta w}\right] \tag{13b}$$

$$1 = \frac{C_M}{C_{MT}} + \frac{X_T}{C_{MT}}\left[\frac{2K_M C_M + 4KCK_M C_M + 2(K_M C_M)^2}{E^2}\right]\frac{w}{1 - \delta w} \tag{13c}$$

For a given set of binding parameters (K, K_x, J_1, J_2, z) and initial concentrations (X_T, C_T, C_{MT}), we can solve Eqs. (13a), (13b) and (13c) simultaneously and determine the free concentrations w, C and C_M. With these we can calculate the concentration of any size aggregate from Eqs. (5) and (7).

In the fluorescence quenching experiments of Erickson et al. (1986) (also see the chapter by Baird et al.), the quantity that is determined is the fraction of sites bound, q. Since $q =$ (the concentration of IgE sites bound)$/(2X_T)$, it follows that

$$q = \frac{1}{2E^2}\left[\sum_{n=1}^{\infty}(2n-2)w_n + (4KC+2K_MC_M)\sum_{n=1}^{\infty}(2n-1)w_n + (2KC+K_MC_M)^2\sum_{n=1}^{\infty}2nw_n\right] + \sum_{n=1}^{\infty}nr_n \tag{14}$$

When we use Eq. (8a) to eliminate the sum over rings in Eq. (14), and use Eqs. (9a) and (9b) to evaluate the remaining sums, we find that

$$q = 1 - \frac{1+2KC+K_MC_M}{E^2}\frac{w}{1-\delta w} \tag{15}$$

We (Erickson, Goldstein, Holowka and Baird) are now in the process of fitting Eq. (15) to bivalent hapten-cell surface IgE binding data, in order to determine the equilibrium binding parameters. The precision of the values that we will obtain for the equilibrium crosslinking constant, K_x, and equilibrium ring-closure constants, J_1 and J_2, will depend in part on how good the equivalent site and random walk approximations are. For short bivalent haptens ($J_1 = 0$), however, we can determine the equilibrium crosslinking constant for the formation of dimers of IgE (K_{2x} in Figure 3) without making either of these assumptions.

When short bivalent haptens that cannot form intermolecular bridges ($J_1 = 0$) bind to IgE at very low bivalent hapten concentrations, almost all the IgE will be in one of three states: both sites free; one site free and one site bound to an uncrosslinked bivalent hapten; or one site free and one site bound to a crosslinked ligand (see Figure 7). This means that if we plot q, the fraction of sites bound, versus C_T, the concentration of bivalent hapten, the slope of this curve as C_T goes to zero will depend only on K, K_{2x} and X_T. In particular, one can show that

$$\lim_{C_T \to 0} q = s\, C_T \tag{16}$$

where the slope s is

$$s = \frac{2K(1+2K_{2x}X_T)}{1+4KX_T+4KK_{2x}X_T^2} \tag{17}$$

To use Eq. (17) to determine K_{2x}, we need to determine the limiting slopes from binding curves, i.e, determine s for different values of X_T and fit Eq. (17) to these values to determine K and K_{2x}. If the values of K and K_{2x} that we obtain in this way agree with the values of K and K_x determined by fitting Eq. (15) to the entire binding curve, then for the range of concentrations studied, the equivalent site approximation is reasonable. This work is now in progress.

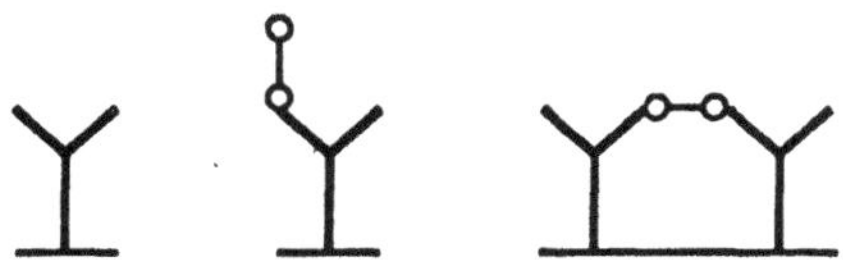

FIGURE 7 The states involving at most one bivalent hapten when the bivalent hapten is short and cannot bridge two binding sites on the same IgE.

BIVALENT HAPTENS AND HISTAMINE RELEASE CURVES

Even when we do not know the values of the bivalent hapten-IgE equilibrium constants, the theory of the previous section makes testable predictions. These predictions follow from the observation that the concentration of all aggregates depends on the free bivalent and monovalent ligand concentrations only through the parameter δ. Eq. (13a) shows that this is true for w. From Eqs. (5) and (7), it then follows that this is also true for the concentration of all aggregates.

From the definition of δ, Eq. (4), it follows that $d\delta/dC = 0$ at $C = (1 + K_M C_M)/(2K)$, and that δ is a maximum at this bivalent hapten concentration. We call this concentration C_{max}, i.e,

$$C_{\text{max}} = \frac{1 + K_M C_M}{2K} \tag{18a}$$

Since $dw/dC = 0$ when $d\delta/dC = 0$, we have that, at $C = C_{\text{max}}$, $dw_n/dC = 0$ and $dr_n/dC = 0$. It can be shown (Dembo and Goldstein, 1978) that at $C = C_{\text{max}}$, w_n and r_n are either at a maximum or a relative minimum. If we substitute Eq. (18a) into Eq. (13b), we find that at $C = C_{\text{max}}$

$$C_T = C_{\text{max}} + X_T \tag{18b}$$

This means that when crosslinking is maximum, there is one bivalent hapten bound per IgE, i.e, $C_{\text{bound}} \equiv C_T - C = X_T$ when $C = C_{\text{max}}$.

It also follows from Eq. (4) that δ is symmetric about C_{max} when it is plotted against $\log(C)$. If we let $y = \log(C/C_{\text{max}})$, then we can rewrite δ as follows:

$$\delta = \frac{K_x X_T}{K C_{\text{max}}} \frac{1}{\left[\exp(-y/2) + \exp(y/2)\right]^2} \tag{19}$$

Thus, we see that $\delta(y) = \delta(-y)$. Since the concentrations of all aggregates are functions of C only through δ, it follows that w_n and r_n have this same symmetry.

(This means, for example, that if $C_{max} = 10^{-7}M$, the concentration of linear chains containing 2 IgEs, or any other number of IgEs, is the same at $C = 10^{-8}M$ and $C = 10^{-6}M$. This would also be true for rings containing 2 IgEs, or any other number of IgEs.)

We have used these results to make predictions about the properties of histamine release curves generated with bivalent haptens (Dembo and Goldstein, 1978). A *histamine release curve* is a plot of the percent histamine release versus the log of the ligand concentration. The ligand can be any crosslinking agent, e.g., a bivalent hapten, an antigen, an anti-IgE, etc. To generate a curve, each of the data points is obtained by exposing sensitized basophils to a fixed concentration of ligand in the presence of calcium (calcium is required for histamine release) and then measuring the amount of histamine released during the next t minutes. Usually $t \geq 30$ min. One wants t to be long enough so that histamine release goes to completion.

For rapidly equilibrating bivalent haptens, we expect the histamine release curve to be controlled by the equilibrium distribution of crosslinks. One measure of this distribution is the *crosslinking curve*, which is the fraction of IgEs in aggregates versus the log of the ligand concentration. If we define S_2 as:

$$S_2 = \sum_{n=2}^{\infty} n(w_n + r_n) \tag{20}$$

then the crosslinking curve is a plot of S_2 versus $\log(C)$. Dembo and Goldstein (1978) showed that if $J_1 = 0$, then S_2 is always maximum at $C = C_{max}$. If the histamine release curve follows the crosslinking curve in a general way, i.e, rises when crosslinks rise, is a maximum when crosslinking is a maximum, and falls when crosslinks fall, then histamine release should also be a maximum at $C = C_{max}$. Also, since the crosslinking curve is symmetric, we predict that the histamine release curve should also be symmetric.

If we are dealing with cells that do not respond to dimers of IgE, such as RBL cells, then a better measure of the bivalent hapten–IgE aggregate distribution that the cell is responsive to is (Wofsy and Goldstein, 1986)

$$S_3 = \sum_{n=3}^{\infty} n(w_n + r_n) \tag{21}$$

Shown in Figure 8 are plots of S_2 and S_3 for a typical set of parameters. Note that both curves have their maximum at $C = C_{max}$ and are symmetric about C_{max}.

If histamine release is maximum at $C = C_{max}$, where C_{max} is given by Eq. (18a), then additional predictions follow. Note that the IgE concentration does not enter Eq. (18a), nor does any property of the bivalent hapten except its single-site

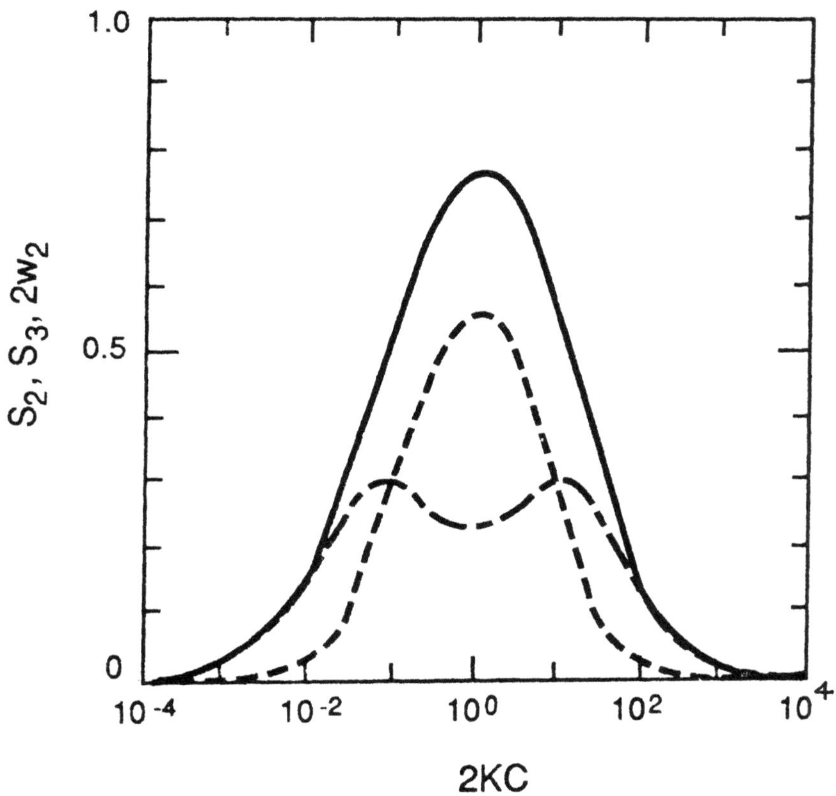

FIGURE 8 Predictions of equilibrium crosslink formation as a function of the bivalent hapten concentration C. S_2 (——) is the fraction of IgE involved in crosslinks, S_3 (-----) is the fraction of IgE involved in aggregates containing three or more IgEs, and $2w_2$ (- - - - -) is the fraction of IgE involved in aggregates containing two IgEs. K is the single site equilibrium binding constant. When solving Eq. (13a), we took $K_x X_T = 5$, and $J_1 = J_2 = 0$.

equilibrium binding constant, K, nor does any property of the basophil. This implies that the bivalent hapten concentration at which histamine release is maximum will not depend on the concentration of IgE on the basophil surface; nor on the properties of the spacer separating the two haptenic groups; nor on the histamine-releasing properties of the basophil; nor on the cell donor (basophils from different donors sensitized with the same monoclonal IgE will have the same C_{max}). Although the theory was developed for monoclonal IgE, when we first tested these predictions, there was no specific monoclonal IgE available. Instead we used human basophils

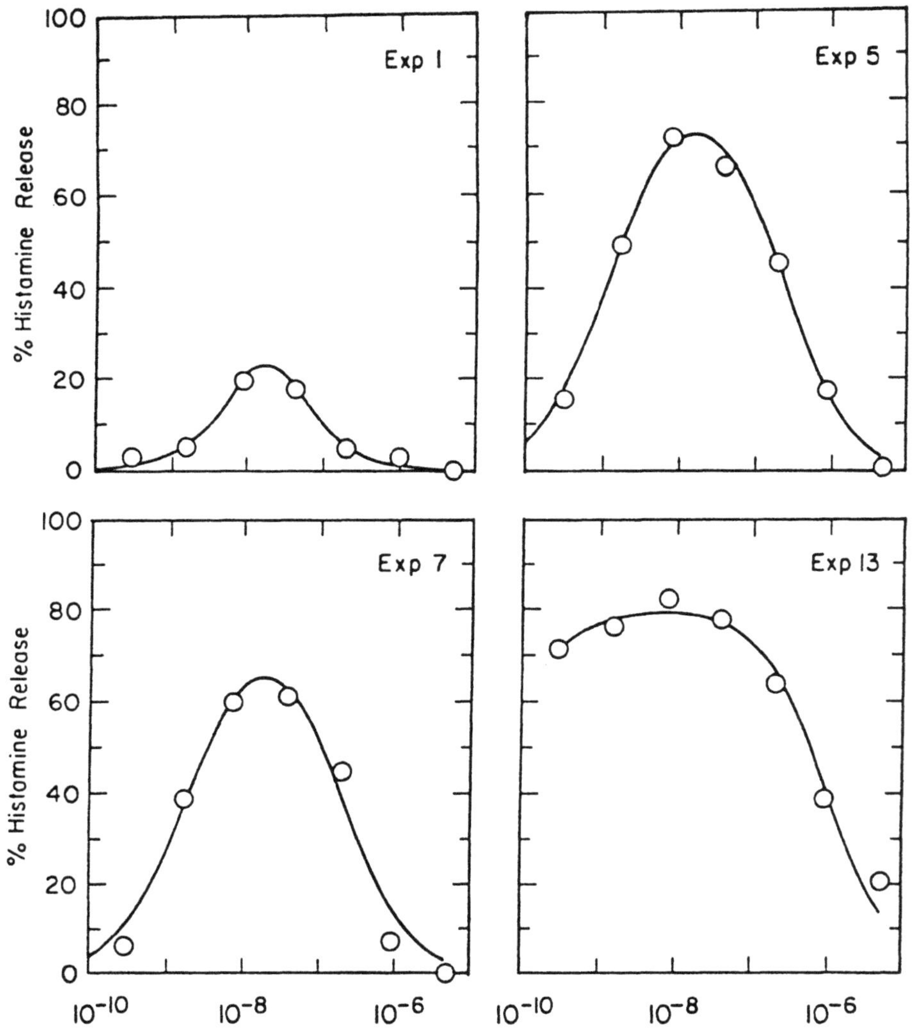

FIGURE 9 Four histamine release curves obtained with the bivalent asymmetric hapten α,ϵ-DNP-lysine. Fifteen experiments were carried out in all. No monovalent hapten was present. The solid curves are the best fits of an empirical symmetric bell-shaped curve to the data. Typical are the fits shown for experiments 1, 5 and 7. In experiment 13, the histamine release curve is much broader than we normally see. (MacGlashan et al., 1985).

from nonallergic donors passively sensitized with a polyclonal IgE. Even so, we obtained good agreement between theory and experiment (Dembo et al., 1978).

More recently, using a monoclonal anti-DNP IgE, we have directly tested Eq. (18a) and the prediction that the histamine release curve is symmetric (MacGlashan et al., 1985). We passively sensitized human basophils with an anti-DNP monoclonal

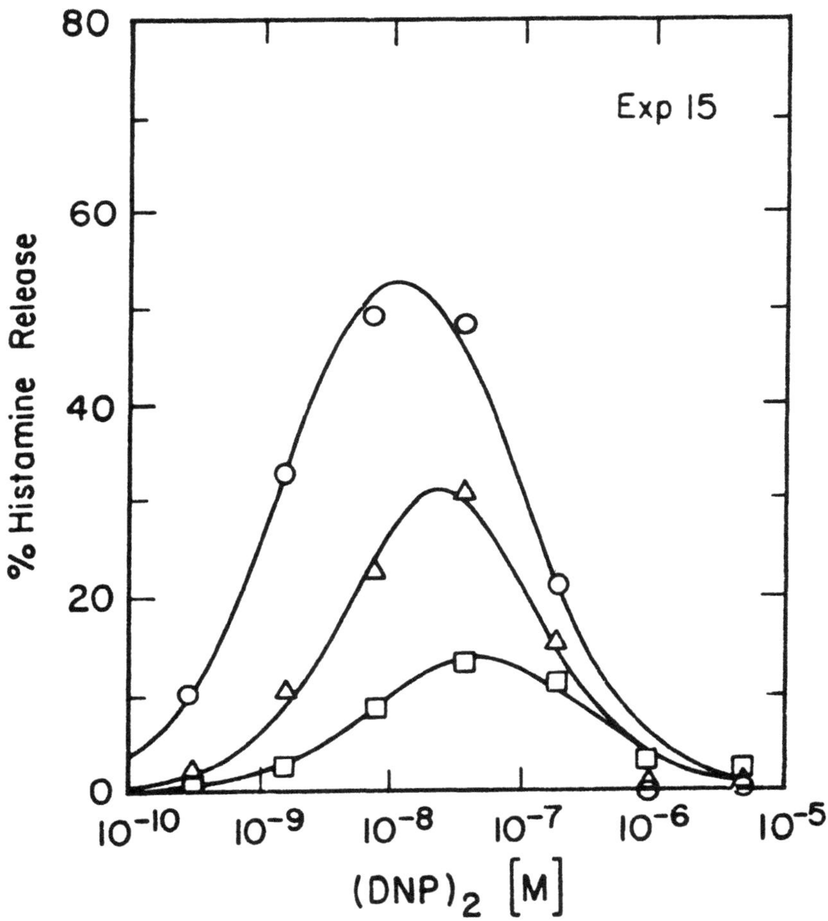

FIGURE 10 The effect on the histamine release curve of adding the monovalent hapten ϵ-DNP-lysine. There was no ϵ-DNP-lysine present in the highest curve (○), $1 \times 10^{-8} M^{-1}$ was present in the middle curve (△), and $5 \times 10^{-8} M^{-1}$ was present in the lowest curve (□). As in Figure 9, the solid curves are best fits of an empirical symmetric bell-shaped curve to the data (MacGlashan et al., 1985).

IgE whose single-site affinity constant with ϵ-DNP-lysine had been determined by equilibrium dialysis to be $7.1 \times 10^7\ M^{-1}$ at 37°C (Liu et al., 1980). We then used the asymmetric bivalent hapten α,ϵ-DNP-lysine, to generate histamine release curves, as shown in Figure 9. Some curves were generated in the presence of both bivalent and monovalent hapten (see Figure 10). We then fit symmetric curves to the data and determined the value of $C_{\max}$ for each histamine release curve. For an asymmetric hapten such as α,ϵ-DNP-lysine, all the results we derived hold provided we replace K by the average affinity constant $(K_\alpha + K_\epsilon)/2$, where K_α and K_ϵ are the single-site equilibrium binding constants respectively for the α hapten moiety and ϵ hapten moiety, and replace $K_M C_M$ by $K_\alpha C_\alpha + K_\epsilon C_\epsilon$, where C_α and C_ϵ are the concentrations of monovalent α and monovalent ϵ hapten (Wofsy et al., 1978). For example, Eq. (18a) becomes

$$C_{\max} = \frac{1 + K_\alpha C_\alpha + K_\epsilon C_\epsilon}{K_\alpha + K_\epsilon} \tag{22}$$

To determine K_α and K_ϵ, we fit Eq. (22) to the values of $C_{\max}$ that we had determined from the histamine release curves (see Figures 8 and 9). The value for K_ϵ that we obtained in this way from histamine release measurements, $K_\epsilon = 6.9 \pm 0.5 \times 10^7 M^{-1}$ at 37°C, agrees with the value of $7.1 \times 10^7 M^{-1}$ determined by Liu et al. (1980) from equilibrium dialysis. The agreement between the two estimates of the ϵ-DNP-lysine affinity constant is consistent with a central prediction of the crosslinking theory, Eq. (22). It also indicates that the hapten binding properties of monoclonal IgE are not altered when the IgE is bound to Fc_ϵ receptors on basophil surfaces. We have also confirmed this latter result from direct binding studies with IgE in solution and on cell surfaces (Erickson et al., 1986). The prediction of the theory that the histamine release curves should be symmetric was also consistent with experiment.

These experiments demonstrate that histamine release is a monotonically increasing function of the number of IgEs crosslinked on the cell surface, i.e., the histamine release curve rises when crosslinks rise, is maximum when crosslinks are maximum, and decrease when crosslinks decrease. When this is so, adding monovalent hapten, as in Figure 10, always decreases histamine release. In general, this correlation between crosslinking and histamine release will only hold when the amount of specific IgE on the basophil surface is small, as was the case in these experiments where basophils from nonallergic donors were used. These cells had relatively few free Fc_ϵ receptors on their surface.

GENERAL FEATURES OF HISTAMINE RELEASE CURVES

The shapes of the histamine release curves in Figures 9 and 10 raise the following question: why do fewer crosslinks yield less histamine release even though histamine

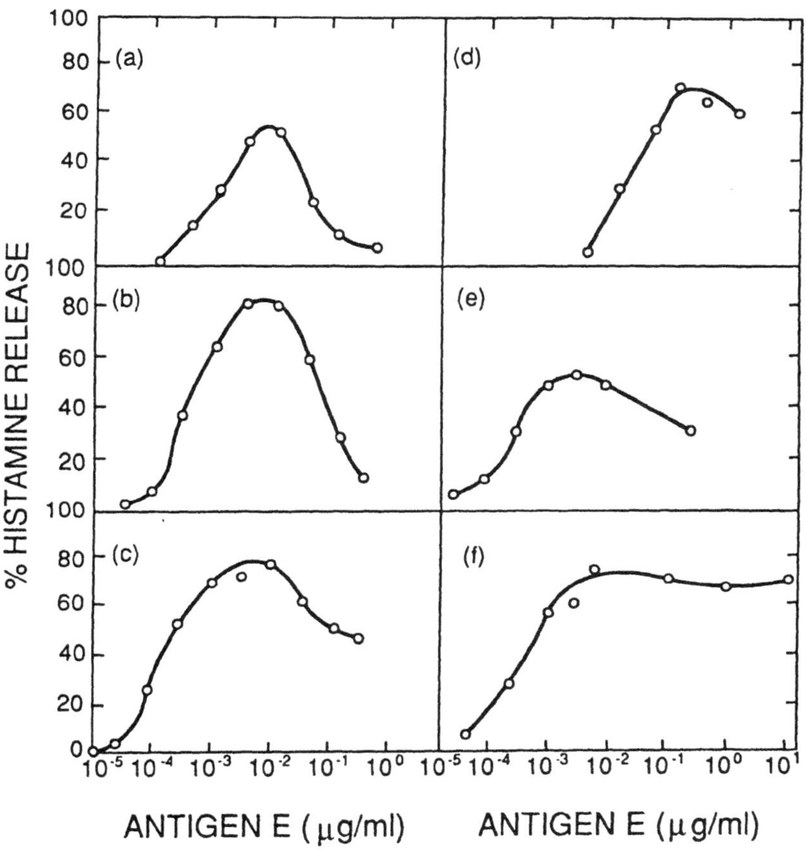

FIGURE 11 Examples of histamine release curves with basophils from ragweed allergic donors. Antigen E is the major antigen isolated from ragweed pollen. Curves (a), (b), and (c) are from Drobis and Siraganian, 1978; curves (c), (d) and (e) from Marone et al., 1981.

release is allowed to go to completion? The bivalent hapten used in these experiments, α,ϵ-DNP-lysine, rapidly equilibrates so that during the time degranulation is occurring, there should be a constant concentration of crosslinks on the basophil surface. If crosslinks signal release, why isn't the same amount of histamine released no matter what the concentration of crosslinks, provided we wait long enough? At nonoptimal antigen concentrations, why does histamine release turn off before total degranulation is achieved? This could come about if all the IgE were internalized, but on human basophils there is still specific IgE on the basophil surface after histamine release has gone to completion (MacGlashan et al., 1983a). The shapes of the histamine release curves appear to imply that aggregates of IgE not only generate "on" signals, but also "off" signals. These latter signals are said to desensitize the cell. A major role of desensitization may be to regulate the release process (Baxter and Adamik, 1975; Kazimierczak et al., 1984).

It is worthwhile, despite the complexity of the reagents used, to go back to the literature and study the shapes of published histamine release curves. Since the cells are usually sensitized with polyclonal IgE, or come from allergic donors, and the valences of the crosslinking agents are usually unknown, a rigorous comparison with a theoretical model is not possible. Still, the patterns that one sees indicates what qualitative features any theory of histamine release must be capable of generating. In Figure 11 are four such curves. The basophils are from ragweed allergic donors and the crosslinking agent is antigen E, the major ragweed antigen. In general, histamine release curves exhibit the following patterns. At low concentrations of the crosslinking agent, histamine release rises. Ultimately the curve levels off and goes through a maximum. In many cases, the maximum is quite broad. As the concentration is further increased the curve either (a) falls to zero, (b) levels off, or (c) decreases, then rises again and goes through a second maximum (see Figure 12).

The falling off of the histamine release curve at supraoptimal antigen concentrations is known as *antigen excess inhibition*. We saw in the previous section that when there are small amounts of specific IgE on the basophil surface crosslinking is a decreasing function of the bivalent hapten concentration when the concentration is above the optimal value, $C_{\max}$. Here antigen excess inhibition comes about because at high enough hapten concentrations the binding sites on IgE become blocked with hapten, preventing crosslink formation. At high enough concentrations, all phenomena triggered by crosslinking must exhibit this type of inhibition. We have called this type I antigen excess inhibition (Goldstein and Dembo, 1984).

The second type of antigen excess inhibition (type II) is characterized by a decreasing histamine release curve over a range of antigen concentrations where crosslinking is increasing. This type of inhibition occurs because the desensitization signals generated by aggregated IgE dominates the release signals generated by these aggregates. Type II inhibition was first detected by Becker et al. (1973) using anti-IgE and human basophils from allergic donors. They showed that anti-IgE produced patching of IgE, and that the onset of patching occurred at the same concentration as the onset of antigen excess inhibition. Recently Robertson et al. (1986) detected type II inhibition with RBL cells while studying the relationship between IgE aggregation, cytoskeletal interactions and degranulation. They observed that crosslinking of IgE with either a polyclonal anti-IgE or a highly haptenated DNP antigen, DNP_n-bovine serum albumin (BSA), induced a detergent-resistant association between IgE aggregates and the cytoskeleton. This association increased at anti-IgE and DNP_n-BSA concentrations that were supraoptimal for histamine release.

There is a simple way to determine what type of inhibition is responsible for the fall off in histamine release at high concentrations, provided that a monomer is available to compete for IgE binding sites with the crosslinking agent being used. Adding monomer always decreases crosslinking. Thus, if excess inhibition is of type I, adding monomer will decrease histamine release in all parts of the histamine release curve as in Figure 10. However, if excess inhibition is of type II, adding monomer will cause histamine release to decrease in the rising portion of the curve and increase in the region of antigen excess inhibition. At supraoptimal antigen concentrations, reducing crosslinks reduces the inhibition signal. Type II antigen excess inhibition was detected in this way by Magro et al. (Magro and

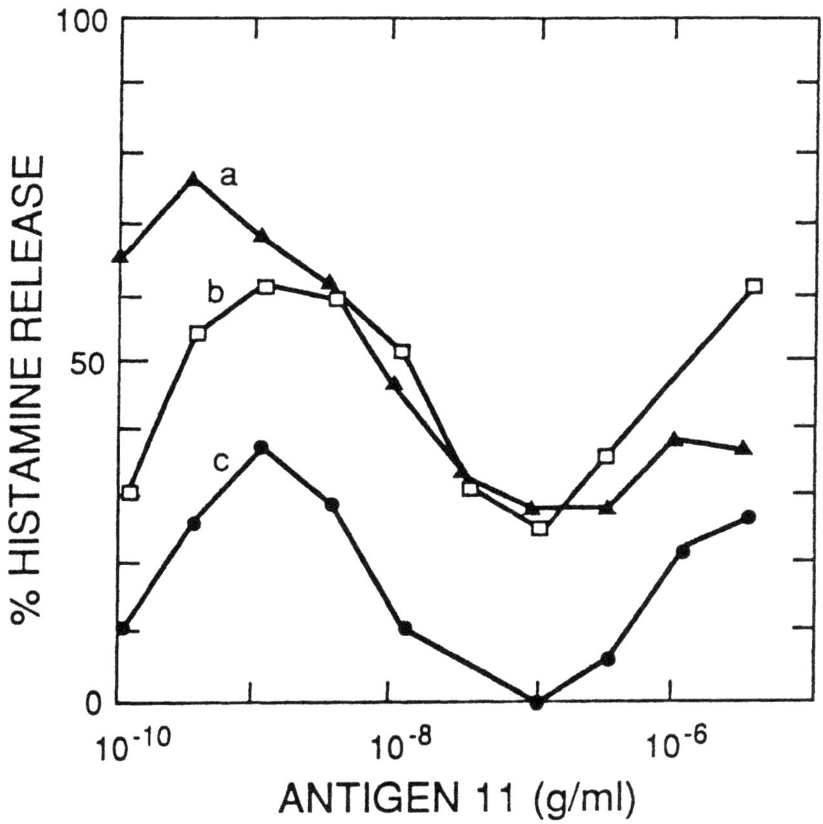

FIGURE 12 Histamine release curves from basophils from three donors who were allergic to the dust mite *Dermatophagoides farinae* (Weyer et al., 1982). Antigen 11 is a highly purified component of *D. farinae*. DeLisi and Siraganian (1979) observed similar double-humped histamine release curves with rabbit basophils.

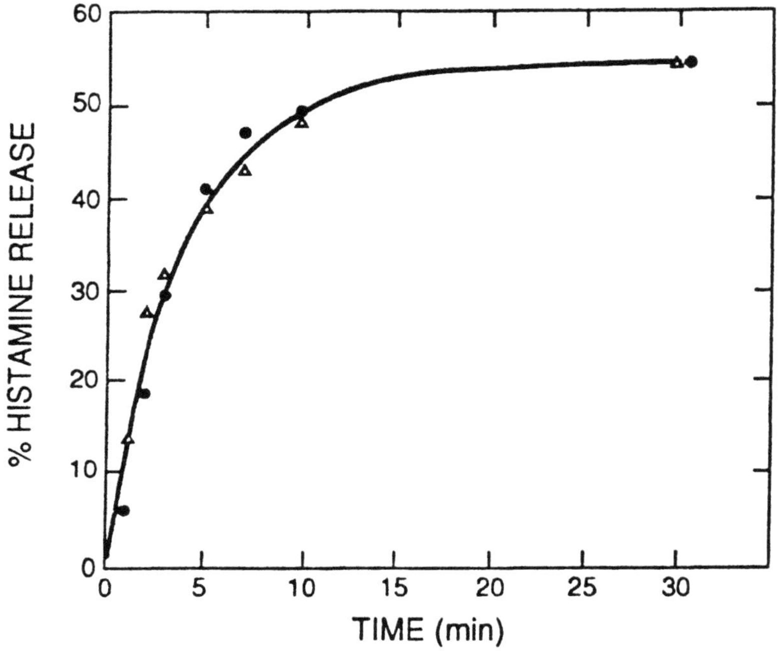

FIGURE 13 The kinetics of histamine release. Human basophils were passively sensitized with antiserum from a penicillin allergic donor and then challenged with $10^{-7}M$ $(BPO)_2$ at 37°C in the presence of Ca^{++}. At the times indicated histamine release was stopped by adding EDTA (•) or $(BPO)_1$ at a final concentration of $3 \times 10^{-5}M$ (△) (Sobotka et al., 1979).

Alexander, 1974; Magro and Bennich, 1977) who showed that adding Fab fragments to basophils exposed to anti-IgE caused an increase in histamine release in the region of excess inhibition and a decrease in histamine release in the rising part of the histamine release curve. Recently, Oliver et al. (1987) observed type II antigen excess inhibition with RBL cells sensitized with monoclonal anti-DNP IgE. When DNP-lysine was added in the presence of DNP-BSA at supraoptimal concentrations, histamine release increased.

THE KINETICS OF RELEASE AND DESENSITIZATION

A necessary requirement for histamine release from human basophils is the presence of calcium in the external medium. Crosslinking of IgE, at least in moderation, leads to an influx of calcium which serves as an essential signal for histamine release. A working hypothesis is that crosslinking leads to the opening of calcium channels or activates calcium carriers that allow calcium to be transported into the cell (Foreman and Mongar, 1973).

Adding a calcium chelating agent such as ethylene diaminetetraacetic acid (EDTA) to the external medium stops histamine release in seconds (Lichtenstein and Osler, 1964). This allows one to study the kinetics of histamine release in the following way. At time $t = 0$, sensitized basophils are exposed to a crosslinking agent and calcium. At a time t later, EDTA is added to the medium to stop histamine release. The amount of histamine released up to time t is then measured. Such a kinetic curve is shown in Figure 13. In these experiments basophils from a nonallergic donor were sensitized with antiserum from an individual who was allergic to penicillin. A bivalent hapten, $(BPO)_2$ (bis benzylpenicilloyl 1,6 diaminohexane), was used to trigger release. With this system, we could compete off the crosslinks formed by $(BPO)_2$ by adding large concentrations of the monovalent hapten $(BPO)_1$ (benzylpenicilloyl formyl-L-lysine). Indeed, it seemed that the $(BPO)_2$ bound so weakly that we could simply wash it off, since washing these cells also stopped histamine release. However, we did not directly assay for bound $(BPO)_2$ after $(BPO)_1$ was added. MacGlashan and Lichtenstein (1983), using a radioactively labeled multivalent BPO antigen, showed that adding $(BPO)_1$ stopped histamine release even though all the antigen was not eluted from the basophil surface. Seagrave et al. (1987) obtained similar results using a fluorescent antigen, DNP-B-phycoerythrin (DNP-B-phE) and RBL cells sensitized with monoclonal anti-DNP IgE. Adding DNP-lysine stopped histamine release, but the longer the time before DNP-lysine was added, the less effective the monomer was in removing the antigen. As can be seen in Figure 13, adding $(BPO)_1$ stopped histamine release as fast as adding EDTA (Sobotka et al., 1979). Crosslinks do not activate a long-lived calcium channel, rather the constant presence of crosslinked IgE is required for activation of human basophils.

In the previous section, we indicated that crosslinking of IgE can generate inhibitory signals as well as signals for release. We can see this directly in the following type of experiment (Lichtenstein, 1971; Foreman and Garland, 1974). At time $t = 0$, sensitized basophils are exposed to antigen (or any other crosslinking agent) in the absence of calcium. At various times after the start of the experiment, calcium is added and histamine release is allowed to go to completion. As shown in Figure 14, the longer the basophils are exposed to antigen, the less able they are to release histamine when calcium is added.

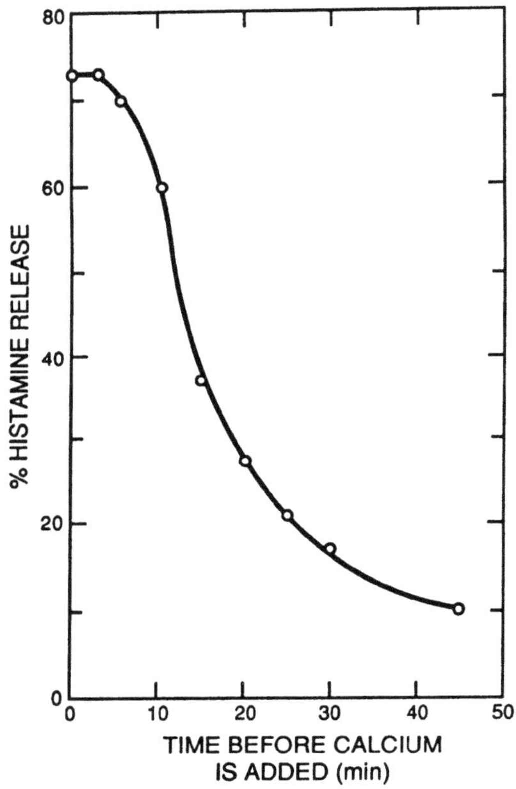

FIGURE 14 The kinetics of desensitization. Basophils from a donor allergic to ragweed pollen were exposed to antigen E at a final concentration of 6.7×10^{-2} $\mu g/ml$ at time $t = 0$ in the absence of calcium. At the times indicated, calcium was added at a final concentration of 0.6mM. Histamine release was then allowed to proceed for 45 min. after calcium was added.

The histamine release dose response curves of the previous section suggested that in the presence of calcium, crosslinking generates both release and desensitization signals. The final amount of histamine that is released is determined by a competition between these signals. The desensitization experiment in Figure 14 shows that in the absence of calcium the release signal is blocked but the desensitization signal in unimpaired. For $(BPO)_2$ interacting with passively sensitized human basophils, we showed that the rate of desensitization is independent of the external calcium concentration (Goldstein et al., 1979).

THE TYPES OF DESENSITIZATION

IgE-mediated desensitization of human basophils can be either antigen specific (Sobotka et al., 1978) or antigen nonspecific (Lichtenstein, 1971; Drobis and Siraganian, 1976). The difference can be understood by considering basophils that are sensitized with two monoclonal IgEs that are specific for two non-crossreacting antigenic determinants. After sensitization, the basophils are incubated in the absence of calcium with an antigen that is specific for one of the monoclonals. If the basophils desensitize specifically, they progressively lose their ability to respond to

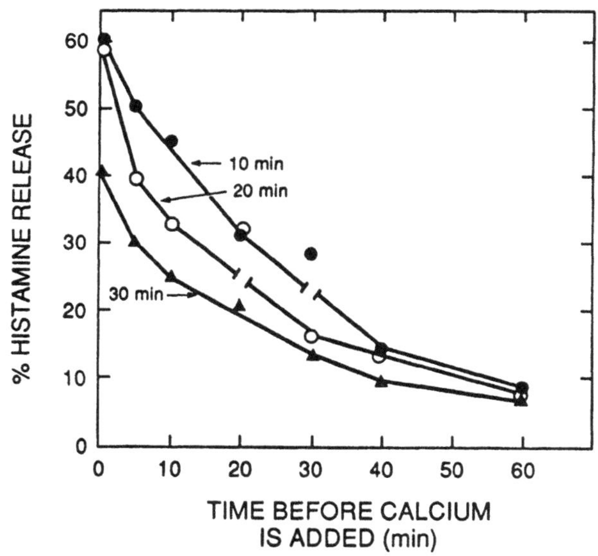

FIGURE 15 The unit signal for specific desensitization is an aggregate of 2 IgEs. Human basophils were passively sensitized with preformed dimers of IgE (10 $\mu g/ml$) in the absence of calcium for 10, 20 or 30 min. At time $t = 0$ further binding of the IgE dimers was stopped by adding 100 $\mu g/ml$ of monomeric IgE. At the times indicated calcium in the presence of D_2O was added and histamine release was allowed to continue for 45 min. The final calcium concentration was 0.6mM and the final concentration of D_2O 44%. When these cells were challenged with anti-IgE, they released histamine normally (data not shown) indicating the desensitization was specific (Kagey-Sobotka et al., 1981).

the desensitizing antigen when calcium is added, but are still capable of responding to the non-crossreacting antigen or anti-IgE. If the basophils desensitize nonspecifically, they progressively lose their ability to respond to any ligand that crosslinks IgE. Both types of desensitization have recently been observed on human lung mast cells (MacGlashan and Lichtenstein, 1987) and on B lymphocytes that express both IgM and IgD (Cambier et al., 1987). Antigen-specific desensitization has been seen with mouse mast cells (Siraganian and Hazard, 1979) and RBL cells (Fewtrell, 1985).

As with release, the minimum-sized aggregate required to induce specific desensitization is two crosslinked IgEs (Kagey-Sobotka et al., 1981). We demonstrated this by first passively sensitizing human basophils with preformed IgE dimers for a fixed period of time (10, 20 or 30 min) in the absence of calcium. We then added a large excess of monomeric IgE to rapidly fill the remaining free Fc_ϵ receptors. This stopped any further crosslink formation. At various times after the addition of monomeric IgE, we added back calcium and measured the subsequent histamine release. The results of the experiment are shown in Figure 15. The longer the IgE dimers remained on the basophil surface in the absence of calcium, the less responsive the cells were when calcium was added. However, when these cells were challenged with anti-IgE, they released histamine normally. Thus, the desensitization was specific.

As specific desensitization progresses, more and more IgE dimers are transformed from an active to an inactive state. Still, even after 40 min. there is significant histamine release, indicating that some IgE dimers have remained active for this entire period. From studies on cells from two different individuals we estimated that the average time τ to inactivate an IgE dimer is 10^2–$10^3 s$ (Dembo et al., 1982). This was considerably slower than the inactivation time we determined for crosslinks formed by the rapidly equilibrating bivalent hapten, $(BPO)_2$. For $(BPO)_2$, we found $\tau = 25s$ (Dembo et al., 1979b). However, it should be remembered that we were not using a cell line, but basophils from a few donors. There is insufficient data to tell whether the result that IgE dimers are more resistant to specific desensitization than $(BPO)_2$ is true for all human basophils.

With a ligand like $(BPO)_2$, we expect crosslinks to constantly form and break up. In the experiment with IgE-dimers outlined in Figure 15, no crosslinks formed during the desensitization period because we blocked all free Fc_ϵ receptors. This experiment, therefore, demonstrates that the constant formation of crosslinks is not required to generate either a signal for release or for specific desensitization.

The type of desensitization a basophil will undergo depends, at least in part, on the surface concentration of IgE (Dembo and Goldstein, 1980). MacGlashan and Lichtenstein (1981) showed that as the amount of specific IgE on the basophil surface is increased, the type of desensitization goes through a smooth transition from specific to nonspecific.

The minimum size IgE aggregate that is required to induce nonspecific desensitization is unknown. As yet, nonspecific desensitization has not been achieved with

IgE dimers. This may be because high enough concentrations of cell surface dimers have not been reached through passive sensitization, or it may be that aggregates larger than dimers are required to induced nonspecific desensitization.

The following suggests that the two types of desensitization turn off different parts of the calcium transport mechanism that is activated when IgE is crosslinked. Specifically desensitized human basophils release normally to non-crossreacting antigens or anti-IgE (Sobotka et al., 1979). Nonspecifically desensitized human basophils release normally when exposed to the calcium ionophore A23187 (Lichtenstein, 1975) or C5a (Siraganian and Hook, 1976). Thus, the degranulation apparatus of the cell is not impaired by desensitization. Ishizaka et al. (1985) have shown that the calcium influx that is normally triggered by crosslinking IgE is blocked in mouse mast cells that have been desensitized.

MODELS FOR DESENSITIZATION

As we have seen, crosslinking is required to trigger degranulation, as well as specific and nonspecific desensitization. Both types of desensitization involve the blocking of calcium transport in different ways. Specific desensitization is a local phenomenon rendering nonfunctional only those Fc_ϵ receptors that have been crosslinked. Nonspecific desensitization is a global phenomenon rendering all Fc_ϵ receptors incapable of participating in the transport of calcium. The mechanisms for both types of desensitization are unknown. Two obvious possibilities that can explain nonspecific desensitization are that (a) crosslinking causes an essential intermediate for calcium transport to be used up, or (b) crosslinking causes an inhibitor of calcium transport to be produced. An inhibitor model has recently been proposed by Oliver et al. (1978). We have developed a mathematical model that can predict the time course of histamine release when basophils with any surface IgE concentration are exposed to antigen (Dembo and Goldstein, 1980; Goldstein and Dembo, 1984). In the model, nonspecific desensitization occurs because an essential intermediate for calcium transport is irreversibly degraded. We proposed that that the influx of calcium is mediated by the production of short-lived channels formed from an inactive precursor, which we called a calcium-gating factor, interacting with crosslinked Fc_ϵ receptors, i.e., IgE aggregates. Once these channels form, they spontaneously decay, using up the available precursor.

Specific desensitization functionally removes crosslinked IgE from the cell surface, although physically they seem to remain (MacGlashan et al., 1983). To explain specific desensitization, we considered two models. In the first model (Dembo et al., 1979a; Dembo and Goldstein, 1980; Dembo et al., 1982), we assumed that the specific deactivation of crosslinked IgE-Fc_ϵ receptor complexes was independent of the decay of the gating factor, i.e., the mechanisms of specific and nonspecific desensitization were uncoupled. We modeled the process as if an enzyme existed that recognized crosslinked IgE-Fc_ϵ receptor complexes and rendered them incapable of

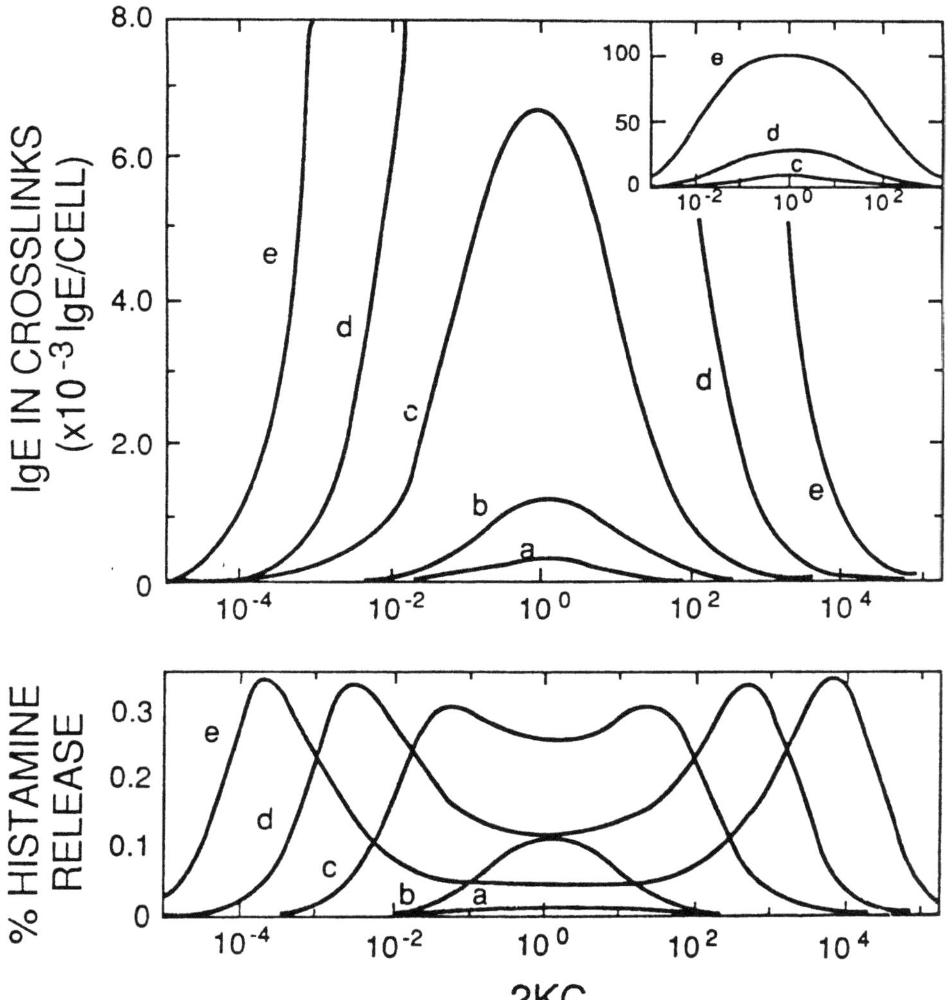

FIGURE 16 The predicted relation of histamine release to crosslinking for sensitized basophils exposed to a bivalent hapten with equilibrium constant K and concentration C (Dembo and Goldstein, 1980). Curves a, b, c, d and e correspond to basophils with 1×10^3, 3×10^3, 1×10^4, 3×10^4 and 1×10^5 IgE/cell.

forming calcium channels. In the second model, we assumed specific and nonspecific desensitization were coupled in such a way that when a calcium channel decays, both the calcium-gating factor and the Fc_ϵ receptors involved in the complex are inactivated (Goldstein and Dembo, 1984).

In both models of specific desensitization, when there are small amounts of specific IgE on the cell surface, negligible amounts of gating factor will be used up during histamine release. The final amount of histamine release, if the reaction is allowed to go to completion, will be determined by the competition between release and specific desensitization. If calcium is not present, but added at a later time, the model predicts that the cells will exhibit specific desensitization. Also, in this low IgE surface concentration limit, the histamine release curve will have a very similar shape to the crosslinking curve, as has been seen (Dembo et al., 1978; MacGlashan et al., 1985).

To understand the role of specific desensitization, at least in the model, it is useful to consider a hypothetical mutant basophil that lacks the ability to specifically desensitize. Such a cell would be unstable when there was a small amount of specific IgE on its surface in the sense that any amount of crosslinking would cause the cell to totally degranulate. If a desensitization experiment were done under these conditions, and one waited long enough, nonspecific desensitization would be observed. Interestingly, if there were large amounts of specific IgE on the cell surface, the cell would behave normally.

In the model, the rate of decay of the calcium-gating factor is proportional to the amount of crosslinked IgE on the cell surface. When this becomes large, the decay will become very rapid and release will become negligible. This dependence of the decay rate on the crosslinked IgE concentration accounts for the transition from specific to nonspecific desensitization.

In Figure 15 are the predicted relations between histamine release and crosslinking. The model predicts that at low specific IgE concentrations ($\leq$ 3000 IgE/cell) the histamine release curve has roughly the same shape as the crosslinking curve. However, as the specific IgE concentration is increased, a threshold is reached beyond which further increases in crosslinking reduces histamine release. This is because in the model, the rate of release saturates with increasing crosslinks, but the rate of nonspecific desensitization does not. This causes a depression in the histamine release curve centered about the hapten concentration at which crosslinking is optimal. As the amount of specific IgE is increased further, we predict broader regions of reduced histamine release. For curves such as d and e, the flat central region of the curves may be so broad that it becomes impossible to observe the predicted rise in the histamine release curve at very high hapten concentrations. Note how all the shapes of histamine release curves seen in Figures 11 and 12 are predicted in Figure 15.

From Figure 15, it can be seen that the model predicts the two types of antigen excess inhibition that have been observed. The model predicts that they should be correlated with the two types of desensitization. It predicts that when the amount of specific IgE on the cell surface is small, cells will desensitize specifically and exhibit type I antigen excess inhibition, while when the amount of specific IgE on the cell surface is large cells will desensitize nonspecifically and exhibit type II desensitization. Recently type II antigen excess inhibition has been observed with multivalent antigens on RBL cells (Robertson et al., 1986; Oliver et al., 1987). Our model leads us to predict that under these same conditions these multivalent antigens will induce the nonspecific desensitization of RBL cells.

CONCLUSION

The existence of basophils and mast cells, with specific IgE bound to their Fc_ϵ receptors, has allowed both experimental and theoretical biologists to test their ideas about the role of antibody aggregation in cell activation and desensitization. Because, for these cells, activation and desensitization are rapid (minutes) and require no accessory cells or growth factors, this is one of the simplest systems for studying such processes. For the theorist, there is more detailed data on these cells than on any other cells in the immune system. The models that have been developed so far are still quite naive. The dramatic progress made by experimentalist in unraveling this system serves as a challenge to theorists to develop realistic models that can explain what has been observed and make new testable predictions.

ACKNOWLEDGEMENTS

I would like to thank Carla Wofsy for her critical comments concerning the manuscript and her helpful discussions.

This work was supported by National Institutes of Health Grant GM35556 and by the United States Department of Energy.

REFERENCES

1. Basciano, L. K., E. H. Berenstein, L. Kmak and R. P. Siraganian (1986), "Monoclonal Antibodies that Inhibit IgE Binding," *J. Biol. Chem.* **261**, 1823–1831.
2. Baxter, J. H. and R. Adamik (1975), "Control of Histamine Release: Effects of Various Conditions on Rate of Release and Rate of Cell Desensitization," *J. Immunol.* **114**, 1034–1041.
3. Becker, K. E., T. Ishizaka, H. Metzger, K. Ishizaka and P. M. Grimley (1973), "Surface IgE on Human Basophils during Histamine Release," *J. Exp. Med.* **138**, 394–409.
4. Cambier, J. C., L. B. Justement, M. K. Newell, Z. Z. Chen, L. K. Harris, V. M. Sandoval, M. J. Klemsz and J. T. Ranson (1987), "Transmembrane Signals and Intracellular 'Second Messengers' in the Regulation of Quiescent B-Lymphocyte Activation, *Immunol. Rev.* **95**, 37–57.
5. Conrad, D. H., H. Bazin, A. H. Sehon and A. Froese (1975), "Binding Parameters of the Interaction between Rat IgE and Rat Mast Cell Receptors," *J. Immunol.* **114**, 1688–1691.
6. Conroy, M. C., N. F. Adkinson, Jr. and L. M. Lichtenstein (1977), "Measurement of IgE on Human Basophils: Relation to Serum IgE and Anti-IgE Induced Histamine Release," *J. Immunol.* **118**, 1317–1321.
7. DeLisi, C., and R. P. Siraganian (1979), "Receptor Cross-Linking and Histamine Release. II. Interpretation and Analysis of Anomalous Dose Response Patterns," *J. Immunol.* **22**, 2293–2299.
8. Dembo, M., and B. Goldstein (1978), "Theory of Equilibrium Binding of Symmetric Bivalent Haptens to Cell Surface Antibody: Application to Histamine Release from Basophils," *J. Immunol.* **121**, 345–353.
9. Dembo, M., and B. Goldstein (1980), "A Model of Cell Activation and Desensitization by Surface Immunoglobin: the Case of Histamine Release from Human Basophils," it Cell **22**, 59–67.
10. Dembo, M., B. Goldstein, A. K. Sobotka and L. M. Lichtenstein (1978), "Histamine Release Due to Bivalent Penicilloyl Haptens: Control by the Number of Cross-Linked IgE Antibodies on the Basophil Plasma Membrane," *J. Immunol.* **121**, 354–358.
11. Dembo, M., B. Goldstein, A. K. Sobotka and L. M. Lichtenstein (1979a), "Degranulation of Human Basophils: Quantitative Analysis of Histamine Release and Desensitization, Due to a Bivalent Penicilloyl Hapten," *J. Immunol.* **123**, 1864–1872.
12. Dembo, M., B. Goldstein, A. K. Sobotka and L. M. Lichtenstein (1979b), "Histamine Release due Bivalent Penicilloyl Haptens: the Relation of Activation and Desensitization of Basophils to Dynamic Aspects of Ligand Binding to Cell Surface Antibody," *J. Immunol.* **122**, 518–528.

13. Dembo, M., A. Kagey-Sobotka, L. M. Lichtenstein and B. Goldstein (1982), "Kinetic Analysis of Histamine Release due to Covalently Linked IgE Dimers," *Mol. Immunol.* **19**, 421–434.
14. Drobis, J. D. and R. P. Siraganian (1976), "Histamine Release from Cultured Human Basophils: Lack of Histamine Resynthesis after Antigen Release," *J. Immunol.* **117**, 1049–1053.
15. Erickson, J., P. Kane, B. Goldstein, D. Holowka and B. Baird (1986), "Cross-Linking of IgE-Receptor Complexes at the Cell Surface: a Fluorescence Method for Studying the Binding of Monovalent and Bivalent Haptens to IgE," *Mol. Immunol.* **23**, 769–781.
16. Fewtrell, C. (1985), "Activation and Desensitization of Receptors for IgE on Tumor Basophils," *Calcium in Biological Systems*, Eds. R. P. Rubin, G. B. Weiss and J. W. Putney, Jr. (New York: Plenum Press), 129–136.
17. Fewtrell, C., and H. Metzger (1980), "Larger Oligomers of IgE are More Effective than Dimers in Stimulating Rat Basophilic Leukemia Cells," *J. Immunol.* **125**, 701–710.
18. Foreman, J. C. and L. G. Garland (1974), "Desensitization in the Process of Histamine Secretion Induced by Antigen and Dextran," *J. Physiol.* **239**, 381–391.
19. Foreman, J. C. and J. L. Mongar (1973), "The Role of the Alkaline Earth Ions in Anaphylactic Histamine Secretion," *J. Physiol.* **224**, 753–769.
20. Furuichi, K., J. Rivera and C. Isersky (1985), "The Receptor for Immunoglobulin E on Rat Basophilic Leukemia Cells: Effect of Ligand Binding on Receptor Expression," *Proc. Natl. Acad. Sci. USA* **82**, 1522–1525.
21. Garcia, R., V. Ureña, C. Lahoz and F. Ortiz (1978), "Quantitation of IgE on the Basophil Surface in Atopic Patients," *Int. Arch. Allergy Appl. Immunol.* **56**, 463–469.
22. Goldstein, B. and M. Dembo (1984), "The IgE Mediated Activation and Desensitization of Human Basophils," *Cell Surface Dynamics: Concepts and Models*, Eds. A. S. Perelson, C. DeLisi and F. W. Wiegel (New York: Marcel Dekker), 331–382s.
23. Goldstein, B., M. Dembo and F. J. Malveaux (1979), "Some Quantitative Sspects of the Passive Sensitization of Human Basophils," *J. Immunol.* **122**, 830–833.
24. Goldstein, B., M. Dembo, A. K. Sobotka and L. M. Lichtenstein (1979), "Some Invariant Properties of IgE-Mediated Basophil Activation and Desensitization," *J. Immunol.* **123**, 1873–1882.
25. Goldstein, B., and A. S. Perelson (1984),. "Equilibrium Theory for the Clustering of Bivalent Cell Surface Receptors by Trivalent Ligands: Application to Histamine Release from Basophils," *Biophys. J.* **45**, 1109–1123.
26. Grant, J. A. and L. M. Lichtenstein (1974),. "Release of Slow-Reacting Substance of Anaphylaxis from Human Leukocytes," *J. Immunol.* **112**, 897–904.

27. Hill, T. L. (1983), "Length Dependence of Rate Constants for End-to-End Association and Dissociation of Equilibrium Linear Aggregates," *Biophys. J.* **44**, 285–288.
28. Isersky, C., J. Rivera, S. Mims and T. J. Triche (1979), "The Fate of IgE Bound to Rat Basophilic Leukemia Cells," *J. Immunol.* **122**, 1926–1936.
29. Isersky, C., J. Rivera, D. M. Segal and T. Triche (1983), "The Fate of IgE Bound to Rat Basophilic Leukemia Cells. II. Endocytosis of IgE Oligomers and Effect on Receptor Turnover," *J. Immunol.* **131**, 388–396.
30. Ishizaka, T., K. Ishizaka, S. G. O. Johansson and H. Bennich (1969), "Histamine Release from Human Leukocytes by Anti-γE Antibodies," *J. Immunol.* **102**, 884–892.
31. Ishizaka, T., A. R. Sterk, M. Daeron, E. L. Becker and K. Ishizaka (1985), "Biochemical Analysis of Desensitization of Mouse Mast Cells," *J. Immunol.* **135**, 492–501.
32. Kagey-Sobotka, A. K., M. Dembo, B. Goldstein, H. Metzger and L. M. Lichtenstein (1981), "Qualitative Characteristics of Histamine Release from Human Basophils by Covalently Cross-linked IgE," *J. Immunol.* **127**, 2285–2286.
33. Kane, P., J. Erickson, C. Fewtrell, B. Baird and D. Holowka (1986), "Cross-Linking of IgE-Receptor Complexes at the Cell Surface: Synthesis and Characterization of a Long Bivalent Hapten that is Capable of Triggering Mast Cells and Rat Basophilic Leukemia Cells," *Mol. Immunol.* **23**, 783–790.
34. Kazimierczak, W., H. L. Meier, D. W. MacGlashan, Jr. and L. M. Lichtenstein (1984), "An Antigen-Activated DFP-Inhibitable Enzyme Controls Basophil Desensitization," *J. Immunol.* **132**, 399–405.
35. Kulczycki, A., Jr., and Metzger, H. (1974), "The Interaction of IgE with Rat Basophilic Leukemia Cells: II. Quantitative Aspects of the Binding Reaction," *J. Exp. Med.* **140**, 1676–1695.
36. Levy, D. A. and A. G. Osler (1966), "Studies on the Mechanism of Hypersensitivity Phenomena: XIV. Passive Sensitization *In Vitro* of Human Leukocytes to Ragweed Pollen Antigen," *J. Immunol.* **107**, 1122–1130.
37. Lichtenstein, L. M. (1971), "The Immediate Allergic Response: *In Vitro* Separation of Antigen Activation, Decay and Histamine Release," *J. Immunol.* **107**, 1122–1130.
38. Lichtenstein, L. M. (1975), "The Mechanism of Basophil Histamine Release Induced by Antigen and by Calcium Ionophore A23187," *J. Immunol.* **114**, 1692–1699.
39. Lichtenstein, L. M. and A. G. Osler (1964), "Studies of the Mechanisms of Hypersensitivity Phenomena: IX. Histamine Release from Human Leukocytes by Ragweed Pollen Antigen," *J. Exp. Med.* **120**, 507–530.
40. Liu, f. T., J. W. Bohn, E. L. Ferry, H. Yamamoto, C. A. Molinaro, L. A. Sherman, N. R. Klinman and D. H. Katz (1980), "Monoclonal Dinitrophenyl-Specific Murine IgE Antibody: Preparation, Isolation, and Characterization," *J. Immunol.* **124**, 2728–2736.

41. MacGlashan, D. W., Jr., M. Dembo and B. Goldstein (1985), "Test of a Theory Relating to the Cross-Linking of IgE Antibody on the Surface of Human Basophils," *J. Immunol.* **135**, 4129–4134.
42. MacGlashan, D. W., Jr. and L. M. Lichtenstein (1981), "The Transition from Specific to Nonspecific Desensitization in Human Basophils," *J. Immunol.* **127**, 2410–1414.
43. MacGlashan, D. W., Jr. and L. M. Lichtenstein (1983), "Studies of Antigen Binding on Human Basophils I. Antigen Binding and Functional Consequences," *J. Immunol.* **130**, 2330-2336.
44. MacGlashan, D. W., Jr. and L. M. Lichtenstein (1987), "Basic Characteristics of Human Lung Mast Cell Desensitization," *J. Immunol.* **139**, 501–505.
45. MacGashan, D. W., Jr., M. Mogowski and L. M. Lichtenstein (1983a), "Studies of Antigen Binding on Human Basophils II. Continued Expression of Antigen-Specific IgE during Antigen-Induced Desensitization," *J. Immunol.* **130**, 2337–2342.
46. MacGlashan, D. W., Jr., R. P. Schleimer and L. M. Lichtenstein (1983b), "Qualitative Differences between Dimeric and Trimeric Stimulation of Human Basophils," *J. Immunol.* **130**, 4–6.
47. MacGlashan, D. W., Jr., S. P. Peters, J. Warner and L. M. Lichtenstein (1986), "Characteristics of Human Basophil Sulfidopeptide Leukotriene Release: Releasability Defined as the Ability of the Basophil to Respond to Dimeric Cross-Links," *J. Immunol.* **136**, 2231–2239.
48. Magro, A. M. and A. Alexander (1974), "Histamine Release: *In Vitro* Studies of the Inhibitory Region of the Dose-Response Curve," *J. Immunol.* **112**, 1762–1765.
49. Magro, A. M. and H. Bennich (1977), "Concanavalin A Induced Histamine Release from Human Basophils *In Vitro*," *Immunol* **33**, 51–58.
50. Malveaux, F. J., M. C. Conroy, N. F. Adkinson, Jr. and L. M. Lichtenstein (1978), "IgE Receptors on Human Basophils: Relationship to Serum IgE Concentration," *J. Clin. Invest.* **62**, 176–181.
51. Marone, G., A. Kagey-Sobotka and L. M. Lichtenstein (1981), "IgE-Mediated Histamine Release from Human Basophils: Differences between Antigen E- and Anti-IgE Induced Secretion," *Int. Archs. Allergy Appl. Immun.* **65**, 339–348.
52. McKenzie, D. S. (1976), "Polymers and Scaling," *Phys. Reports* **27**, 35–88.
53. Mendoza, G. and H. Metzger (1976), "Disparity of IgE Binding between Normal and Tumor Mouse Mast Cells," *J. Immunol.* **117**, 1573–1578.
54. Menon, A. K., D. Holowka, W. W. Webb and B. Baird (1986a), "Clustering, Mobility, and Triggering Activity of Small Oligomers of Immunoglobulin E on Rat Basophilic Leukemia Cells," *J. Cell Biol.* **102**, 534–540.
55. Menon, A. K., D. Holowka, W. W. Webb and B. Baird (1986b), "Crosslinking of Receptor-Bound Immunoglobulin E to Aggregates Larger than Dimers Leads to Rapid Immobilization," *J. Cell Biol.* **102**, 541–550.

56. Metzger, H., G. Alcaraz, R. Hohman, J.-P. Kinet, V. Pribluda and R. Quarto (1986), "The Receptor with High Affinity for Immunoglobulin E," *Ann. Rev. Immunol.* **4**, 419–470.
57. Oliver, J. M., JC. Seagrave, R. F. Stump, J. R. Pfeiffer and G. G. Deanin (1987), "Signal Transduction and Cellular Response in RBL-2H3 Mast Cells," *Progress in Allergy* (in press).
58. Osler, A. G., L. M. Lichtenstein and D. A. Levy (1968), "*In Vitro* Studies of Human Reaginic Allergy," *Adv. Immunol.* **8**, 183–231.
59. Pecht, I. and A. Corcia (1987), "Stimulus-Secretion Coupling Mechanisms in Mast ells," *Biophys. Chem.* **26**, 291–301.
60. Perelson, A. S. (1980) Receptor Clustering on a Cell Surface II. Theory of Receptor Cross-Linking by Ligands Bearing Two Chemically Distinct Functional Groups," *Math. Biosci.* **49**, 87–110.
61. Perelson, A. S., and C. DeLisi (1980), "Receptor Clustering on a Cell Surface I. Theory of Receptor Cross-Linking by Ligands Bearing Two Chemically Identical Groups," *Math. Biosci.* **48**, 71–110.
62. Pfeiffer, J. R., JC. Seagrave, B. H. Davis, G. G. Deanin and J. M. Oliver (1985), "Membrane and Cytoskeletal Changes Associated with IgE-Mediated Serotonin Release from Rat Basophilic Leukemia Cells," *J. Cell Biol.* **101**, 2145–2155.
63. Pruzansky, J. J., L. C. Grammer, R. Patterson and M. Roberts (1983), "Dissociation of IgE from Receptors on Human Basophils. I. Enhanced Passive Sensitization for Histamine Release," *J. Immunol.* **131**, 1949–1953.
64. Quarto, R., J.-P. Kinet and H. Metzger (1985), "Coordinate Synthesis and Degradation of the α-, β- and γ-Subunits of the Receptor for Immunoglobulin E," *Mol. Immunol.* **22**, 1045–1051.
65. Robertson, D., D. Holowka and B. Baird (1986), "Cross-Linking of Immunoglobulin E-Receptor Complexes Induces Their Interaction with the Cytoskeleton on Rat Basophilic Leukemia Cells," *J. Immunol.* **136**, 4565–4572.
66. Schweitzer-Stenner, R., A. Licht, I. Lüscher and I. Pecht (1987), "Oligomerization and Ring Closure of Immunoglobulin E Class Antibodies by Divalent Haptens," *Biochem.* **26**, 3602–3612.
67. Segal, D. M., J. D. Taurog and H. Metzger (1977), "Dimeric Immunoglobulin E Serves as the Unit Signal for Mast Cell Degranulation," *Proc. Natl. Acad. Sci. USA* **74**, 2993–2997.
68. Seagrave, JC., G. G. Deanin, J. C. Martin, B. H. Davis and J. M. Oliver (1987), "DNP-Phycobiliproteins, Fluorescent Antigens to Study Dynamic Properties of Antigen-IgE-Receptor Complexes on RBL-2H3 Rat Mast Cells," *Cytometry* **8**, 287–295.
69. Siraganian, R. P. and K. A. Hazard (1979), "Mechanisms of Mouse Mast Cell Activation and Inactivation for IgE-Mediated Histamine Release," *J. Immunol.* **122**, 1719–1724.
70. Siraganian, R. P. and W. A. Hook (1976), "Complement Induced Histamine Release from Human Basophils: II. Mechanism of Histamine Release Action," *J. Immunol.* **116**, 639–646.

71. Siraganian, R. P., W. A. Hook and B. B. Levine (1975), "Specific *In Vitro* Histamine Release from Basophils by Bivalent Haptens: Evidence for Activation by Simple Bridging of Membrane Bound IgE," *Immunochem.* **12**, 149–157.
72. Sobotka, A. K., M. Dembo, B. Goldstein and L. M. Lichtenstein (1979), "Antigen Specific Desensitization of Human Basophils," *J. Immunol.* **122**, 511–517.
73. Sterk, A. R. and T. Ishizaka (1982), "Binding Properties of IgE Receptors on Normal Mouse Mast Cells," *J. Immunol.* **128**, 838–843.
74. Truesdell, C. (1945), "On a Function which Occurs in the Theory of the Structure of Polymers," *Annals of Math.* **46**, 144–157.
75. Wank, S. A,, C. DeLisi and H. Metzger (1983), "Analysis of the Rate-Limiting Step in a Ligand-Cell Receptor Interaction: the Immunoglobulin E System," *Biochem.* **22**, 954–959.
76. Weyer, A., J.-P. Dandeu, F. Marchand and B. David (1982), "*In Vitro* Histamine Release from Human Basophils Triggered by a Purified Allergen from *Dermatophagoides Farinae*: Bimodal Aspect of the Dose Response Curve," *Ann. Immunol. Inst. Pasteur* **133D**, 87–94.
77. Wofsy, C. (1980), "Analysis of a Molecular Signal for Cell Function in Allergic Reactions," *Math. Biosci.* **49**, 69–86.
78. Wofsy, C. and B. Goldstein (1980), "Theory of Equilibrium Binding of a Bivalent Ligand to Cell Surface Antibody: the Effect of Antibody Heterogeneity on Crosslinking," *J. Math. Biol.* **10**, 347–366.
79. Wofsy, C. and B. Goldstein (1987), "The Effect of Co-operativity on the Equilibrium Binding of Symmetric Bivalent Ligands to Antibodies: Theoretical Results with Application to Histamine Release from Basophils," *Mol. Immunol.* **24**, 151–161.
80. Wofsy, C., B. Goldstein and M. Dembo (1978), "Theory of Equilibrium Binding of Asymmetric Bivalent Haptens to Cell Surface Antibody: Application to Histamine Release from Basophils," *J. Immunol.* **121**, 593–601.

BARBARA BAIRD, JON ERICKSON, BYRON GOLDSTEIN,† PATRICIA KANE, ANANT K. MENON, DEBORAH ROBERTSON, and DAVID HOLOWKA
Cornell University, Baker Laboratory, Ithaca, New York 14853 and †Theoretical Division, University of California, Los Alamos National Laboratory, Los Alamos, New Mexico 87545

Progress Toward Understanding the Molecular Details and Consequences of IgE-Receptor Crosslinking

INTRODUCTION

The high-affinity receptor for immunoglobulin E (IgE) on mast cells and basophils mediates antigen-induced cellular degranulation to release histamine in the allergic response. Aggregation of receptors at the cell surface by multivalent antigen or by other means is known to be necessary to stimulate this process, but little is known about the mechanism by which this aggregation leads to a variety of intracellular changes that eventually result in degranulation.[1] This receptor serves as a model for other types of immunological receptors, and our laboratory's investigation has focused on features that highlight its central role in the signal transduction process. We have used primarily an experimental system of rat basophilic leukemia (RBL) cells[2] and monoclonal IgE of various specificities. Our main areas of investigation have been: (1) the way in which IgE binds to receptor, (2) the way in which IgE-receptor complexes bind multivalent antigen and are crosslinked, and (3) the way in which clustered IgE-receptor complexes interact with other cellular structures to generate a transmembrane signal. This paper will briefly summarize our contributions in the first and third areas and devote somewhat greater attention to our more recent results in the second area. Quite recently, we have begun to characterize

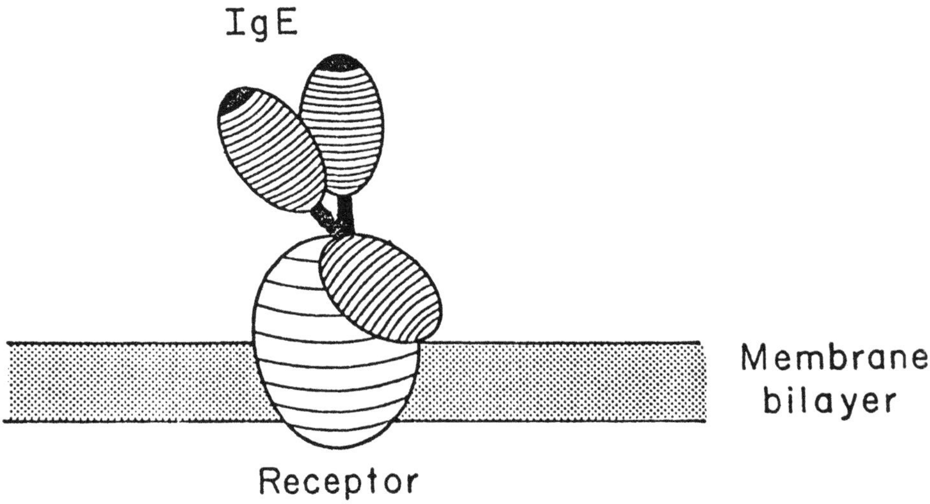

FIGURE 1 Model for the structure of IgE bound to receptor on the plasma membrane surface.

an antigen-induced desensitization phenomenon in this system, and some relevant aspects of these initial results are also mentioned.

STRUCTURAL FEATURES OF THE IgE-RECEPTOR COMPLEX

The receptor is a transmembrane protein containing 3 subunits.[1] The α subunit is exposed primarily on the outside of the cell and binds IgE in a tight monovalent interaction ($K_a > 10^{10} M^{-1}$). The β and disulfide-bonded γ_2 subunits are mostly inside the membrane and are exposed cytoplasmically. IgE has the typical Y-shaped structure of immunoglobulin with antigen binding sites at the ends of the two Fab segments and an Fc segment (containing $C_\epsilon 2$, $C_\epsilon 3$ and $C_\epsilon 4$ domains) that binds to receptor. As shown in Figure, 1 we have proposed that IgE assumes a bent conformation when it binds to receptors on the membrane surface. This model is consistent with a series of resonance energy-transfer measurements of distances between donor probes on receptor-bound IgE and acceptor probes at the membrane surface.[3-6] The model is also consistent with biochemical evidence that a crucial site for interaction is between the $C_\epsilon 2$ and $C_\epsilon 3$ domains.[7] Recently we used genetic engineering techniques to prepare a recombinant IgE with the $C_\epsilon 4$ domain replaced by the corresponding C-terminal domain of a human IgG_1 which does not bind to the high-affinity receptors on RBL cells. Initial studies show that this recombinant IgE binds as tightly as the native IgE,[8] providing further evidence that the $C_\epsilon 4$

domain does not interact appreciably with the receptor as suggested by the model. Substantial pharmaceutical interest in the critical components of the IgE-receptor interaction springs from ongoing efforts to design a small molecule with structural features similar to the IgE Fc binding domain to act as an antagonist that could be used in the treatment of allergies. Of more basic interest is the possibility that the hypothesized asymmetric structure of the IgE-receptor complex influences the way receptors come together in a triggering configuration.

INTERACTIONS OF CLUSTERED RECEPTORS ACCOMPANYING CELL TRIGGERING

The structural interactions that are most pertinent to the physiological response are those that occur after the IgE-receptors have been crosslinked. The minimal amount of crosslinking necessary was indicated by observations that small oligomers of IgE but not IgE monomers are capable of triggering cellular degranulation.[9] We now have several lines of evidence showing that other cellular components become associated with small clusters of receptors and may thereby affect their subsequent interactions. Initial observations were made with the use of fluorescently labeled, covalent oligomers of IgE (IgE_n, where $n \leq 6$)and fluorescence microscopy.[10] Fluorescent IgE monomers bound to RBL cell receptors appear uniformly distributed over the cell surface, and this appearance persists indefinitely. Small IgE oligomers bound to receptors also appear uniformly distributed initially; small clusters of ≤ 6 cannot be resolved at this level of microscopy. However, after a few hours at 4°C and faster at higher temperatures, patches of fluorescence form on the cells representing clusters of 1,000–10,000 IgE receptor complexes. Internalization also occurs at the higher temperatures. Since no additional external crosslinker was added, these results suggest that the cell recognizes oligomeric receptors as being different than monomeric receptors and responds by pulling these small clusters together. It is unlikely that this dramatic patching phenomenon is necessary for the signal transduction leading to degranulation since the time period is too long, however, it may represent a long-term manifestation of an earlier recognition step that is critical for triggering.

The observed redistribution of receptors suggests engagement with cytoskeletal components. We could monitor these interactions more sensitively with the method of fluorescence photobleaching recovery which allows measurement of changes in lateral diffusion of membrane proteins. Monomeric IgE-receptor complexes have mobility properties typical for membrane proteins: a diffusion coefficient of about $3 \times 10^{-10} cm^2 s^{-1}$ and a mobile fraction of about 80%. We obtained the striking result that when IgE oligomers were bound to receptors, the mobile fraction became greatly reduced, although those receptors remaining mobile had about the same diffusion coefficient ([11]; Figure 2). The immobilization did not require visible

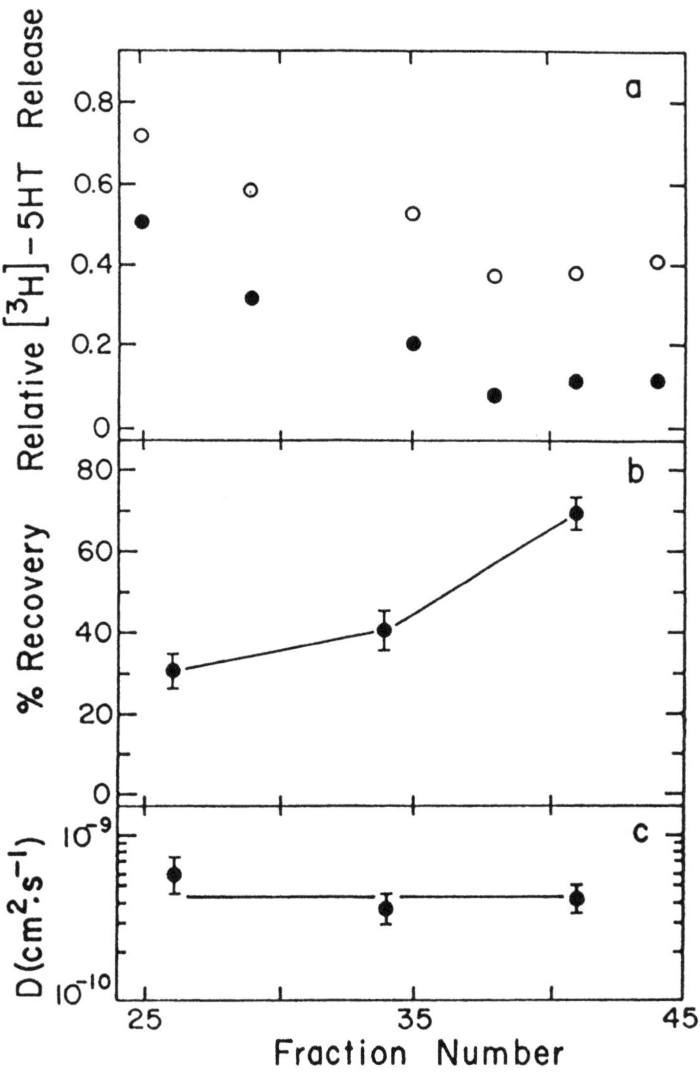

FIGURE 2 (a) [^{3}H]-5HT ([^{3}H]-serotonin) release induced by 4 μg/ml (○) or 0.4 μg/ml (●) of oligomeric IgE fractions after gel filtration chromatography. Fraction 26 corresponds roughly to pentamers–hexamers and fraction 42 corresponds roughly to dimers–trimers. The data are expressed relative to release induced by unlabeled monomeric IgE + DNP_{16}-BGG. This degranulation activity can be compared to fluorescence photobleaching recovery measurements of lateral diffusion: (b) mean percent recovery (R) and (c) diffusion coefficient (D) values for the same fractions. Bars represent SEM. Figure taken from ref. [11].

patching, and our interpretation of these results was that the receptors that had become crosslinked into some oligomeric state became anchored while those less crosslinked maintained normal mobility. The dramatic reduction in mobility cannot be simply explained by size increase according to hydrodynamic theory which predicts that the diffusion coefficient of membrane proteins has a weak logarithmic dependence on molecular size.[12] Of primary interest is the relationship between this induced immobilization and the signal for degranulation. Figure 2 shows fractions from a gel filtration column with the size of the oligomeric IgE increasing with decreasing fraction number (roughly dimers to hexamers). As shown, there is a good correlation between reduced mobility and increased degranulation.

We could assess the generality of this correlation and apply more stringent tests on the basis of two known features of IgE receptor-mediated degranulation of RBL cells: (i) compared to trimers and higher oligomers, IgE dimers cause a poor response that is enhanced in the presence of D_2O,[13] and (ii) degranulation is initiated within minutes by multivalent antigen, and it is stopped immediately upon addition of excess monovalent ligand that acts to break up the crosslinked IgE-receptors or prevent new crosslinks from forming. We could distinguish and compare the properties of monomers, dimers, and higher oligomers with the monoclonal anti-IgE(Fc) antibody, A2.[14] Binding studies showed a single tight site and at least one weak site for A2 on receptor-bound IgE, such that low concentrations of A2 caused the formation of receptor dimers on the cell surface, and further addition of A2 caused crosslinking to form higher oligomers. As before, oligomers were immobilized and caused substantial degranulation. Dimers were nearly as mobile as monomers and caused very little degranulation, but in the presence of D_2O, dimer mobility was reduced and degranulation was doubled.[15] The rate and reversibility of the immobilization process could be assessed with photobleaching measurements of monoclonal anti-DNP IgE bound to receptors on RBL cells. At the earliest time measurable (2 min.) after addition of the multivalent antigen, $DNP_{16}BGG$, the crosslinked receptors were almost totally immobilized, and the receptors were completely re-mobilized equally rapidly after the subsequent addition of monovalent DNP-lysine.[15]

Other experiments showed that after a certain degree of external crosslinking, the receptors become detergent insoluble, further evidence that anchoring to the insoluble cytoskeleton does occur under these conditions. When cells bound with ^{125}I-IgE are treated with buffers containing nonionic detergents, only 5% of the radioactive counts sediment with the insoluble cytoskeleton, while after crosslinking externally with a polyclonal anti-IgE antibody, 80% co-sediment.[16] Multivalent antigens prepared by random conjugation of proteins with DNP (e.g., $DNP_{16}BGG$ and $DNP_{23}BSA$) can also cause detergent insolubility, although with some preparations it is necessary to add chemical crosslinkers, such as dithiobisproprionimidate, to stabilize the cytoskeletal interactions. Our preparation of $DNP_{23}BSA$ required no chemical crosslinkers, and we tested the relationship between insolubility and cell triggering. They are not parallel as shown in Figure 3. Insolubility apparently increases with the degree of crosslinking: it goes up with addition of more antigen and then starts to go down as added antigen begins to compete and break up the aggregates on the cell surface. The striking result with this antigen is that degranulation peaks before insolubility is detectable and goes down as crosslinking and

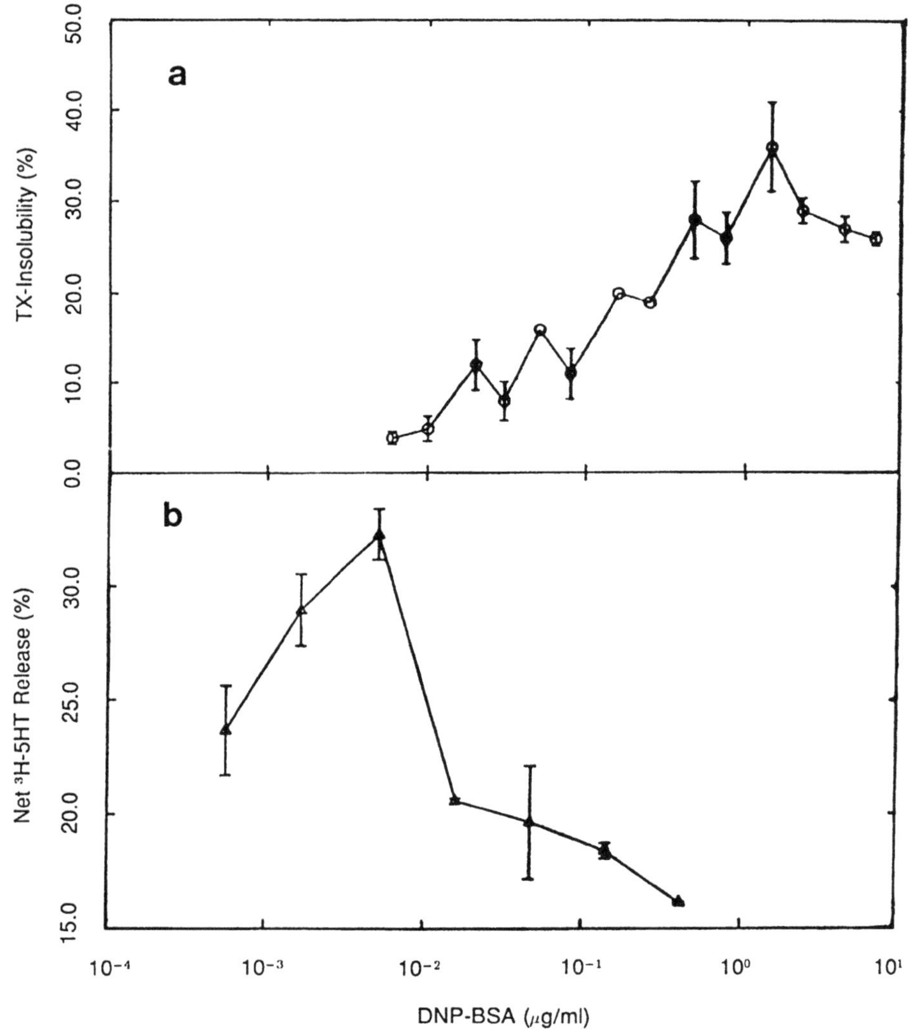

FIGURE 3 Dose-response curves for DNP_{23}-BSA comparing detergent (Triton X-100) insolubility (a) with ^{3}H-5HT release (b). Experiments were carried out as described in ref. [16].

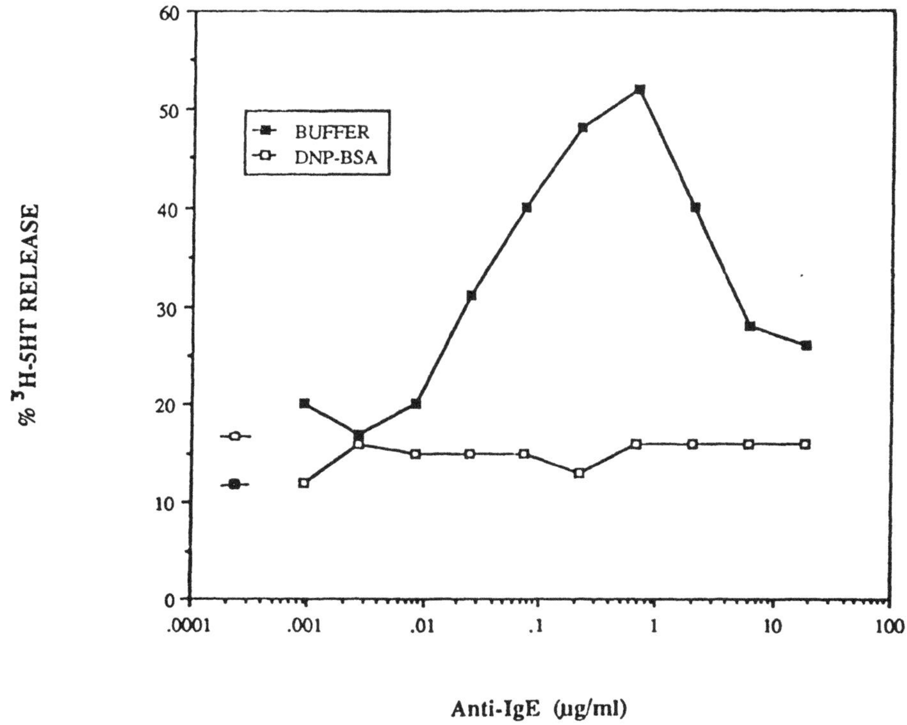

FIGURE 4 ^{3}H-5HT response to polyclonal anti-IgE after pre-incubation for 1 hr. at room temperature with 4 μg/ml DNP_{23}-BSA (□) or with buffer only (■). Isolated points to the left indicate the release in samples with no anti-IgE added. The release assays were carried out as described in ref. [36].

insolubility increases. This result is consistent with previous suggestions of others that excessive crosslinking can lead to a turn-off or desensitization in these cells.[17,18]

We have now been able to detect desensitization directly in a degranulation assay using a supra-optimal dose of DNP_{23}BSA. For the experiment shown in Figure 4, the cells were pre-incubated with a high dose of DNP_{23}BSA or with buffer only at room temperature. Then anti-IgE was added to trigger cells, and the cells were tested for degranulation at 37°C. Cells pre-incubated in buffer show a dose-dependent degranulation with anti-IgE, and with no anti-IgE, there is a low level of spontaneous release. For cells pre-incubated in DNP_{23}BSA with no anti-IgE added, there is some higher level of release than spontaneous release in the other samples. This is presumably due to triggering by DNP_{23}BSA during the pre-incubation. None of the doses of anti-IgE cause appreciable degranulation above background levels.

This and similar experiments provide strong evidence that high doses of this multivalent antigen turn off or desensitize these cells in their ability to respond to further crosslinking. We have some evidence that the desensitization can affect receptors that were not involved in the original crosslinks, but these experiments as well as those testing Ca^{2+}—and temperature—dependence are still in progress.

In the picture that is emerging from these studies, crosslinking of receptors causes interactions with other cellular components and leads to transmembrane signals that turn on as well as turn off the degranulation response. The possibly competing activities of crosslinked receptors in an immobilized state are probably complicated and may depend on other conditions such as time, temperature, and the supply of co-factors. Our results with a variety of crosslinking ligands have shown that freshly—and not excessively–crosslinked receptors are immobilized (as detected by photobleaching) under conditions where there is an active signal for degranulation. Crosslinking of receptors that is more advanced also corresponds to an immobilized state that is more likely to be detergent insoluble, and desensitization probably occurs in this state. It will be an interesting challenge to unravel the interplay between the turn-on and turn-off signals and the structural interactions that mediate them.

LIGAND BINDING AND CROSSLINKING

Most fundamental to initiating these transmembrane signals is the crosslinking of receptors. Hence, basic understanding of the system requires investigation of the critical molecular features of crosslink formation that control the nature and the magnitude of this signal. Constraints may be structural as well as kinetic and thermodynamic. Since it has been shown that $< 10\%$ of the receptors need to be bridged in an appropriate manner to trigger maximal degranulation,[19] the use of highly conjugated multivalent antigens, polyclonal anti-IgE, and other heterogeneous reagents might obscure rather specific requirements for productive crosslinking. More precise information can come from structurally defined, low-valency ligands that can be tested in correlated binding measurements and degranulation assays and can be analyzed with relatively simple theoretical treatments that exist and continue to be developed.[18]

The binding method we have developed utilizes fluorescence quenching to monitor the occupancy of DNP-containing ligands in the antibody combining sites of fluorescein isothiocyanate(FITC)-modified monoclonal anti-DNP IgE.[20] FITC-IgE in solution or bound to receptors on RBL cells is monitored in the fluorimeter cuvette, and after addition of the DNP-ligand, there is a drop in fluorescence that levels off. By comparison with the maximal quenching obtained with excess DNP-ligand at equilibrium, the fraction of antibody-combining sites occupied as a function of DNP concentration and other variables can be determined. Initially we tested several different bivalent DNP-ligands that were incapable of eliciting a degranulation response, even though they crosslinked anti-DNP IgE very efficiently in solution.[20] Since all of these were less than 30 Å end-to-end length, we suspected that crosslinking of cell surface receptor-bound IgE could not occur sufficiently. Therefore, we synthesized a bivalent ligand, N,N′-bis(DNP-aminocaproyl-L-tyrosine)-L-cystine ($(DCT)_2$-cys), that is greater than 40 Å in length, and we found it to be capable of triggering cellular degranulation with a dose-response curve similar in shape to that predicted by the equilibrium theory of Dembo and Goldstein.[21,22] However, the position of the maximum was shifted to higher concentrations than expected on the basis of purely equilibrium considerations, and the magnitude of the response was low and sometimes undetectable. Figure 5 shows the equilibrium data for binding between the monovalent DCT or the bivalent $(DCT)_2$-cys and monoclonal anti-DNP IgE bound to receptors on RBL cells.[20] The fraction of IgE sites that are bound is plotted against the total concentration of DNP ligand sites. The DCT curve can be fit to obtain the mono-

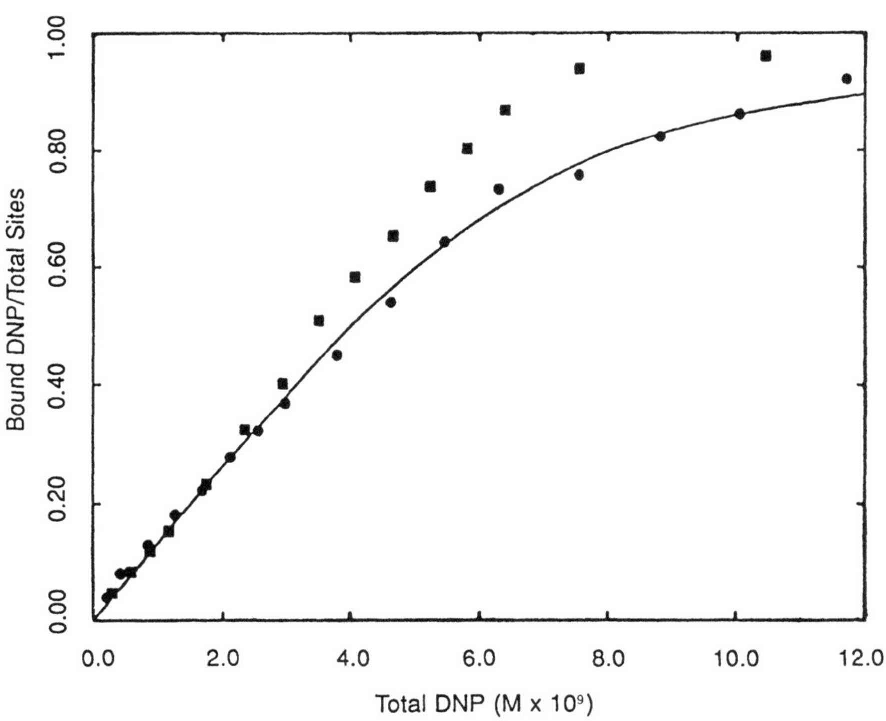

FIGURE 5 Equilibrium of DCT (•) and $(DCT)_2$-cys (▪) to FITC-anti-DNP IgE bound to receptors on RBL cells. Total IgE combining sites were 6.7 nM in both experiments. Figure taken from ref. [20].

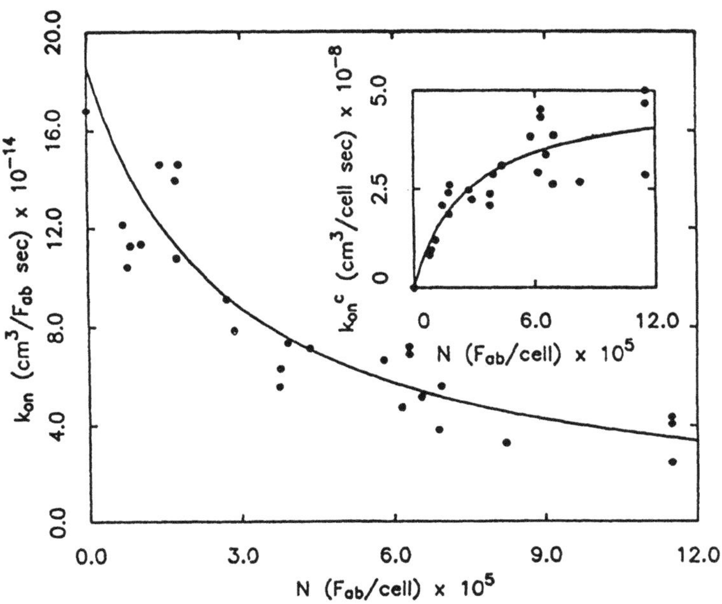

FIGURE 6 Accumulated k_{on} data and best fit on Eq. 1. *Inset*: same data plotted on a per cell basis. Solid line represents best fit of Eq. 1 multiplied by N. Figure taken from ref. [24].

valent intrinsic affinity constant of $K = 1.5 \times 10^9 M^{-1}$. The $(DCT)_2$-cys curve is different from the DCT curve. For both curves at low ligand concentrations, essentially all the DNP groups are bound. Therefore, at low concentrations of $(DCT)_2$-cys, crosslinking occurs readily since both ends of the ligand tend to be bound. Derivation of both K and a crosslinking affinity constant, K_x, from these data is possible in principle, and a simple treatment yields the same value as DCT for K and a lower bound for a bulk $K_x > 2 \times 10^9 M^{-1}$ (conversion of K_x to molecular and two-dimensional units requires knowledge of the surface density of IgE-receptor complexes). An exact analysis of the equilibrium binding is complicated because the bivalent ligand can form dimers, trimers, and longer linear chains as well as rings of all sizes, and consequently, more affinity constants are involved.[22] Numerical analyses of detailed $(DCT)_2$-cys equilibrium binding data similar to that shown in Figure 5 are in progress to obtain these constants,[23] but the basic result is clear that this bivalent ligand which does not trigger well is very efficient at crosslinking IgE-receptor complexes on the cell surface. Thus, it appears that there are additional requirements of the crosslinked species for effective delivery of a transmembrane signal.

A reasonable explanation for the poor triggering by $(DCT)_2$-cys is kinetic, i.e., that the binding is on and off and that the receptors do not remain crosslinked for some necessary period of time. In general, kinetic parameters may be more meaningful than equilibrium parameters in the regulation of RBL cell triggering, particularly since we know that a small amount of clustering results in interactions with other cellular components. We have been able to determine bimolecular rate constants as small as $10^8 M^{-1} sec^{-1}$ in the fluorimeter since we can easily measure the fluorescence signal of nanomolar concentrations of FITC-IgE. In initial studies, we measured the forward rate of binding of the monovalent ligand, DCT, as a function of IgE density on the cell surface.[24] Since IgE binds so tightly to receptors, this density was easily varied over a range of 10^4–10^6 Fab sites/cell by incubating the RBL cells with different concentrations of IgE or by using the same saturating concentration of IgE, but changing the ratio of anti-DNP IgE to IgE of a different specificity in the incubation mixture. As shown in Figure 6, the forward rate constant decreases as IgE density increases reflecting increasing competition amongst these surface-bound ligand receptors. The data are well fit by an equation derived from the theory of Berg and Purcell:[25]

$$k_{on}\left(\frac{cm^3}{receptor \cdot sec}\right) = \frac{4\pi D a \kappa_{on}}{4\pi D a + N\kappa_{on}} \qquad (1)$$

where D is the ligand diffusion coefficient, a is the cell radius, N is the number of receptors per cell and κ_{on} is the intrinsic "reaction-limited" ligand-receptor forward rate constant. At the zero density limit, $N = 0$, the extrapolated $k_{on} = \kappa_{on} = 19 \times 10^{-14} cm^3 site^{-1} sec^{-1} = 1.1 \times 10^8 M^{-1} sec^{-1}$ is the same as that measured for DCT and IgE in solution. The insert shows the same data expressed as k_{on} per cell (multiplied by N), and this reveals the other limit: at very high density k_{on} approaches the diffusion limited value of $4\pi D a = 5 \times 10^{-8} cm^3 cell^{-1} sec^{-1}$. If we assume $D = 10^{-5} cm^2 sec^{-1}$, then $a = 4 \mu m$, and both of these are quite reasonable values for this system. These data provide the first experimental verification of the Berg-Purcell theory which says that there is little additional kinetic advantage gained by a cell expressing more than $4\pi Da/\kappa_{on}$ receptors, typically less than 1% of its total surface area.[25]

We are in the process of measuring the rates of binding and crosslinking of cell-bound FITC-IgE by the bivalent ligand $(DCT)_2$-cys for comparison with DCT. Both the monovalent and the bivalent ligands show a similar initial rapid phase of fluorescence, quenching corresponding monovalent attachment. The $(DCT)_2$-cys shows a slower second phase of quenching, not observed with DCT, corresponding to subsequent crosslinking events.[26] A detailed analysis to determine rate constants is underway and takes advantage of the K and K_x affinity constants derived in the equilibrium studies described above.[23]

The dissociation kinetics of the DNP-ligands are being measured directly by adding an excess of soluble unlabeled anti-DNP as a competitive "sink" for the ligand as it dissociates from cell-bound fluorescein-IgE.[26,27] This approach appears to be valid since: (1) the rate of dissociation of DCT measured by this method is identical for ligand to solution IgE and cell-bound IgE (at low densities), and (2) the rate obtained for this case ($0.05\,sec^{-1}$) is the same as that calculated from the equilibrium and forward rate constants. The data for $(DCT)_2$-cys can be fit with an empirical equation containing parameters for the initial distribution of crosslinking vs. monovalently bound ligand and two dissociation rate constants. The faster rate constant is characteristic for the rate of monovalent ligand dissociation ($\leq 0.05\,sec^{-1}$), and the slower one represents a rate constant for the breaking of crosslinks ($\leq 5 \times 10^{-4} sec^{-1}$). As expected, the dissociation rates for both DCT and $(DCT)_2$-cys decrease with surface density of IgE-receptors, but the bivalent ligand is much more sensitive to this effect, since bivalent binding is possible and two bonds must break before it can escape from the crosslinked complex at the surface.

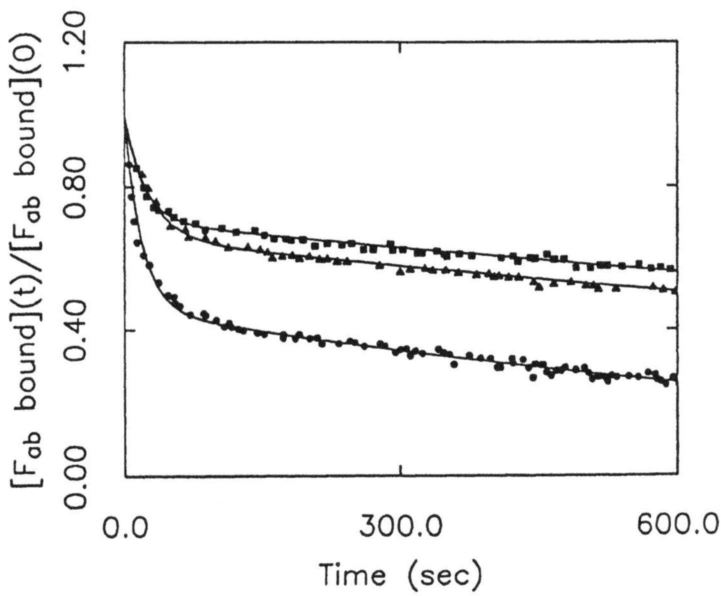

FIGURE 7 Dissociation rate of $(DCT)_2$-cys from cell-bound FITC-IgE after pre-incubation for 11 (•), 65 (▲), or 160 (■) min. and then addition of excess unlabeled anti-DNP IgE. The concentration of cell-bound FITC-Fab sites was 0.41 nM, added $(DCT)_2$-cys was 0.45 nM, and the added unlabeled Fab sites was 1.3 μM. The solid curves are theoretical predictions of a two-step dissociation reaction using the following best-fit parameters: the fraction of ligand bound bivalently (increases with pre-incubation time from .23 to .33 to .35), the reverse crosslinking constant (decreases with pre-incubation time from $5.0 \times 10^{-4} sec^{-1}$ to $2.2 \times 10^{-4} sec^{-1}$ to $2.0 \times 10^{-4} sec^{-1}$), and a reverse monovalent binding constant (remains about the same at $\sim .05\,sec^{-1}$). A saturation effect is observed.

The most interesting finding in these kinetic experiments is that pre-incubation of bivalent ligands with the cells for increasingly longer periods of time before the addition of the sink (5 minutes–2 hours) results in a significant decrease in the dissociation rate (Figure 7). This appears to be due to both an increase in the fraction of $(DCT)_2$-cys attached bivalently and a decrease in the crosslink-breaking rate constant. Notably, this time-dependent effect is not due to receptor internalization, and it does not occur on membrane vesicle preparations from RBL cells that are devoid of an intact cytoskeleton found in whole cells.[28] An attractive explanation is that the reduction of dissociation rate with time is due to the ability of the cells to recognize crosslinked receptors and to gather them into larger clusters such that the local density of those receptors is increased with time, progressively enhancing rebinding over dissociation. Support for this explanation comes from experiments where anti-IgE and anti-receptor antibodies as well as the lectin concanavalin A were used to crosslink IgE-receptor complexes at sites other that the antibody-combining site. This way of stabilizing receptor clustering caused a similar reduction in the dissociation rate of $(DCT)_2$-cys while the dissociation of the monovalent DCT did not change detectably. The cell-induced receptor clustering revealed by the dissociation experiments is consistent with that observed with IgE oligomers by fluorescence microscopy after long times. In general, dissociation kinetics should provide a very sensitive monitor of receptor distributions on a short time scale with a variety of ligands and under different conditions.

The thermodynamic- and kinetic-binding studies with $(DCT)_2$-cys imply that not only does crosslinking occur readily on the cell surface, but the crosslink lifetime is sufficiently long ($\geq 1000\, sec$) to initiate a triggering signal if crosslinking alone were required. An intriguing possibility is, under the particular conditions used during crosslink formation with this bivalent ligand, a desensitization signal competes with and often dominates the activation signal. Preliminary experiments have not yet revealed desensitization caused by $(DCT)_2$-cys, but more sensitive tests are underway.

Structural inadequacy is another possible explanation for the poor triggering ability of $(DCT)_2$-cys. Since the anti-DNP IgE (H1.26.82; [29]) we have used in these experiments has a very high affinity of binding corresponding to a deep antibody-combining site,[19] engulfment of a large part of the bivalent ligand may result in a receptor bridging distance that is shorter or more rigid than optimal for effective delivery of a transmembrane signal. We can address this possibility with monoclonal anti-DNP IgE of lower affinities.[29].

We have taken another approach to investigate distance constraints on productively crosslinked IgE-receptor complexes by making rigid bivalent ligands of different lengths.[31,32] These ligands were made by taking advantage of the tight binding between avidin and biotin ($K > 10^{16} M^{-1}$) and making complexes of 1-dinitropenyl-amino-12-biotinamidododecane (DNP-C_{12}-biotin) and avidin.[33] We found that, although avidin is tetravalent for biotin, two of the DNP-C_{12}-biotin derivatives bind relatively weakly and dissociate during dialysis. Several lines of evidence indicate that the remaining two DNP-C_{12}-biotin are on opposite faces of the avidin protein, approximately 41 Å apart. Thus, we could prepare (DNP-C_{12}-biotin$)_2$-avidin, and

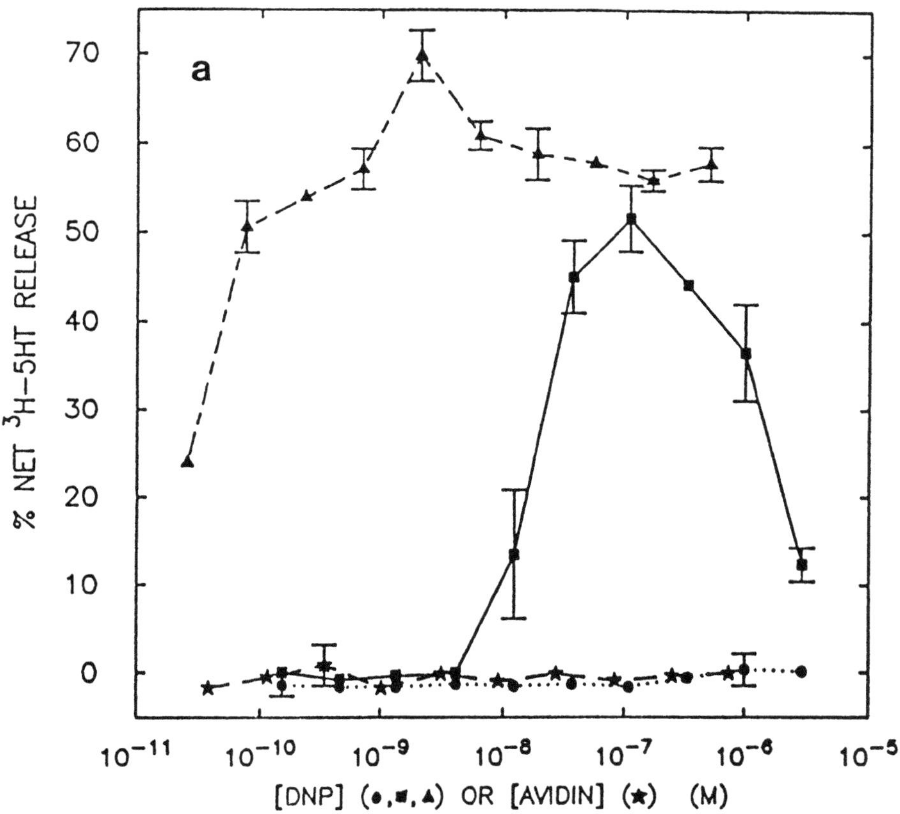

FIGURE 8a ^{3}H-5HT release stimulated by DNP-C_{12}-biotin (●) of avidin (⋆) alone compared to that stimulated by a (DNP-C_{12}-biotin$)_2$-avidin complex (■) and DNP_{16}-BGG (▲). 12.8% spontaneous release was subtracted. Figure taken from ref. [31].

ligand proved to be potent in triggering degranulation as shown in Figure 8a.[31] For this ligand, the maximum of the bell-shaped dose-response curve does occur as [DNP] $\sim K^{-1}$ ($K = 10^7 M^{-1}$ is the intrinsic affinity constant for binding to the anti-DNP IgE) as predicted by the theory of Dembo and Goldstein.[22] These ligands could be extended to increasing lengths with the use of bis-biotin-1,2-diaminododecane to crosslink two or more avidin molecules together before capping each end with a DNP-C_{12}-biotin conjugate.[32] A chemical crosslinking reagent was used to stabilize the polymers, and fluorescamine modification was used for detection in preparative polyacrylamide gel electrophoresis. A heterogeneous mixture of these polymers were separated by size to obtain fractions of monomers up to hexamers corresponding to lengths of about 40 to 240 Å between the end DNP groups. When tested in the biological assay, the avidin polymers, dimers through hexamers, all triggered a high level of degranulation. These levels were higher than the degranulation observed with the monomer species from this preparation (Figure 8b). This monomer is less effective than that prepared in the absence of the chemical crosslinker, and the difference may reflect subtle structural changes. In the case of the experiment shown in Figure 8b, it serves as a useful indicator that the potency of the polymers is not due to some artifact of the polymer preparation procedure or a small percentage of ligands within the fractions that have the DNP group on the same side of the avidin. The major result of these experiments is that the DNP

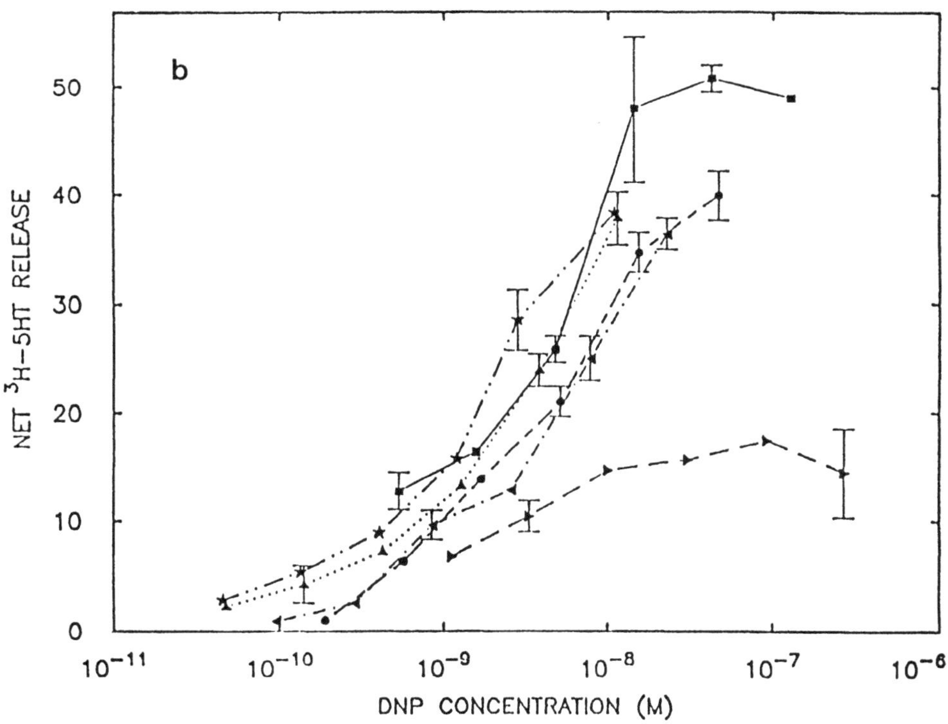

FIGURE 8b ^{3}H-5HT release stimulated by bivalent (DNP-C_{12}-biotin)$_2$-avidin polymers that had been electrophoretically separated according to size: monomers (▸), dimers (■), trimers (•), tetramers (◂), pentamers (⋆), and hexamers (▲). 6% spontaneous release was subtracted, and bars represent the range of duplicate samples. Figure taken from ref. 32.

haptenic groups can be separated by a rigid spacer > 240 Å and still be an effective triggering reagent. Since it now appears that the receptor-bound IgE molecule has only very restricted segmental flexibility,[34,35] our structural mapping studies[3–6] together with this result lead us to conclude that tethered IgE-receptor complexes need not interact directly with each other to deliver a productive transmembrane signal. Hence, the important interactions may occur between crosslinked receptors and other cellular components, such as the immobile components revealed in the fluorescence photobleaching recovery measurements.[10,14] However, since the fluorescence microscopy studies[9] and the bivalent ligand dissociation kinetics[26,27] have indicated that receptors crosslinked externally to a limited extent are recognized by the cell and, subsequently, gathered into larger clusters, the possibility remains that receptors that are not directly crosslinked do come into contact. To address this question, it should be possible to visualize clusters of IgE-receptor complexes that have been crosslinked by these bivalent avidin polymers on the cell surface using gold-labeled IgE and electron microscopy techniques.[17]

SUMMARY

Although crosslinking of IgE-receptors complexes by a variety of multivalent reagents stimulates a cellular response, some critical molecular features have begun to emerge with the use of structurally defined bivalent ligands. It appears that the directly crosslinked receptors do not have to come into physical contact to be productive and that some shorter crosslinks are less effective. The crosslinked receptors attach directly or indirectly with the cytoskeleton, and this is followed by further receptor clustering that is controlled by the cell. These dynamic structural interactions occur concomitantly with the events that activate as well as desensitize the cell. The roles they play in generating these transmembrane signals remain to be elucidated.

ACKNOWLEDGMENTS

We are grated to Mr. Gary Labrecque and Ms. Cheryl DiDomenico for help in carrying out some of these experiments. This research was supported by National Institutes of Health grants AI18306, AI18610, AI22449, GM3556 and by the United States Department of Energy.

REFERENCES

1. Metzger, H., G. Alcaraz, R. Hohman, J.-P. Kinet, V. Pribluda, and R. Quarto (1986), "The Receptor with High Affinity for Immunoglobulin E," *Ann. Rev. Immunol.* **4**, 419–470.
2. Barsumian, E. L., C. Isersky, M. G. Petrino, and R. Siraganian (1981), "IgE-Induced Histamine Release from Rat Basophilic Leukemia Cell Lines: Isolation of Releasing and Nonreleasing Clones," *J. Immunol.* **11**, 317–323.
3. Holowka, D., and B. Baird (1983), "Structural Studies on the Membrane-Bound Immunoglobulin E-Receptor Complex. 2. Mapping of Distances between Sites on IgE and the Membrane Surface," *Biochem.* **22**, 3475.
4. Baird, B., and D. Holowka (1985), "Structural Mapping of Fc Receptor-Bound Immunoglobulin E: Proximity to the Membrane Surface of the Antibody Combining Site and Another Site in the Fab Segments," *Biochemistry* **24**, 6252–6259.
5. Holowka, D., D. A. Conrad, and B. Baird (1985), "Structural Mapping of Membrane-Bound Immunoglobulin E-Receptor Complexes: Use of Monoclonal Anti-IgE Antibodies to Probe the Conformation of Receptor-Bound IgE," *Biochemistry* **24**, 6260–6267.
6. Baird, B., and D. Holowka, "Structural Mapping of Membrane-Associated Proteins: A Case Study of the IgE-Receptor Complex," *Spectroscopic Membrane Probes*, Ed. L. Loew (Boca Raton: CRC Press, Inc.), in press.
7. Perez-Montfort, R., and H. Metzger (1982), "Proteolysis of Soluble IgE-Receptor Complexes: Localization of Sites on IgE which Interact with the Fc Receptor," *Mol. Immunol.* **19**, 1113–1125.
8. Baird, B., R. Shopes, D. Holowka, and V. T. Oi, unpublished results.
9. Fewtrell, C. (1985), "Activation and Desensitization of Receptors for IgE on Tumor Basophils," *Calcium in Biological Systems*, Eds. G. Weiss, J. Putney and R. Rubin (New York: Plenum Press), 129–136.
10. Menon, A. K., D. Holowka, and B. Baird (1984), "Small Oligomers of Immunoglobulin E (IgE) Cause Large-Scale Clustering of IgE Receptors on the Surface of Rat Basophilic Leukemia Cells," *J. Cell Biol.* **98**, 577–583.
11. Menon, A. K., D. Holowka, W. W. Webb, and B. Baird (1986), "Clustering, Mobility, and Triggering Activity of Small Oligomers of Immunoglobulin E on Rat Basophilic Leukemia Cells," *J. Cell Biol.* **102**, 534–540.
12. Saffman, P. G., and M. Delbruck (1975), "Brownian Motion in Biological Membranes," *Proc. Natl. Acad. Sci. USA* **72**, 3111.
13. Fewtrell, C., and H. Metzger (1980), "Larger Oligomers of IgE are More Effective than Dimers in Stimulating Rat Basophilic Leukemia Cells," *J. Immunol.* **125**, 701.
14. Conrad, D., E. Studer, J. Gervasoni, and T. Mohanajumar (1983), "Properties of Two Monoclonal Antibodies Directed against the Fc and Fab Regions of Rat IgE," *Int. Arch. Allergy Appl. Immunol.* **70**, 352–360.

15. Menon, A. K., D. Holowka, W. W. Webb, and B. Baird (1986), "Crosslinking of Receptor-Bound Immunoglobulin E to Aggregates Larger than Dimers Leads to Rapid Immobilization," *J. Cell Biol.* **102**, 541–550.
16. Robertson, D., D. Holowka, and B. Baird (1986), "Crosslinking of Immunoglobulin E-Receptor Complexes Induces Their Interaction with the Cytoskeleton on Rat Basophilic Leukemia Cells," *J. Immunol* **136**, 4565–4572.
17. Oliver, J., J. C. Seagrave, and R. F. Stump (1988), this volume.
18. Goldstein, B. (1988), this volume.
19. Fewtrell, C., A. Kessler and H. Metzger (1979), "Comparative Aspects of Secretion from Tumor and Normal Mast Cells," *Adv. Inflam. Res.* **1**, 205–221.
20. Erickson, J., P. Kane, B. Goldstein, D. Holowka, and B. Baird (1986), "Crosslinking of IgE-Receptor Complexes at the Cell Surface: A Fluorescence Method for Studying the Binding of Monovalent and Bivalent Haptens to IgE," *Molec. Immunol.* **23**, 769–781.
21. Kane, P., J. Erickson, C. Fewtrell, B. Baird and D. Holowka (1986), "Crosslinking of IgE-Receptor Complexes at the Cell Surface: Synthesis and Characterization of a Long Bivalent Hapten that is Capable of Triggering Mast Cells and Rat Basohpilic Leukemia Cells," *Molec. Immunol.* **23**, 783–790.
22. Dembo, M., and B. Goldstein (1978), "Theory of Equilibrium Binding of Symmetric Bivalent Haptens to Cell Surface Antibody: Application of Histamine Release from Basophils," *J. Immunol.* **121**, 345–353.
23. Erickson, J., B. Goldstein, D. Holowka, and B. Baird, unpublished results.
24. Erickson, J., B. Goldstein, D. Holowka, and B. Baird (1988), "The Effect of Receptor Density on the Forward Rate Constant for Binding of Ligands to Cell Surface Receptors," *Biophys. J.* **52**, 657–662.
25. Berg, H. C., and E. M. Purcell (1977), "Physics of Chemoreception," *Biophys. J.* **20**, 193–219.
26. Erickson, J., B. Goldstein, D. Holowka, and B. Baird (1988), "Bivalent Ligand-Surface Immunoglobulin Surface Kinetics as a Probe of Cell Surface Dynamics," *Biophys. J. Abstracts*, in press.
27. Erickson, J., B. Goldstein, D. Holowka, and B. Baird, manuscript in preparation.
28. Holowka, D., and B. Baird (1983), "Structural Studies of the Membrane-Bound Immunoglobulin E-Receptor Complex. 1. Characterization of Large Plasma Membrane Vesicles from Rat Basophilic Leukemia Cells and Insertion of Amphipathic Fluorescent Probes," *Biochem.* **22**, 3466–3474.
29. Liu, F. T., J. W. Bohn, E. L. Ferry, H. Yamamoto, C. A. Molinaro, L. A. Sherman, N. R. Klinman, and D. Katz (1980), "Monoclonal Dinitrophenyl-Specific Murine IgE Antibody: Preparation, Isolation, and Characterization," *J. Immunol.* **124**, 2728–2736.

30. Schweitzer-Stenner, R., A. Licht, I. Luscher, and I. Pecht (1987), "Oligomerization and Ring Closure of Immunoglobulin E Class Antibodies by Divalent Haptens," *Biochem.* **26**, 3602–3612.
31. Kane, P. M., D. Holowka, and B. Baird (1988), "Crosslinking of IgE-Receptor Complexes by Rigid Bivalent Antigens > 200 Å in Length Triggers Cellular Degranulation," submitted for publication.
32. Kane, P., D. Holowka, and B. Baird (1988), "Characterization of Model Antigens Composed of Biotinylated Haptens Bound to Avidin," submitted for publication.
33. Green, N. M. (1975), "Avidin," *Adv. in Protein Chem.* **29**, 85.
34. Slattery, J., D. Holowka, and B. Baird (1985), "Segmental Flexibility of Receptor-Bound Immunoglobulin E," *Biochem.* **24**, 7810–7820.
35. Holowka, D., T. Wenzel, B. Baird, and L. Stryer, manuscript in preparation.
36. Baird, B., D. Sajewski, and S. Mazlin (1983), "A Microtiter Plate Assay using Cellulose Acetate Filters for Measuring Cellular [^{3}H]-Serotonin Release," *J. Immunol. Meth.* **64**, 365–375.

J. M. OLIVER, J. C. SEAGRAVE and R. F. STUMP
Department of Pathology, University of New Mexico, School of Medicine, Albuquerque, New Mexico

Two Distinct States of Crosslinked IgE Receptors That May Trigger and Terminate Secretion from RBL-2H3 Mast Cells

1. INTRODUCTION

In mast cells and basophils, the crosslinking of cell surface IgE receptors is the first of a series of events leading to the release of inflammatory mediators and vasoactive amines from granules. The recent development of several permanent mast cell lines has provided investigators with large numbers of cells to analyze the molecular mechanisms of this secretory response. One of the best characterized is RBL-2H3, a line of rat leukemia cells that provides an *in vitro* model for the study of mucosal mast cells.

The experiments described here address the mechanisms that couple (and uncouple) IgE receptor crosslinking to the activation of biochemical, biophysical and functional responses in RBL-2H3 cells. These data form the basis of a new model for the control of secretion from mast cells. Its central hypothesis is that crosslinked IgE receptors can exist in two distinct states at the cell surface, one that can interact with transducer and effector moieties to trigger secretion, and one that is either inactive or can inhibit secretion.

2. THE ACTIVATION OF RBL-2H3 CELLS

RBL-2H3 cells are a cultured line developed by Metzger, Siraganian and colleagues from a basophil-like solid tumor that occurred in a β-chloroethylamine-treated rat.[1–3] By electron microscopy, they appear as rather undistinguished mononuclear cells containing relatively small, randomly dispersed, more or less electron-dense granules.[4] They have been classified as derivatives of mucosal mast cells based on studies of the biochemical composition of these granules.[5,6]

The biochemical properties of the cell surface Fc_ϵ receptors (IgE receptors) of RBL-2H3 cells have been described in detail by Metzger and colleagues (reviewed in 7). The receptor consists of 1 α, 1 β, and 2 γ chains, where only the glycosylated chain, that contains the binding site for IgE, faces the external milieu. It binds hapten-specific IgE molecules with a remarkably high affinity ($K_d \cong 10^{-10}$M for most IgE's that have been studied). This binding occurs via only one face of the Fc region and it is associated with a specific bending of the IgE molecule (reviewed in 8).

The IgE receptor-mediated activation of RBL-2H3 cells can occur either directly or in a two-step reaction (Figure 1). Under physiological conditions, a two-step process controls secretion. The first event is the binding of IgE to IgE receptors that

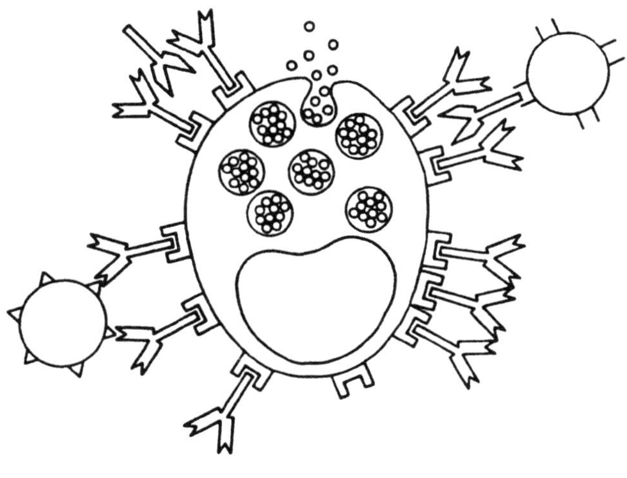

FIGURE 1 Steps in the activation of RBL-2H3 cells. Under physiological conditions, cells are prepared for secretion by binding IgE ( ) to their cell surface Fc_ϵ receptors ( ). Receptor crosslinking and secretion can then be induced by the addition of multivalent antigen ( ) or by the addition of polyclonal anti-IgE ( ). Anti-IgE bound to fixed cells can be labelled with protein A-gold particles ( ). Cells can also be activated directly by the addition of IgE multimers ( ).

exist at high concentration (around 3×10^5 receptors/cell) on the cell surface.[3] Most laboratories now use the dinitrophenyl (DNP)-specific monoclonal antibody, HI-DNP-ϵ-26.82 (anti-DNP-IgE) of F.-T. Liu and colleagues.[9] The second step, which causes secretion, is the crosslinking of these IgE-receptor complexes. Anti-DNP-IgE-receptor complexes can be crosslinked with a wide range of multivalent DNP-proteins,[10,11] including DNP-BSA (for biochemical assays), DNP-phycoerythrins (DNP-PhE; for studies by fluorescence microscopy and flow cytometry of the binding, distribution and mobility of antigen) and DNP-BSA-gold particles (for electron microscopy). Polyclonal (and some monoclonal) anti-IgE antibodies are also valuable crosslinking agents for *in vitro* studies. As shown in Figure 1 and below, these antibodies can be labelled after fixation with protein A-conjugated gold colloids. Alternatively, chemically crosslinked IgE multimers or anti-receptor antibodies that directly crosslink IgE receptors can stimulate secretion in a single-step reaction.

IgE receptor crosslinking causes the margination of granules and a decrease in granule number due to exocytosis. Other responses to receptor crosslinking include the release of vasoactive amines (serotonin, histamine) and lysosomal enzymes (e.g., β-hexosaminidase) and a series of membrane and cytoskeletal changes including the transformation of the cell surface from a microvillous to a lamellar topography, the increased assembly of F-actin, stimulated cell spreading and adhesion to substrates, increased fluid pinocytosis, and an increase in cortical stiffness. These responses are described in [4], [10] and [12].

The biochemical basis of these functional responses is fairly well understood. RBL-2H3 cells are one of many systems in which ligand-receptor interaction stimulates a membrane-associated phospholipase C that, in turn, enhances the turnover of inositol phospholipids.[13] The hydrolysis of these phospholipids produces diacylglycerol (DAG), that activates protein kinase C, plus a family of inositol phosphates, including inositol trisphosphate (IP_3). Work in this and other laboratories (reviewed in [4]) has indicated that both products of inositol phospholipid hydrolysis appear to be required for optimal secretion: IP_3 to mobilize the cytoplasmic Ca^{2+} stores that initiate secretion and diacylglycerol to active protein kinase C and initiate the Na^+-dependent influx of Ca^{2+} that maintains secretion. The membrane and cytoskeletal responses, on the other hand, appear to depend on events that are initiated by the activation of protein kinase C and do not depend on extracellular Ca^{2+}.[4,10] There is limited evidence that a GTP-binding transducer protein(s) is involved in the stimulation of phospholipase C activity in RBL-2H3 cells.[4]

One missing element in the analysis of signal transduction in RBL-2H3 cells has been insight into the molecular mechanisms that couple crosslinked IgE receptors to transducer and effector proteins.

3. DYNAMIC PROPERTIES OF THE IgE RECEPTOR

A. INTRODUCTION

By virtue of its abundance, biochemical characterization, and extra-membrane crosslinking, the IgE receptor system of RBL-2H3 cells is ideally suited for studies of dynamic receptor properties regulating signal transduction. This was first recognized by Baird, Holowka and colleagues, who published two particularly important studies in RBL-2H3 cells. In one of these studies, Menon and colleagues[14,15] tested the possibility that the lateral mobility of IgE receptors may be modulated by crosslinking agents. Using the technique of fluorescence recovery after photobleaching (FRAP), they first established that non-crosslinked fluorescein-IgE-receptor complexes are able to diffuse relatively freely in the RBL-2H3 membrane ($D \cong 2 \times 10^{-10}$ cm/sec.). They then demonstrated that the addition of antigen or anti-IgE antibody causes a rapid and substantial increase in the fraction of immobile receptors ($D \leq 5 \times 10^{-12}$ cm/sec.). Receptors were also immobilized by chemically crosslinked fluorescent IgE multimers. In antigen-treated cells, receptor mobility could be restored by the addition of an excess of monovalent hapten. These data raised the possibility that receptor immobilization plays a role in triggering secretion. Curiously, it was found by FRAP analysis that receptor immobilization occurs at 4°C and in the presence of cytochalasin B. As discussed below, crosslinked IgE-receptor complexes can cap under these two conditions, suggesting that receptors that appear to be immobile on the time scale of FRAP analysis may nevertheless retain a limited degree of long-range mobility.

In a related study, Robertson et al.[16] showed that incubation of IgE-primed cells for 30 min. with polyclonal anti-IgE antibody can cause a transition of the IgE receptor from a Triton-X-100-soluble form to a detergent-resistant form. This transition was also caused by less active crosslinking agents (antigen and IgE trimers), provided chemical crosslinking agents were added prior to detergent extraction. It was suggested that the interaction of crosslinked receptors with the cytoskeleton may be a necessary step for the delivery of a transmembrane signal. However, in cells incubated with a polyclonal anti-IgE antibody, the extent of receptor association with the cytoskeleton continued to increase beyond the anti-IgE concentration that gave maximal degranulation. This result, as well as the relatively slow rate of conversion to detergent insolubility, is consistent with our hypothesis (below) that receptor-cytoskeleton interaction is involved in the formation of stable, highly crosslinked receptor complexes that may terminate secretion.

B. THE KINETICS OF ANTIGEN BINDING TO IgE-RECEPTOR COMPLEXES

An experiment performed during the characterization by flow cytometry of DNP-B-phycoerythrin (DNP-B-PhE) binding to anti-DNP-IgE-receptor complexes provided direct evidence for the formation of highly crosslinked, cytoskeleton-associated, inactive or inhibitory IgE-receptor complexes on the surface of RBL-2H3 cells.[11] Having demonstrated that DNP_{42}-B-PhE binds rapidly and saturably to IgE-receptor complexes, we carried out experiments to quantify the displacement of antigen by the monovalent, competing hapten, DNP-lysine. DNP-lysine in a 5000X molar excess immediately blocks the secretion of [^{3}H]serotonin when added to RBL-2H3 cells at any time after the addition of antigen (Figure 2). We expected that it would displace surface-bound antigen with the same kinetics.

The experimental result, given in Figure 3A (solid lines), was considerably different from the predicted outcome. It was found that DNP-lysine added shortly after DNP_{42}-B-PhE displaces the majority of surface-bound multivalent antigen. However, with increasing incubation time, the rate of antigen displacement by monovalent ligand becomes progressively slower. The conversion of antigen-IgE-receptor complexes to a DNP-lysine-resistant form that does not trigger secretion also occurred on cells treated with phenylglyoxal, a reagent that inhibits endocytosis

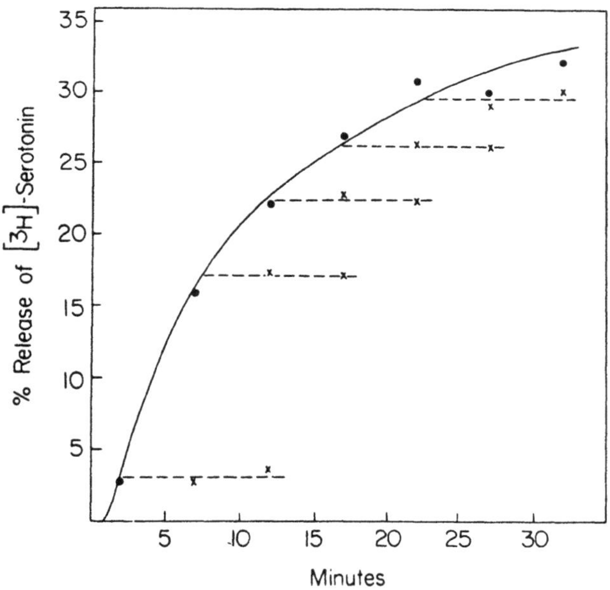

FIGURE 2 The inhibition of antigen-induced secretion by competing monovalent hapten. Cells were primed with anti-DNP IgE (3 μg/ml, 2 hours at 37°C) then incubated with 1.9 nM DNP_{42}-B-PhE in Hanks' medium. The solid line shows the time course of antigen-stimulated release. The dashed lines show that DNP-lysine (10 μM) causes the rapid and complete inhibition of secretion when added at any time during the process of secretion. Results are corrected for basal release (approx. 0.5% of the total [^{3}H]serotonin/min). From Seagrave et al.[11]

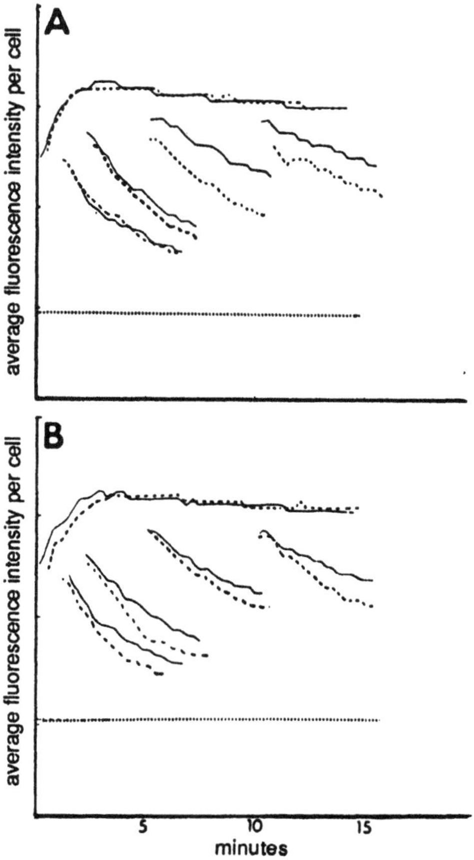

FIGURE 3 The binding of DNP-B-PhE to IgE-receptor complexes and its displacement by DNP-lysine. In panel A, IgE-primed RBL-2H3 cells were incubated with DNP_{42}-B-PhE (1 nM; 0.25 μg/ml) in Hanks' medium (solid lines) or in Hanks' medium with 5 μg/ml dihydrocytochalasin B (dashed lines). In panel B, 0.75 mM phenylglyoxal, an inhibitor of endocytosis, was added to the same incubation mixtures. Binding of the fluorescent antigen to IgE-receptor complexes was measured at 37°C in a flow cytometer. Binding was followed continuously for 15 min. to give the upper curves. The lower curves were obtained by incubating separate samples for 1, 2, 5 or 10 min. with DNP-PhE before adding DNP-lysine (5 μM) and measuring the resulting displacement of the multivalent antigen. Fluorescence is reported in arbitrary units on a log. scale. Neither dihydrocytochalasin nor phenylglyoxal alters the kinetics of multivalent antigen binding. However, antigen bound to cytochalasin-treated cells remains consistently more displaceable by DNP-lysine. The dotted line represents the fluorescence of cells that were incubated with 1 nM DNP-B-PhE without prior exposure to IgE. From Seagrave et al.[11]

(Figure 3B, solid lines; Figure 5D). This shows that the changes in rate of antigen displacement cannot be explained solely on the basis of antigen internalization.

Further experiments indicated a role for antigen-stimulated F-actin assembly in the transition to DNP-lysine-resistant and biologically inactive complexes. It was already known that both the rate and extent of [^{3}H]serotonin secretion are increased by cytochalasin treatment (Figure 4). We found that the time course of antigen binding is unaltered by cytochalasin B and its analogs. However, treatment with cytochalasins delays the shift from readily displaceable to poorly displaceable antigen at every time point (Figures 3A and B, dashed lines).

This experiment raised the possibility that crosslinked IgE receptors may exist in two forms. The first is a short-lived excited state whose formation is inhibited by DNP-lysine. The second is a more highly crosslinked (DNP-lysine-resistant) state that does not trigger secretion. The enhancement of secretion and concomitant increased lifetime of DNP-lysine-sensitive complexes in cytochalasin-treated cells suggests a role for F-actin in the conversion of receptors to the stably crosslinked, monovalent hapten-resistant, inactive form.

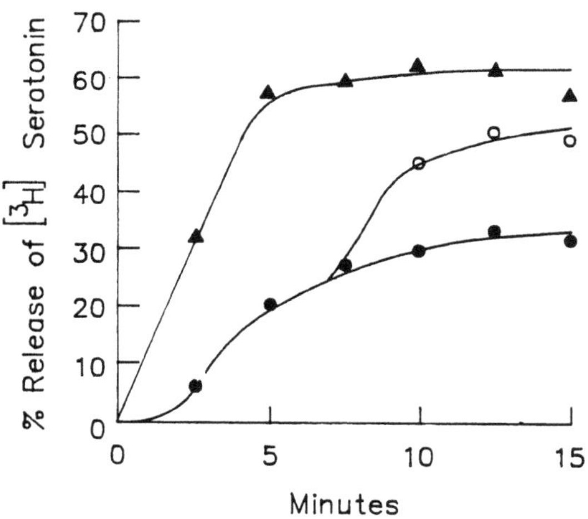

FIGURE 4 Dihydrocytochalasin B enhances antigen-induced secretion from RBL-2H3 cells. Anti-DNP IgE-primed (3 μg/ml, 2 hrs.) cells were incubated with DNP-B-PhE (1 μg/ml) in the presence or absence of dihydrocytochalasin B (5 μg/ml). Cells in Hanks' medium alone released around 30% of their total [^{3}H]serotonin in 15 min. (solid circles). Cytochalasin treatment increased secretion when it was present throughout the experiment (triangles) and when it was added 7 min. after the addition of antigen (open circles). Results are corrected for spontaneous release, that was unaffected by cytochalasin. From Seagrave et al.[11]

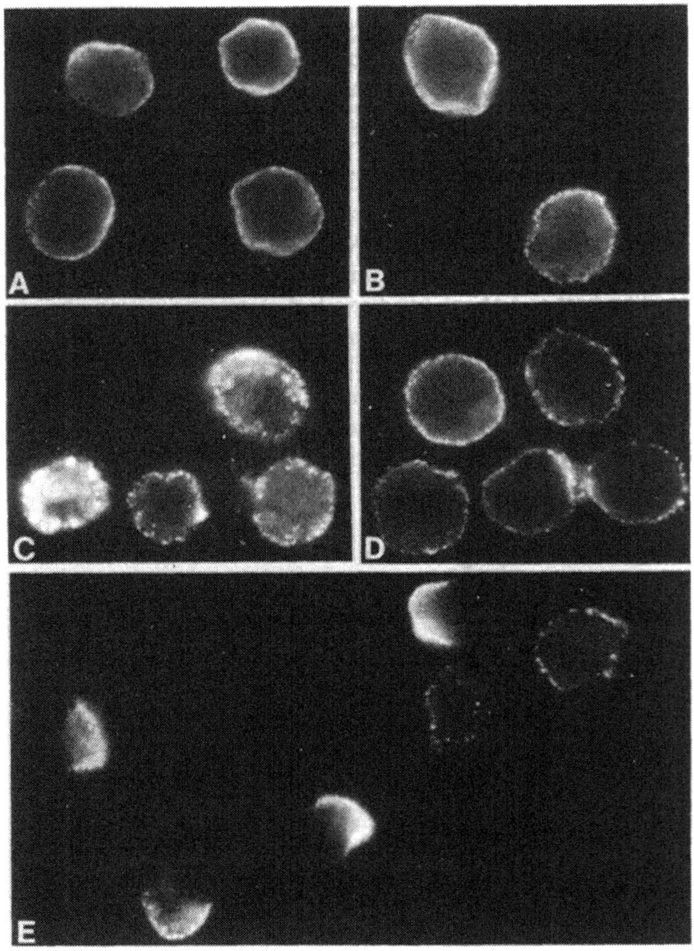

FIGURE 5 The distribution of antigen-IgE-receptor complexes on RBL-2H3 cells. IgE-primed cells were incubated under various conditions with DNP-B-PhE (1 μg/ml), then fixed with 2% paraformaldehyde and examined by fluorescence microscopy. At 4° C, antigen remains uniformly distributed on the cell surface (A). Cells incubated at 37° C develop surface clusters of antigen-IgE-receptor complexes after 5 min. incubation (B). The clusters are more prominent and internalization is evident after 10 min. (C). Cells incubated with antigen for 10 min. at 37° C in the presence of phenylglyoxal form surface clusters that are not internalized (D). Up to 60% of cells can redistribute antigen-IgE-receptor complexes into caps in the presence of dihydrocytochalasin B (0.25 μg/ml, 10 min.: E).

C. IgE RECEPTOR TOPOGRAPHY

In RBL-2H3 cells, it has been known for at least 10 years that the crosslinking of fluorescein-IgE-receptor complexes with antigen, anti-fluorescein or anti-IgE is followed by the clustering and subsequent internalization of the complexes.[4,7] The same pattern of clustering and internalization is seen when unlabelled IgE-receptor complexes are crosslinked with the fluorescent antigen, DNP-B-PhE (Figure 5A-C), and when receptors are directly crosslinked with fluorescent IgE trimers or higher oligomers or by fluorescent anti-receptor antibodies.[14,15,17] Additional studies with fluorescent probes have shown that IgE-receptor complexes can form surface caps on RBL-2H3 cells under two exceptional circumstances. The first is during prolonged (4–12 hrs.) incubation of IgE-primed cells with antigen at 4°C, when large clusters form on the cell surface and may eventually coalesce into caps.[14,15] Capping can also be induced in 2–5 min. at 37°C by incubating IgE-primed cells with cytochalasin B plus antigen (Figure 5E; 11).

The resolution of these fluorescence studies was insufficient to draw conclusions about the relationship of receptor topography to cell activation. This critical information has now been obtained by labelling IgE receptors with gold probes that can be observed at high magnification in the scanning electron microscope. Our studies ([18]; and R.F. Stump, J.R. Pfeiffer, M.C. Schneebeck, E. Collins and J.M. Oliver, unpublished) have focused on the two gold probes indicated in cartoon form in Figure 1. IgE-receptor complexes can be crosslinked directly with DNP-BSA-gold particles. Alternatively, IgE-primed cells can be crosslinked with anti-IgE antibodies and these crosslinking antibodies can be labelled with protein A-gold particles. Silver-enhancement techniques are used to enlarge the particles and to 'roughen' their surfaces. Silver enhancement makes particles as small as 5 nm in diameter easily visible in the backscattered electron imaging (BEI) mode of the scanning electron microscope (SEM). Surface topography is observed in the secondary electron imaging (SEI) mode of the microscope. A series of results is given in Figures 6–11. In each case, panel A is the SEI image and panel B is the BEI image of the same field of view.

In Figure 6, cells were fixed, then labelled sequentially with IgE, with anti-IgE and with protein A-gold. The majority (around 96%) of such cells maintain a microvillous surface topography (A) and an apparently random distribution of membrane-bound gold particles marking IgE receptors (B). The cell in Figure 7 was treated similarly, except that it was fixed after IgE labelling. This cell was selected for its spontaneously lamellar topography. There is a small apparent increase in the density of gold particles marking IgE-receptor complexes on the lamellar surfaces, as expected due to the increase in the amount of membrane being examined. Nevertheless, gold particles are still distributed over the whole cell surface. To a first approximation, it thus appears that non-crosslinked IgE receptors and IgE-receptor complexes are distributed at random on the RBL-2H3 cell surface.

After 2 min. of crosslinking with anti-IgE at 37°C (Figure 8) essentially all the cells have assumed a lamellar topography. This is accompanied by a striking change in the apparent distribution of gold particles marking anti-IgE-IgE-receptor

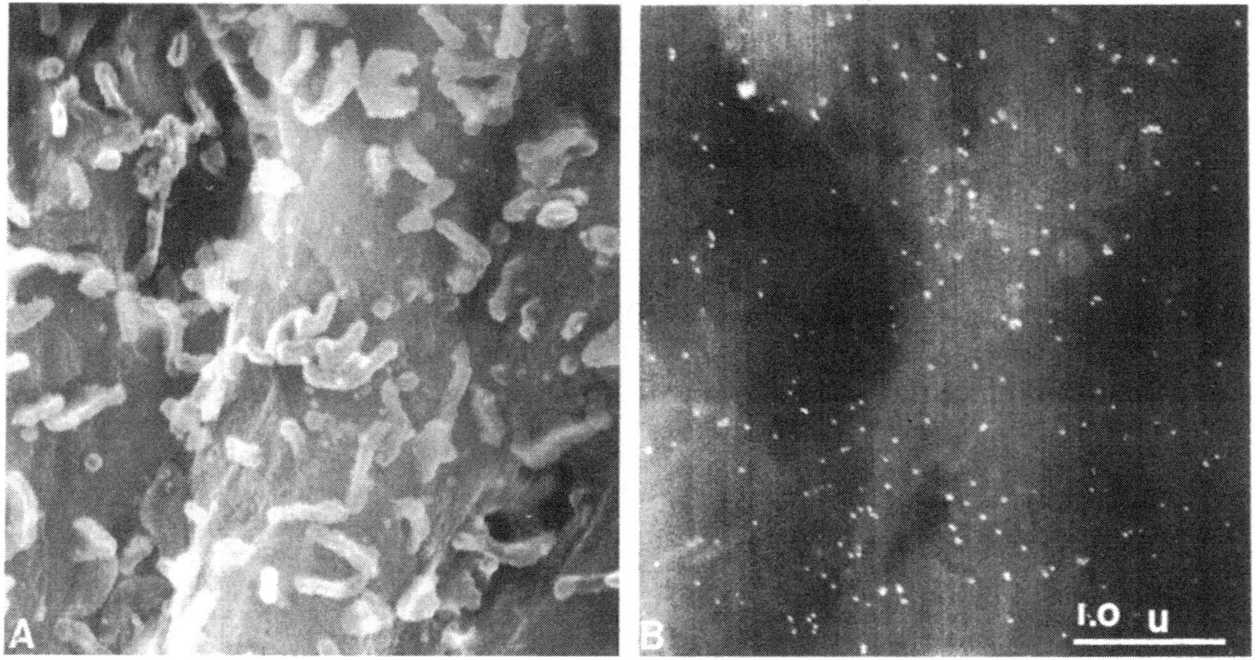

FIGURE 6 The distribution of IgE receptors on fixed cells. Cells were fixed in 10% paraformaldehyde, 0.075% glutaraldehyde, 0.2% picric acid in 0.1 M phosphate, pH 7.2 (a new fixative developed in this laboratory by M. C. Schneebeck that preserves the IgE-binding activity of fixed receptors), then incubated sequentially with IgE, anti-IgE and 15 nm protein A-gold. They were silver-enhanced and processed for SEM without the use of metals (that would generate an unwanted signal in the backscattered imaging mode). The cells were examined by sequential secondary and backscattered electron imaging (SEI/BEI) as described in [18]. An apparently random distribution of gold particles marking anti-IgE-IgE-receptor complexes is observed.

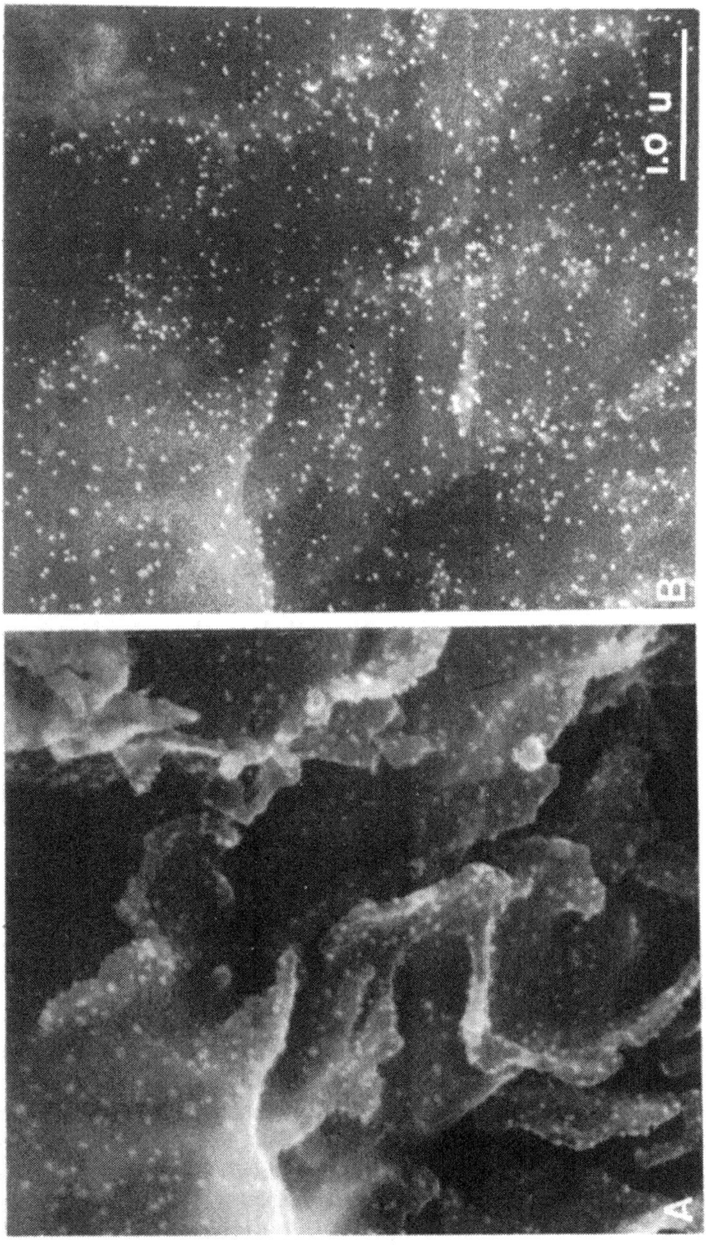

FIGURE 7 The distribution of IgE-receptor complexes on fixed cells. Cells were processed as described in Figure 6 except that they were incubated to saturation with anti-DNP IgE before fixation. An apparently random distribution of gold particles is again observed.

complexes. Specifically, gold particles are concentrated as singlets and small clusters on lamellar surfaces, while the interlamellar membranes carry a relatively lower density of gold particles as singlets and small clusters. We recognize the possibility that this apparent redistribution of small clusters onto lamellar surfaces is an optical artifact due to our 2-dimensional view of the 3-dimensional cell surface. With E. Collins and C. Wofsy, we are digitizing and analyzing stereo pairs of SEI/BEI images in order to quantify receptor density on lamellar and non-lamellar membranes. However, the reality of this redistribution is indicated by its reproducibility on briefly crosslinked samples at the level of both the light and scanning electron microscope. It is reinforced by comparison with the image in Figure 7 in which a spontaneously lamellar cell that was fixed after IgE labelling showed little difference in the apparent density of gold particles between its lamellar and non-lamellar surfaces.

Lamella-associated particles are still prominent after 5 min. of incubation (Figure 9). In addition, cells incubated for 5 min. at 37°C show the first evidence of a new population of gold particles that exist as intermediate-sized clusters (up to 20 or more particles) on interlamellar membrane.

With further incubation, the clusters of gold particles marking anti-IgE-IgE-receptor complexes on interlamellar membrane become progressively fewer in number and larger in size. By 10 min., clusters of up to 50 or more particles are visible (Figure 10), while cells incubated with anti-IgE at 37°C for 15 min. show rafts of up to 100 or more particles on interlamellar membrane (Figure 11). At these later times, there is also a decrease in the total number of surface-associated gold particles, presumably due to internalization. Nevertheless, there is always a coexisting population of lamella-bound gold particles that remain in the form of singlets or small clusters.

We emphasize that the preferential association of small clusters on lamellar surfaces is probable, but not definitively proven. In contrast, the exclusion of the larger complexes from lamellae is absolute: no large clusters of gold particles have ever been observed on lamellae.

4. THE COUPLING AND UNCOUPLING OF RECEPTORS AND SIGNALLING ELEMENTS

An hypothesis to explain the association between IgE receptors and signalling elements has been developed from the dynamic biochemical, biophysical and topographical studies reviewed above. Specifically, we propose that crosslinked IgE receptors can exist in two states at the RBL-2H3 cell surface, one that is competent to trigger secretion and one that is either inactive or directly inhibitory to secretion.

Some properties of the triggering-competent (active) receptor aggregates are listed in Table I, Column I. We postulate that the lifetime of active complexes is brief. This is based on the immediate inhibition of secretion by monovalent hapten

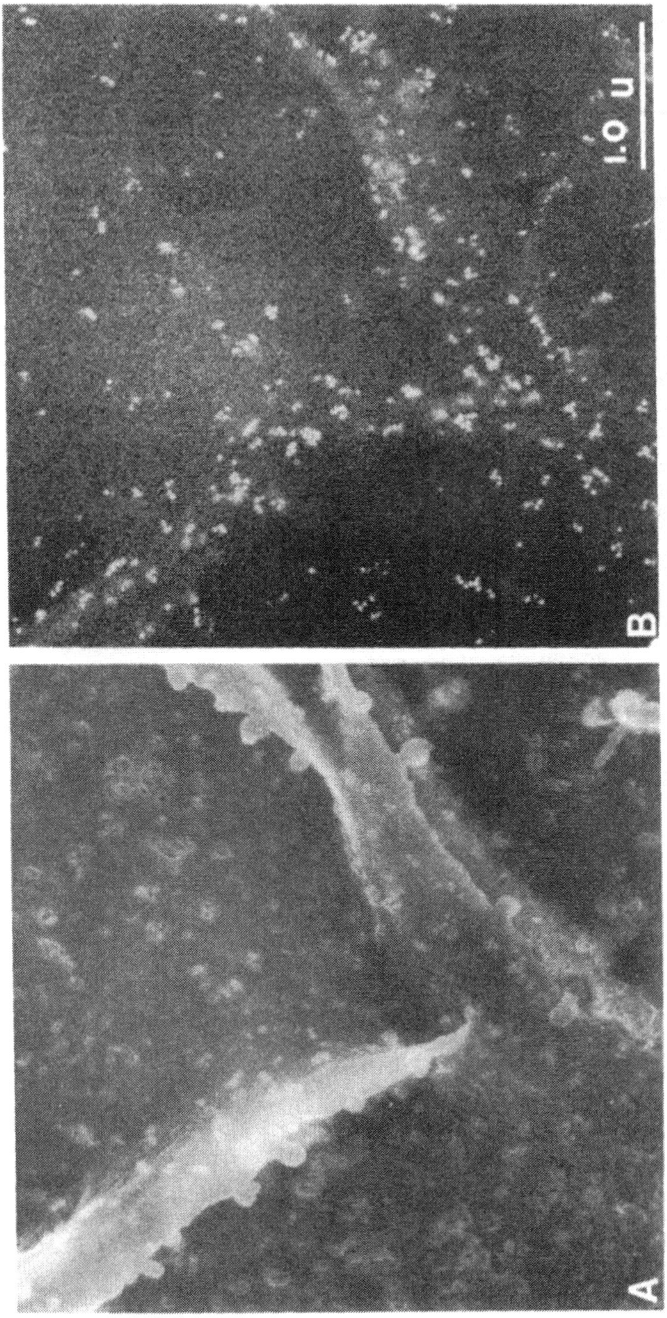

FIGURE 8 The effect of crosslinking for 2 min. on IgE receptor distribution. The IgE-primed cell illustrated here was incubated with anti-IgE at 37°C for 2 min. It was then fixed and processed as in Figure 6. The cell has assumed a lamellar topography and there is a striking redistribution of gold particles marking anti-IgE-IgE-receptor complexes onto lamellar surfaces. Small clusters are also present at lower density on the interlamellar membrane.

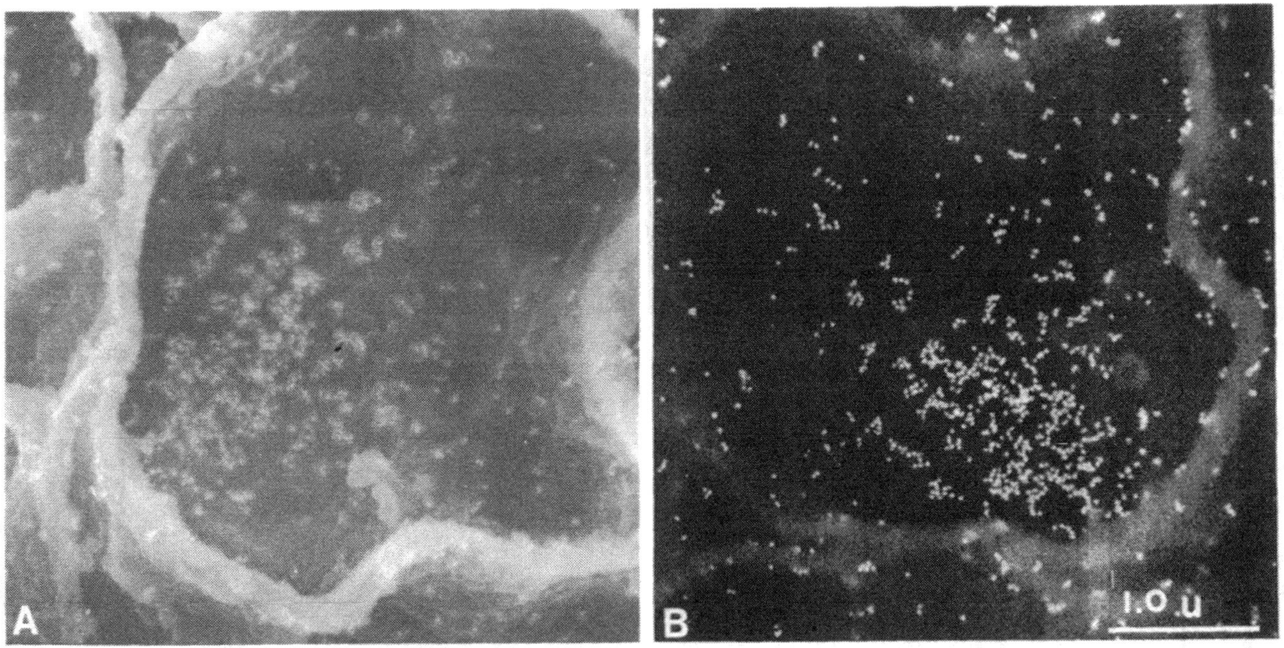

FIGURE 9 The effect of crosslinking for 5 min. on IgE receptor distribution. This IgE-primed cell was incubated for 5 min. with anti-IgE before fixation. Small aggregates of gold particles are still associated with lamellae. There is an additional population of larger clusters on the interlamellar membranes.

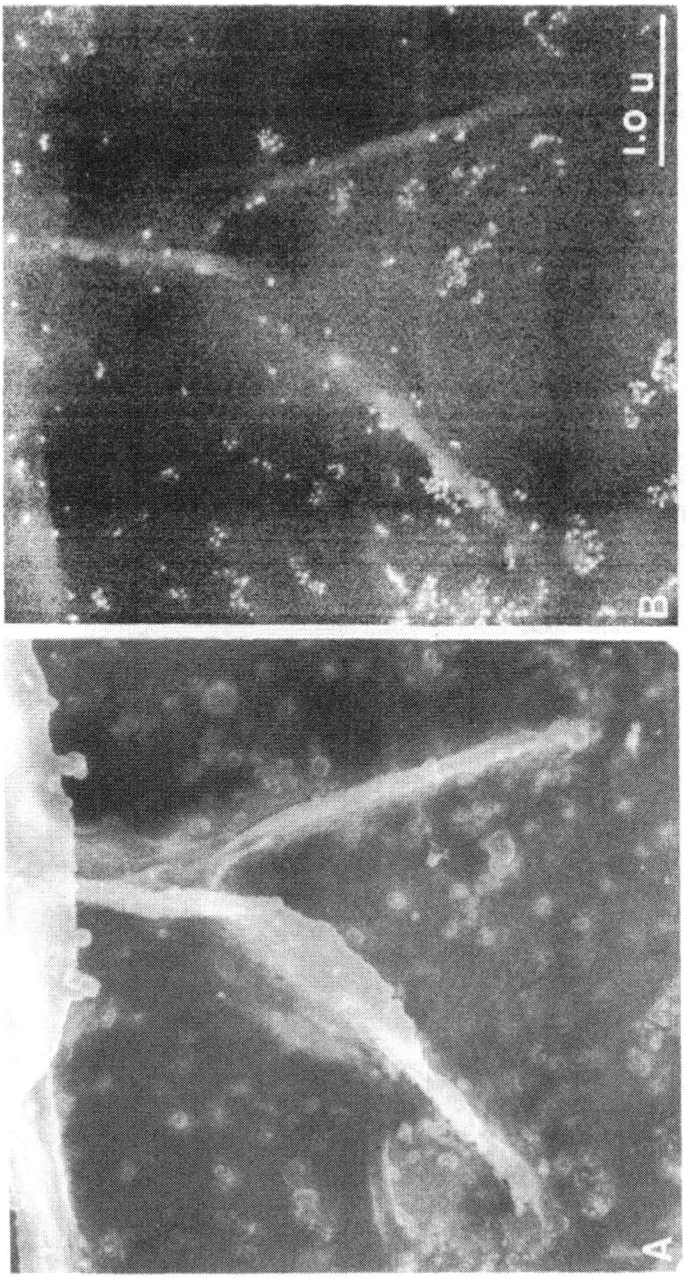

FIGURE 10 The effect of crosslinking for 10 min. on IgE receptor distribution. This IgE-primed cell was incubated with anti-IgE for 10 min. at 37°C before fixation. Small clusters of gold particles marking crosslinked IgE-receptor complexes persist on the lamellar membranes. There is a decrease in the number and increase in the size of the interlamellar clusters.

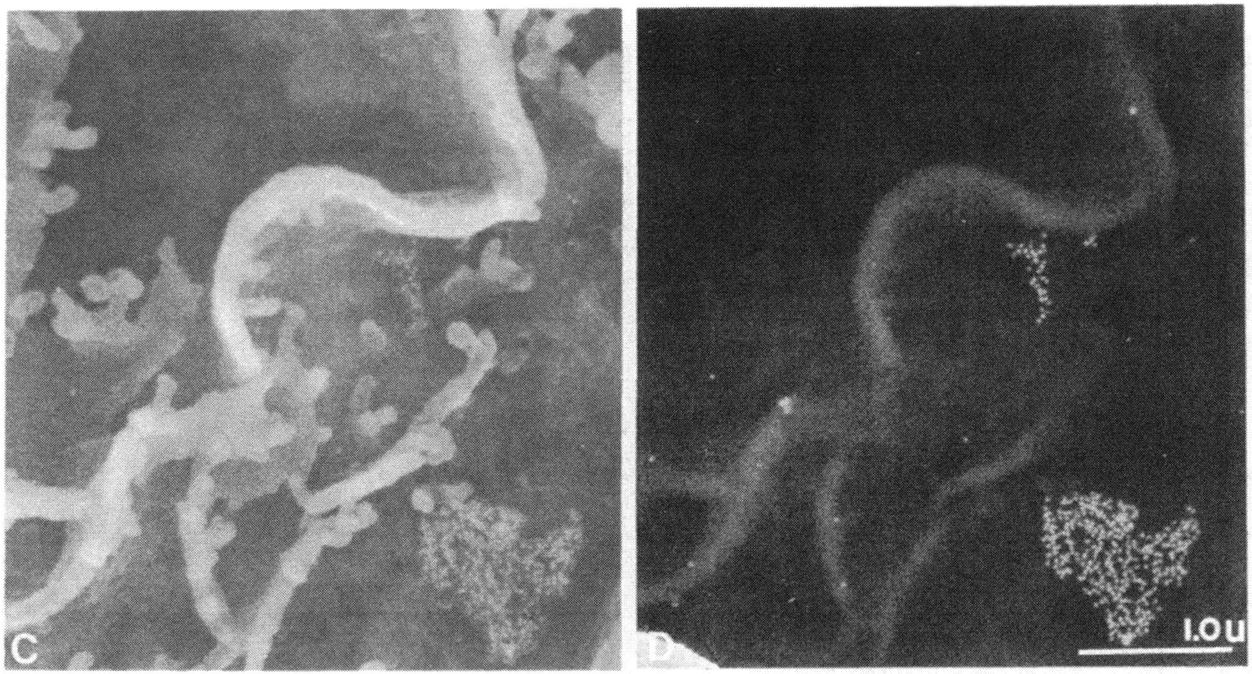

FIGURE 11 The effect of crosslinking for 15 min. on IgE receptor distribution. After 15 min. at 37°C, anti-DNP-IgE-IgE-receptor complexes have two distinct distributions: as small clusters on lamellar membranes and as larger rafts on interlamellar membranes.

even though crosslinked receptor aggregates persist at the cell surface.[11] We postulate further that these active complexes are not linked directly to F-actin and are not extensively crosslinked. These predictions are based on the effects of cytochalasin treatment on receptor properties and triggering activity. As described,[11] the cytochalasins increase the rate and extent of secretion, indicating that antigen-stimulated F-actin assembly is not needed to generate active crosslinked receptor complexes. The cytochalasins also inhibit the formation of extensively crosslinked (monovalent hapten-resistant) receptor complexes, indicating that active complexes may consist of relatively small groups of receptors. We suppose that active complexes remain susceptible to detergent extraction. This is based on the evidence of Robertson et al.[16] that the transition of crosslinked IgE receptors to detergent insolubility occurs relatively slowly and requires extensive crosslinking.

Active complexes are proposed to be relatively immobile. The capping of crosslinked receptors on cytochalasin-treated cells[11] shows that active complexes preserve at least a limited capacity for long-range movement. The FRAP analyses of Menon et al.[14,15] indicate that, nevertheless, the mobility of a large fraction of receptors is reduced by several orders of magnitude within minutes of crosslinking. This immobilization occurs at 4°C and in the presence of cytochalasin, indicating its independence of active metabolism or F-actin assembly. However, no immobilization is measured by FRAP analysis on membrane blebs, indicating that other membrane-associated cytoskeletal proteins, talin or fodrin, for example (see [19,20]), may play a role in this approach to receptor immobility.

Finally, we propose that active receptor complexes may be subject to topographical restraints. Specifically, it is possible that the small receptor aggregates that appear to associate preferentially with membrane lamellae throughout the entire time course of secretion may represent active complexes. We note that this may be a preferred, not an absolute, topographical pattern. Some or all of the small receptor clusters on interlamellar membrane may be relatively immobile, not attached directly to the F-actin cytoskeleton and competent to trigger secretion. Receptors that crosslink at the periphery of the large receptor clusters could also share these properties.

The proposed properties of inactive complexes are given in the right column of Table I. We postulate that signalling is terminated by the movement of crosslinked receptors into larger aggregates that interact with and are stabilized by F-actin. These aggregates are highly crosslinked, detergent-resistant and immobile based on both FRAP and capping assays (the fully immobile population). They are restricted to the interlamellar membranes of RBL-2H3 cells.

Figure 12 incorporates these properties into a working model for the control of IgE receptor-mediated secretion in RBL-2H3 cells. Panel A shows that non-crosslinked IgE-receptor complexes maintain an apparently random distribution that is unaffected by the variable geometry of the cell surface. IgE receptor crosslinking leads to the initial formation of small receptor clusters that interact with transducer and effector moieties to stimulate inositol phospholipid turnover and the recruitment of protein kinase C to the membrane. One consequence of this is a

TABLE 1 Proposed Properties of Active and Inactive Crosslinked IgE Receptor Complexes on RBL-2H3 Cells

Active	Inactive/Inhibitor
Short-lived	Relatively long-lived
Not linked to F-actin (but may be linked to other membrane-associated cytoskeletal elements)	Linked to F-actin
Not extensively crosslinked	Extensively crosslinked
Sensitive to detergent solubilization	Insoluble to detergent
Relatively immobile	Fully immobile
May associate preferentially with lamellar membranes	Accumulate in clusters on non-lamellar membrane and are removed by endocytosis through coated pits
Interact productively with transducer and effector proteins	Do not interact productively with with transducer and effector moieties

change in the cell surface from a microvillous to a lamellar topography and an accompanying or subsequent redistribution of a portion of the crosslinked receptors (and the putative receptor-associated cytoskeletal proteins) onto lamellar membranes (Panel B). The reason for this redistribution is not known with certainty. However, we postulate that the geometry of lamellar membrane makes it a region of minimal potential energy (maximum interaction) for receptor aggregates within a particular size range. That is, the small, relatively immobile, biologically active receptor aggregates may follow the predictions of the surf-riding model that was developed by Hewitt[21] and expanded by Berlin and Oliver[22] to explain the association of particles (e.g., carbon) and crosslinked lectin receptors with lamellae during capping in lymphocytes and macrophages.

We postulate that continued receptor crosslinking generates aggregates that become too large to remain associated with lamellae. This redistribution is probably the result of steric interference: as aggregates become larger, they can no longer bend to conform to the shape of the lamella and so they are displaced once again to regions of reduced curvature. Once on the interlamellar membrane, the aggregates associate with cortical F-actin, becoming detergent-resistant and susceptible to internalization through coated pits (Panel C). This association with F-actin may occur directly or via the hypothetical receptor-associated cytoskeletal (talin or fodrin-like) proteins. In this final form, receptors cannot redistribute into caps,

A
B
C
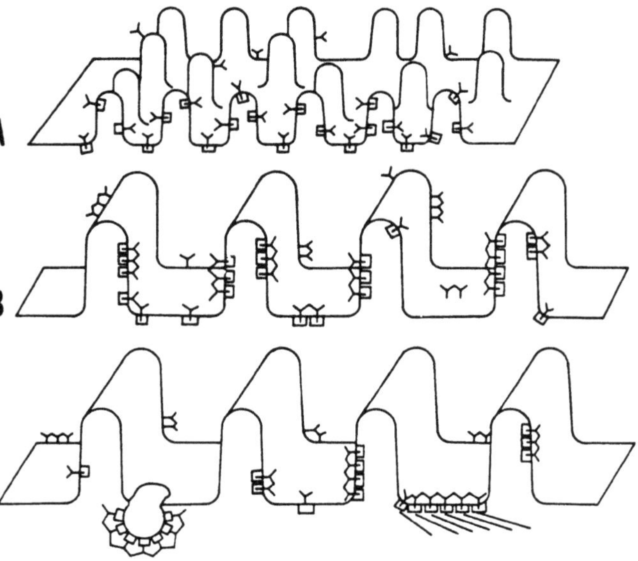

FIGURE 12 A model for the control of IgE-mediated secretion from RBL-2H3 cells. Panel A shows that non-crosslinked IgE-receptor complexes maintain an apparently random distribution that is unaffected by the variable topography of the cell surface. Panel B illustrates the preferential (but probably not obligatory) redistribution of active, crosslinked receptor complexes onto lamellar membranes. Panel C shows the displacement of larger, inactive (or inhibitory) complexes onto interlamellar membranes, their interaction with F-actin and their eventual internalization through coated pits. Additional properties of the 2 forms of the crosslinked receptors are given in the text and in Table I.

indicating a further level of immobilization, and they are incapable of triggering secretion.

We are exploring two possible explanations for the apparent inability of highly crosslinked receptors to trigger secretion. The first is that receptor immobilization simply limits the association of crosslinked IgE-receptor complexes with elements of the signal transduction pathway (G proteins, phospholipase C) whose activation is required to cause secretion. For example, if only the most peripheral receptors in a raft of 100 or more receptors are accessible to transducers, the resulting signal will be small or absent. In this case, secretion would be triggered effectively by the small, relatively immobile, crosslinked aggregates that associate with lamellae, but only poorly by the large, fully immobile interlamellar receptor aggregates.

An alternative explanation is that highly crosslinked receptors generate a Ca^{2+}-independent, F-actin-dependent termination signal that can co-exist with (and oppose) the Ca^{2+}-dependent cell activation signal. This hypothesis can explain two experiments in which the rate and extent of secretion were dissociated from the extent of antigen-stimulated inositolphospholipid turnover and Ca^{2+} mobilization. In the first case, Maeyama et al.[23] demonstrated a predictable relationship between the concentration of crosslinker, the size of the biochemical response (PI turnover or Ca^{2+} mobilization) and the amount of secretion when RBL-2H3 cells were incubated with relatively low concentrations of antigen. However, they reported that the PI and Ca^{2+}-Quin-2 responses continue to increase beyond the concentration of antigen that causes maximal secretion. Conversely, we have found that cytochalasin treatment causes a 2–3 fold increase in antigen-induced secretion with no corresponding increase in [$^{45}Ca^{2+}$] uptake or in the size of the Ca^{2+}-Quin-2 spike, and with only a slight increase in the Ca^{2+}-Quin-2 plateau (G. G. Deanin, J. M. Oliver, and J. C. Seagrave, unpublished). To explain the high dose inhibition of secretion but not of Ca^{2+} mobilization, we propose that high antigen concentrations may accelerate the conversion of a portion of the IgE receptors to their highly crosslinked, inhibitory form. This would reduce the secretory response to Ca^{2+} mobilization triggered by co-existing active, receptor complexes. Conversely, cytochalasin treatment may enhance secretion by limiting the formation of inhibitory complexes.

The nature of this proposed inhibitory signal is completely unknown. It is possible that inhibitory receptor complexes could cause the synthesis or stabilization of chemical messengers (e.g., 3′5′ cyclic AMP) that antagonize the effects of messengers generated by phosphatidyl inositol turnover. However, we favor the hypothesis that the extensively crosslinked, inhibitory complexes may block the fusion of secretory granules with membranes by decreasing membrane fluidity or by stabilizing cortical microfilaments so that they form a physical barrier to the fusion of granules with the plasma membrane.

ACKNOWLEDGMENT

We thank J. R. Pfeiffer for excellent technical assistance and M. Guerrero for manuscript preparation. The original work described here was supported in part by NIH grants GM37202, GM35811, RR02700, by NSF grant PCM–850552 and by CTR grant 1649.

REFERENCES

1. Eccleston, E., B. J. Leonard, J. S. Lowe, and H. J. Welford (1973), "Basophilic Leukemia in the Albino Rat and a Demonstration of the Basopoetin," *Nature New Biol.* **24**, 73–76.
2. Kulczycki, Jr., A., C. Isersky, and H. Metzger (1974), "The Interaction of IgE with Rat Basophilic Leukemia Cells. 1. Evidence for Specific Binding of IgE," *J. Exp. Med.* **13**, 600–616.
3. Barsumian, E. L., C. Isersky, M. G. Petrino, and R. P. Siraganian (1981), "IgE-Induced Histamine Release from Rat Basophilic Cell Lines: Isolation of Releasing and Non-releasing Clones," *Eur. J. Immunol.* **1**, 317–323.
4. Oliver, J. M., J. C. Seagrave, R. F. Stump, J. R. Pfeiffer, and G. G. Deanin (1987), "Signal Transduction and Cellular Response in RBL-2H3 Mast Cells," *Progress in Allergy*, Eds. E. L. Becker and S. Karger (in press).
5. Seldin, D. C., S. Adelman, K. F. Austen, R. L. Stevens, A. Hein, J. P. Caulfield, and R. G. Woodbury (1985), "Homology of the Rat Basophilic Leukemia Cell and the Rat Mucosal Mast Cell," *Proc. Natl. Acad. Sci. USA* **8**, 3871–3875.
6. Seldin, D. C., K. R. Austen, and L. Stevens (1985b), "Purification and Characterization of Protease-Resistant Secretory Granule Proteoglycans Containing Chondroitin Sulfate Di-B and Heparin-Like Glycosaminoglycans from Rat Basophilic Leukemia Cells," *J. Biol. Chem.* **26**, 11131–11139.
7. Metzger, H., G. Alcaraz, R. Hohman, J.-P. Kinet, V. Pribluda, and R. Quarto (1986), "The Receptor with High Affinity for Immunoglobulin E," *Ann. Rev. Immunol.* **4**, 419–470.
8. Baird, B., and D. Holowka (1986), "Structural Mapping of Membrane-Associated Proteins: A Case Study of the IgE-Receptor Complex," *Spectroscopic Membrane Probes*, Ed. L. Loew (Boca Raton, FL: CRC Press).
9. Liu, F.-T., J. W. Bohn, E. L. Ferry, H. Yamamoto, C. A. Molinaro, L. A. Sherman, N. R. Klinman, and D. H. Katz (1980), "Monoclonal Dinitrophenyl-Specific Murine IgE Antibody: Preparation, Isolation and Characterization," *J. Immunol.* **12**, 2728–2736.
10. Pfeiffer, J. R., J. C. Seagrave, B. H. Davis, G. G. Deanin, and J. M. Oliver (1985), "Membrane and Cytoskeletal Changes Associated with IgE-Mediated Serotonin Release from Rat Basophilic Leukemia Cells," *J. Cell Biol.* **10**, 2145–2155.
11. Seagrave, J. C., G. G. Deanin, J. C. Martin, B. H. Davis, and J. M. Oliver (1987), "DNP-Phycobiliproteins, Fluorescent Antigens to Study Dynamic Properties of Antigen-IgE-Receptor Complexes on RBL-2H3 Rat Mast Cells," *Cytometry* **8**, 287–295.
12. Liu, Z.-Y., J.-I. Young, and E. L. Elson (1987), "Rat Basophilic Leukemia Cells Stiffen when They Secrete," *J. Cell Biol.* in press.

13. Beaven, M. A., J. P. Moore, T. R. Smith, T. R. Hesketh, and J. C. Metcalfe (1984), "The Calcium Signal and Phosphatidylinositol Breakdown in 2H3 Cells," *J. Biol. Chem.* **25**, 7137–7142.
14. Menon, A. K., D. Holowka, W. W. Webb, and B. Baird (1986a), "Clustering, Mobility and Triggering Activity of Small Oligomers of Immunoglobulin E on Rat Basophilic Leukemia Cells," *J. Biol. Chem.* **10**, 534–540.
15. Menon, A. K., D. Holowka, W. W. Webb, and B. Baird (1986b), "Crosslinking of Receptor-Bound IgE to Aggregates Larger than Dimers Leads to Rapid Immobilization," *J. Cell Biol.* **10**, 541–550.
16. Robertson, D., D. Holowka, and B. Baird (1986), "Crosslinking of Immunoglobulin E-Receptor Complexes Induce Their Interaction with the Cytoskeleton of Rat Basophilic Leukemia Cells," *J. Immunol.* **13:**, 565–4572.
17. Menon, A. K., D. Holowka, and B. Baird (1984), "Small Oligomers of Immunoglobulin E (IgE) Cause Large-Scale Clustering of IgE Receptors on the Surface of Rat Basophilic Leukemia Cells," *J. Cell Biol.* **9**, 577–583.
18. Stump, R. F., J. R. Pfeiffer, J. C. Seagrave, and J. M. Oliver (1987), "Mapping Gold-Labelled IgE Receptors on Mast Cells by Scanning Electron Microscopy: Receptor Distributions Revealed by Silver Enhancement, Backscattered Electron Imaging and Digital Image Analysis," *J. Histochem. Cytochem.* in press.
19. Geiger, B (1985), "Microfilament-Membrane Interaction," *Trends in Biochem. Sci.* **1**, 456–460.
20. Jacobson, K., A. Ishihara, and R. Inman (1987), "Lateral Diffusion of Proteins in Membranes," *Ann. Rev. Physiol.* **4**.
21. Hewitt, J. A. (1979), "Surf-Riding Model for Cell Capping," *J. Theor. Biol.* **8**, 115–127.
22. Berlin, R. D., and J. M. Oliver (1982), "The Movement of Bound Ligands over Cell Surfaces," *J. Theor. Biol.* **9**, 69–80.
23. Maeyama, K., R. J. Hohman, H. Metzger, and M. A. Beaven (1986), "Quantitative Relationships between Aggregation of IgE Receptors, Generation of Intracellular Signals and Histamine Secretion in Rat Basophilic Leukemia (2H3) Cells. Enhanced Responses with Heavy Water," *J. Biol. Chem.* **26**, 2583–2592.

HOWARD M. DINTZIS and RENEE Z. DINTZIS
Department of Biophysics, and Departments of Cell Biology and Anatomy and Biophysics, respectively; Johns Hopkins University, School of Medicine, Baltimore, Maryland

A Molecular Basis For Immune Regulation: The Immunon Hypothesis

INTRODUCTION

The mechanisms controlling the immune response have had a profound fascination for scientists from many disciplines for decades. Particularly intriguing has been the question of whether or not there might be a central mechanism which plays a controlling role in overall immune regulatory phenomena. It has long been recognized that antigenic material in an "immunogenic form" is important in initiating an immune response, while antigenic material in a "toleragenic," "suppressive" or "inhibitory" form is somehow capable of reducing an immune response. However, the chemical and physical parameters defining these alternative "forms" of antigenic material have not been established.

Factors other than the antigenic material itself have been implicated in immune regulation. Antibody molecules produced during an immune response can interact with the initiating immunogen and affect its subsequent stimulatory behavior. During the past two decades a plethora of additional complex facts have been discovered which make difficult an unambiguous assignment of cause and effect within the multiple interacting phenomena of immune regulation. Numerous types of differentiated cells exist within the lymphoid tissues and these cells produce many products, both on their surfaces and secreted in soluble form. In addition, these cells have different receptors for, and respond to a variety of, molecules in their surroundings. Lastly, antibody molecules can be generated against chemically different antibody molecules within the same animal, raising the possibility of yet another kind of regulatory interaction at this level.

Our experimental results and their interpretation, which we shall discuss in this report, have led us to the conclusion that antigenic material itself in its various physical forms may play a central role, perhaps *the* central role, in the ongoing regulation of the immune response. As we shall show below, this regulatory capability occurs because antigenic material can exist either in a physical state which is intrinsically inhibitory or in a physical state which can be alternatively stimulatory or inhibitory, depending on the amount present. Because of the amphipathic nature of material in the latter state, we propose calling antigenic materials in such a form "amphigens" rather than "antigens." The amphipathic nature of such antigenic material allows it to exert either a positive or negative regulatory effect on the immune response. We shall attempt to demonstrate below that these effects occur through the facilitation of, or the interference with, the formation of multireceptor complexes, which are essential for the stimulation of B cells. Regulatory effects of this type have not yet been sufficiently considered in theoretical models of the immune response.

EXPERIMENTAL OBSERVATIONS

A number of years ago, we became interested in trying to define more clearly the molecular characteristics of an antigenic material which can cause it to act as an effective immunogen. To keep model materials experimentally accessible, we employed as antigens soluble polymeric molecules which are not readily degraded within the mammalian body and which give rise to the predominantly IgM antibody response of the so called "T-cell independent antigens." We avoided using "adjuvants" in the preparation and administration of our antigens. Adjuvants are substances conventionally mixed with antigens to produce or to increase the stimulatory ability of the antigen. However, they are substances which can exert powerful aggregating effects on antigen molecules. Since it is impossible to define the actual molecular state of any antigenic materials after conventional adjuvants are mixed with them, we have not used adjuvants in any of the experiments described below.

We synthesized and size fractionated a series of soluble haptenated polymers so that each fraction consisted of narrowly defined molecular sizes and contained narrowly defined numbers of haptenic epitopes. We then studied the ability of such soluble molecules to stimulate production of anti-hapten antibodies in mice. We measured dose-response curves for these well characterized antigenic preparations, both in whole animals (*in vivo*) and in cultures of cells prepared from spleens (*in vitro*). For *in vivo* studies, measurements were made either of the concentration of serum antibody (IgM) or the number of spleen cells producing IgM molecules

TABLE 1 A Summary of the Molecular Characteristics of the Polymer Molecules Used in the Studies Discussed in this Report[1]

Polymer Molecule	Polymer Backbone	Backbone Composition	Backbone Characteristic	Hapten or Epitope	Hapten or Epitope Characteristic
Dnp-PA	Polyacrylamide	Synthetic polyethylene polymer	Linear homopolymer, uncharged	Dinitrophenyl	Hydrophobic
Flu-PA	Polyacrylamide	Synthetic polyethylene polymer	Linear homopolymer, uncharged	Fluorescein	Hydrophilic
Flu-Fic	Ficoll	Polysaccharide synthesized from sucrose	Three-dimensional, highly crosslinked	Fluorescein	Hydrophilic
Flu-Dex	Dextran	Bacterial polysaccharide of glucose subunits	Predominantly linear, somewhat branched homopolymer	Fluorescein	Hydrophilic
Flu-CMC	Carboxymethyl-cellulose	Carboxy-methylated plant polyglucose	Linear, negatively charged homopolymer	Fluorescein	Hydrophilic
Flu-PVA	Polyvinyl alcohol	Synthetic polyethylene polymer	Linear homopolymer	Fluorescein	Hydrophilic
SIII	Pneumococcal polysaccharide	Bacterial polysaccharide	Linear, negatively charged heteroglycan	Alternating glucose-glucuronic acid	Hydrophilic

[1] The table illustrates the diversity of chemical and physical properties represented by this group of polymeric molecules.

by hemolytic plaque technique, 5 or 6 days after injection of antigen. For *in vitro* studies, measurements were made by plaque technique of the number of cells which were secreting IgM antibody molecules after 3 days of incubation in tissue culture medium containing known concentrations of soluble antigen. Although until now the data we have published[1–8] have been derived from studies using one polymeric carrier, polyacrylamide, and one haptenic epitope, dinitrophenyl (Dnp), we will here present new data concerning studies using a variety of additional carriers and epitopes. These new results confirm and extend our findings with Dnp-polyacrylamide.[9–11]

Table 1 lists the properties of the molecules used in our studies. As can be seen, this list is comprised of polymeric materials which are either natural or synthetic, charged or uncharged, hydrophilic or hydrophobic, linear or branched. The epitopes or haptens are either chemically added onto a carrier polymer "backbone" or naturally built into it as part of the structure. All these materials generated immune responses *in vivo* and *in vitro* which were consistent with those previously generated by Dnp-polyacrylamide. We will here summarize the main points of the Dnp-polyacrylamide studies and the "immunon" theory derived from those studies. We will then discuss how the more recent extended studies relate to these earlier findings.

Uncrosslinked molecules of polyacrylamide are water soluble, uncharged, and have a linear flexible structure which can be expected to approximate a random coil shape in water solution. By appropriate size fractionation and substitution with the chemical group dinitrophenyl (Dnp) as haptenic epitope, we were able to prepare a series of separate molecular forms of the same material. These forms differed from each other in molecular mass and also in the number of haptenic epitopes contained within a single molecule (haptenic valence). Typical *in vivo* dose-response curves for four of these preparations are shown in Figure 1. Note that the two polymers with molecular weights greater than 100 kD and with hapten valences of 53 and 126 are immunogenic while the smaller polymers, of 40 and 60 kD and valences of 11 and 19, are not immunogenic. Later studies showed such smaller molecules to be non-immunogenic at doses even 1000 fold higher than those shown here.

From a number of similar studies, carried out both *in vivo* and *in vitro*, we made some general observations:

OBSERVATION 1. Dnp-polyacrylamide molecules of a size smaller than that corresponding to a molecular weight of approximately 100 kD did not stimulate an anti-Dnp antibody response at any measured dose *in vivo* or *in vitro*, even though the number of Dnp epitopes per molecule was as large as 19.

OBSERVATION 2. Dnp-polyacrylamide molecules larger in size than approximately 100 kD, and substituted with more than approximately 20 Dnp groups per molecule were immunogenic *in vivo* and *in vitro*, giving bell-shaped, dose-response curves in both cases. For each polymer type, the immune response to the Dnp haptenic epitope increased gradually with increasing dose of polymer, peaking for the *in vivo* studies at approximately 1 microgram of polymer per mouse and for the *in vitro* studies at a concentration of approximately 1 picomole polymer per liter, 10^{-12} M. At polymer concentrations above these remarkably low optimum dose levels, the immune response declined rapidly with increasing dose, indicating that the immunogenic polymer was strongly inhibiting *itself* when present above its optimum concentration.

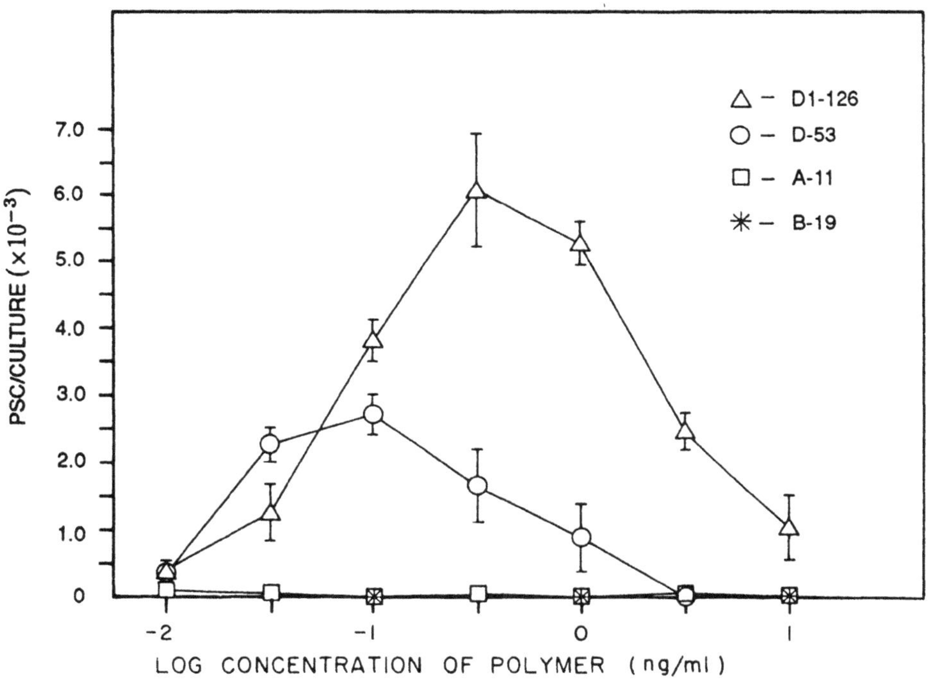

FIGURE 1 *In vitro* dose-response curves to four Dnp-polyacryamide polymers of comparable hapten density, but differing hapten numbers and molecular weights. Each point represents the mean and S.D. of triplicate 3-day cultures. The number of Dnp haptens per molecule is indicated by the subscript or the number after the dash. Molecular weights in kD are as follows: $D1\text{-}126 = 430$, $D\text{-}53 = 145$, $A\text{-}11 = 40$, $B\text{-}19 = 60$.[4]

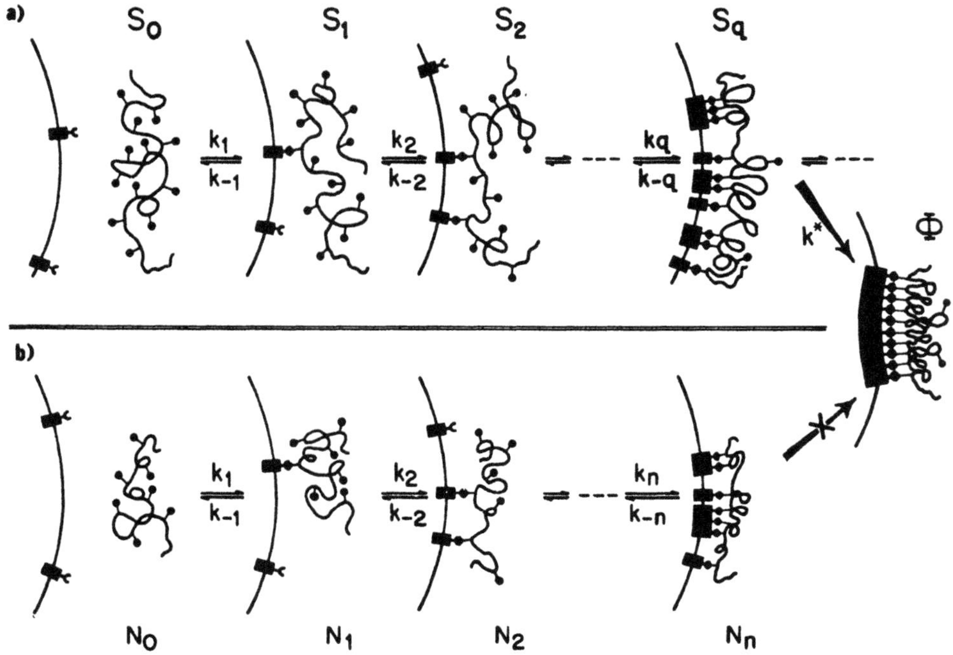

FIGURE 2 A diagrammatic representation of the stepwise assembly of an immunon cluster. The solid rectangles represent B cell surface mobile protein receptor molecules containing binding sites (cups). The highly curved solid lines represent flexible polyacrylamide molecules with the attached Dnp haptenic epitopes shown as solid circles. Receptors are shown as reacting stepwise with epitopes on an immunogenic polymer, S, or on a non-immunogenic polymer, N. Polymer N cannot form immunons, Φ, because it cannot simultaneously bind enough receptors.[2,3]

OBSERVATION 3. The smaller, i.e., under 100 kD, non-immunogenic forms of Dnp-polyacrylamide, although themselves incapable of generating positive immune responses, could strongly inhibit the immune response to the larger immunogenic forms of the same kind of molecule. This inhibition was very potent, with 50% inhibition occurring at the very low dose levels which corresponded to the doses giving the optimum response for the immunogenic forms of polymer. This fact strongly suggests competitive inhibition, but does not indicate its mechanism or site of action.

THE IMMUNON MODEL

Consideration of the three experimental observations above led us to propose a molecular model, which we called the Immunon Model, whose essence is shown diagrammatically in Figure 2. The model proposes that a polymer molecule with properly spaced multiple copies of an epitope can engage multiple receptors on the surface of a B lymphocyte possessing receptors with specific affinity for that epitope. As a result of being simultaneously bound to epitopes on a spatially constrained polymer molecule, the receptors (surface bound IgM or IgD molecules) are concentrated, i.e., brought closer together in space than they would be otherwise. If enough receptors are concentrated sufficiently, for a long enough period of time, the cooperative assembly of a complex molecular signalling structure (an immunon) is facilitated. The formation of enough such immunons on the cell surface triggers a B cell to begin the slow process of cellular division and differentiation which accompanies immune stimulation.

This model is logically consistent with the three experimentally based observations listed above, and to our knowledge is the only simple molecular model which is so consistent. Observation 1 notes that molecules of a molecular weight smaller than approximately 100 kD did not stimulate an immune response. This observation is expected to result from the inability of small molecules to bring together a sufficient number of receptors to form a specific signalling structure. Since each receptor is a globular protein of relatively large physical size, a stimulatory molecule, able to link a sufficient number of them simultaneously, must have considerable linear spread of dimension. To achieve this spatial spread, a stimulatory molecule must have substantial mass, which our experiments indicate is in the range of 100 kD. Molecules with a molecular weight less than 100 kD simply do not have adequate mass to span a sufficient number of membrane receptors. Observation 3 notes that small non-immunogenic molecules can be strongly inhibitory. This observation is to be expected if such small polymer molecules, which cannot by themselves gather enough receptors to form a signalling structure, can steal receptors from other polymer molecules which could do so. Thus, these small, non-immunogenic molecules are competitive inhibitors. Such molecules can be viewed as intrinsically inhibitory. Observation 2 notes that at the high end of the dose-response curve, immunogenic molecules can be self-inhibitory. This observation is to be expected.

If, at sufficiently high concentration, immunogenic polymer molecules can steal receptors from each other, so that the ability of any one molecule to gather enough receptors for a sustained time period is reduced, i.e., competitive inhibition by like molecules. Thus, the particular properties of any one specific type of molecule can cause it to be alternatively immunogenic or inhibitory of immune response to epitopes contained within it, depending on dose. We can describe such a molecule as "amphigenic." It can be expected that both amphigens and intrinsically inhibitory molecules will exert considerable control over the extent of an immune response in a dosage-dependent manner.

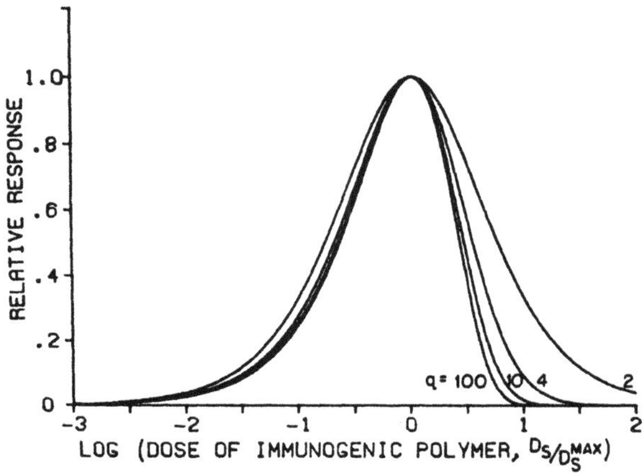

FIGURE 3 Theoretical dose-response curves calculated from Eq. 1 for various values of the immunon number, q.[2]

PREDICTIONS OF THE IMMUNON MODEL

Straightforward mathematical modelling of Figure 2 led us to an equation which predicted the expected dose response and dose inhibition as a function of the amount of antigenic material.[2] Thus, if we consider a group of experimental animals, G, injected with doses D_S of stimulatory (amphigenic) polymer S, and D_N of nonstimulatory polymer N, while a second group of similar animals, G', receives different doses, D'_S and D'_N, the expected ratio of immune responses produced in the two sets of animals, r, has been shown to be given by the equation:[2]

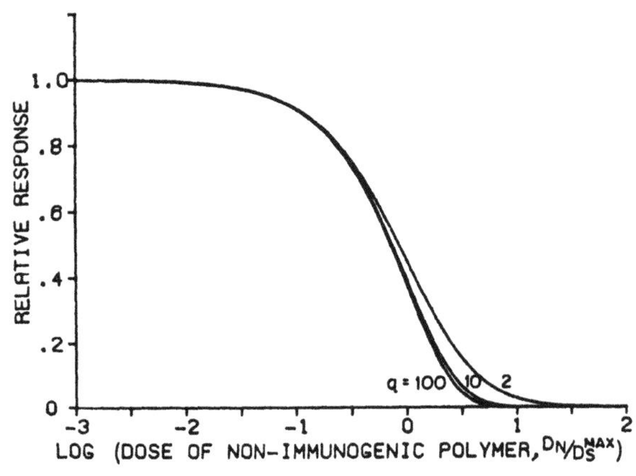

FIGURE 4 Theoretical dose-suppression curves calculated from Eq. 1 for various values of the immunon number, q, at constant value of $D_S = D_S{}^{\max}$.[2]

$$r = \frac{D_S}{D'_S}\left[\frac{(q-1)D_S{}^{\max} + D'_S + D'_N}{(q-1)D_S{}^{\max} + D_S + D_N}\right]^q , \tag{1}$$

where q is the minimum number of receptors which must be closely clustered in order to form an immunon and $D_S{}^{\max}$ is the dose of stimulatory polymer which gives the maximum immune response. Equation 1 can be used to predict the relative extent of immune responses produced by any combination of concentrations of stimulatory and nonstimulatory polymers *in vivo* or *in vitro*, as is illustrated in Figures 3 and 4.

Although the basis for the model is undoubtedly oversimplified, this mathematical modelling led to predictions which proved to be in substantial accord with experimental results as shown in Figures 5 and 6 for *in vivo* measurements. In Figure 6, the theoretical curve was not fitted to the experimental dose-inhibition points, but was calculated from the data of Figure 5 and the theoretical model equations. Measurements of the same type done *in vitro* have given agreement between theory and experiment similar to that obtained when the experiments were done *in vivo*.[3]

Among the simplifications used to derive a workable mathematical model were the assumptions that all responding B cells have the same concentration of receptors per unit surface area and that the receptors on all binding cells have the same binding affinity for epitope. Neither of these assumptions is realistic in considering

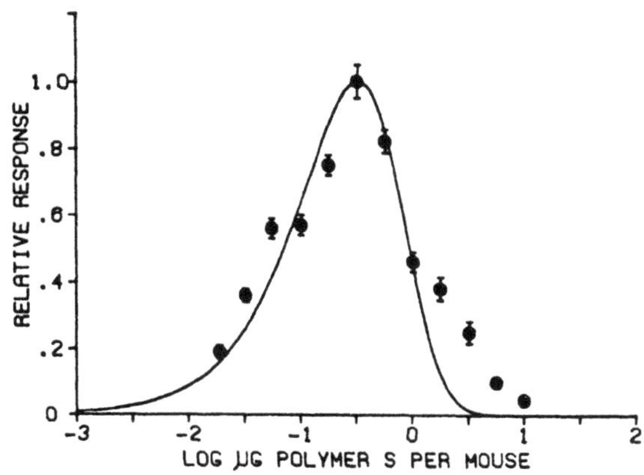

FIGURE 5 Dose-response measurements showing the mean and SEM of the relative concentration, in serum from individual mice, of IgM antibody against Dnp at 6 days after injection of polymer S (polyacrylamide, 130 kD, 43 Dnp groups per molecule) in amounts shown, 10 mice per point. The solid curve shows the theoretical response calculated from Eq. 1 for a peak response at a dose of 0.3 micrograms per mouse and an immunon size, q, of 10.[3]

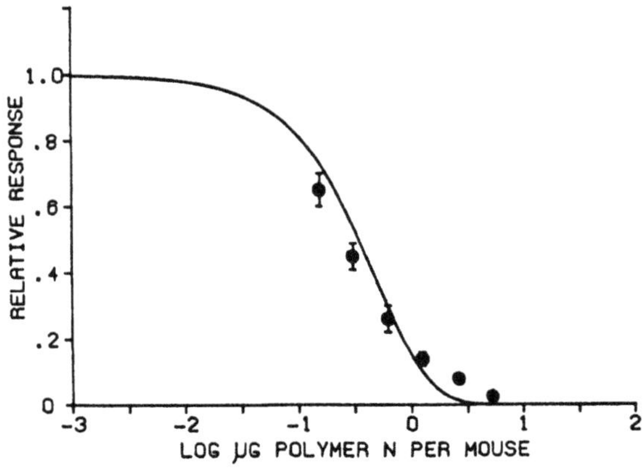

FIGURE 6 Response-inhibition measurements for increasing doses of non-immunogenic polymer N (polyacrylamide, 60 kD, 19 Dnp groups per molecule) injected simultaneously with a constant and optimum dose, 0.3 microgram, of polymer S (Figure 5). Each point shows the mean and SEM for serum measurements on 10 mice at 6 days. The solid curve shows the response expected from Eq. 1, calculated from the data of Figure 5.[3]

a heterogeneous population of B cells. As shown in Figure 7, the experimentally observed broadened dose-response curves are consistent with the cumulative responses of a population of somewhat diverse B cells contributing clonally to the total response. Further simplifications employed in the original immunon model include omission of the effects of some important molecular properties of the polymeric antigen, such as epitope spacing or "density," molecular shape, size, charge, rigidity, etc. There are also many properties of cell surface receptors, such as their number, spatial distribution, mobility, binding constant, etc. which must be brought into a coherent theoretical framework. An attempt has been made recently by Macken and Perelson[12] to define more precisely some of the complexities which need interpretation when a randomly haptenated linear polymeric antigen is used in experiments. Perelson[13] has outlined some of the difficulties in reconciling the results and conclusions discussed above with current theoretical models.

FLUORESCEIN AS A HAPTEN

For more recent experimental studies of immune response, we have used the haptenic epitope fluorescein covalently bonded to a number of different polymeric carriers as shown above in Table 1. Fluorescein is a substantially larger and more complex hapten than Dnp, containing multiple benzenoid rings as well as two phenolic groups and a charged carboxyl group. Nevertheless, the immune response to fluorescein attached to a number of chemically diverse polymers is remarkably similar to the immune response to Dnp on Dnp-polyacrylamide, indicating a high degree of generality to the observations. Both *in vivo* and *in vitro*, all the fluoresceinated polymers with molecular weights greater than approximately 100 kD and hapten valence greater than about 20 haptens per molecule were immunogenic, whereas polymers with molecular weights less than ~ 100 kD were non-immunogenic. As is illustrated in Figure 8 for *in vitro* studies, immunogenic fluoresceinated polymers gave bell-shaped, dose-response curves peaking at concentrations of a few micrograms per liter. *In vivo* studies (not shown) with the immunogenic fluoresceinated polymers gave similar bell-shaped, dose-response curves peaking within the range of 1 μg to 1 mg per mouse.

In analogy to the Dnp-polyacrylamide studies, non-immunogenic fluoresceinated polymers could strongly inhibit the immune response to immunogenic fluoresceinated polymers of the same chemical type (data not shown). The multiplicity of types of fluoresceinated polymer carrier permitted an experimental test of "cross inhibition," i.e., inhibition of immune response to hapten on one type of polymeric carrier by the same hapten on another type of polymeric carrier. As Figure 9 illustrates, such cross inhibition is readily demonstrable and is very potent for a variety of polymeric carriers, provided that they carry the same hapten

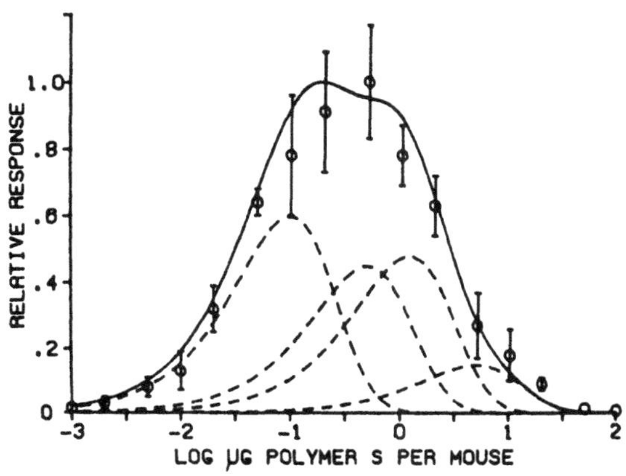

FIGURE 7 Fitting of an experimentally determined set of dose-response measurements by an additive sum of arbitrarily chosen, individual, theoretical cellular clonal responses calculated from Eq. (1). The experimental points are the mean and SEM for 5 mice per point. The solid curve is the sum of the individual dashed curves, each of which is calculated from Eq. 1 for different values of $D_S{}^{max}$ and different values of relative peak height.[2]

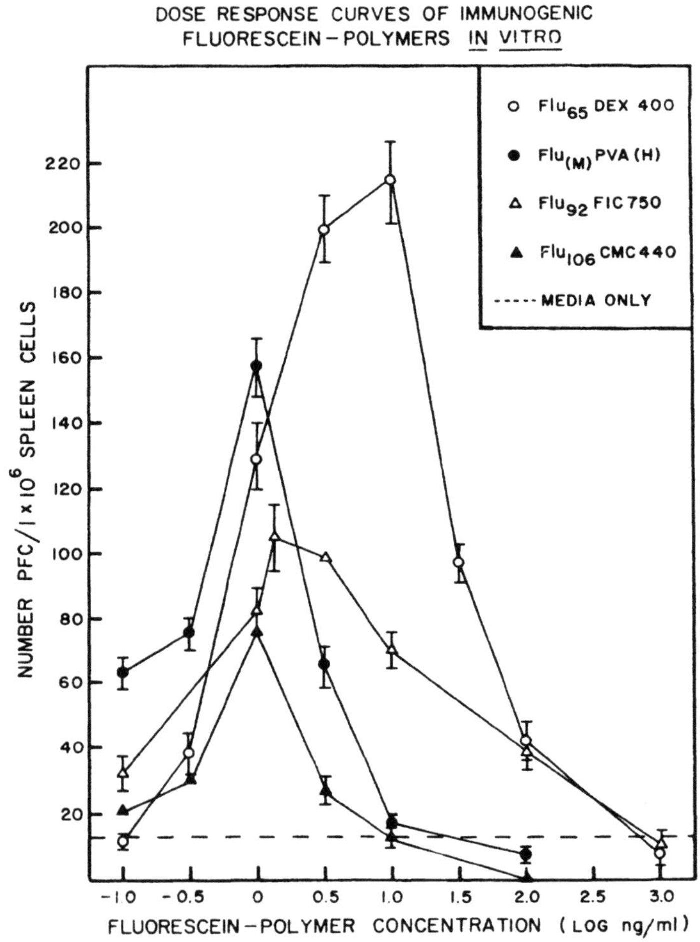

FIGURE 8 Dose-response curves for a number of different fluoresceinated polymers showing similar bell shaped curves. In the polymer description, the subscript after Flu indicates the number of haptenic epitopes per molecule and the number after the polymer abbreviation gives the molecular weight in kilodaltons, kD.[9,10]

(fluorescein). As illustrated in the figure, when the carrier is substituted with an irrelevant hapten, such as Dnp, it cannot inhibit the anti-fluorescein response. These data are in accord with the idea that competitive inhibition of immune response to an immunogenic polymer can be caused by competitive interference with critical receptor clustering. Figure 9 data indicate that the competitive inhibition is relatively independent of the chemical nature of the soluble polymer carrier, but is strongly dependent upon parameters such as molecular size and the number and spacing of haptenic epitopes.

NATURAL POLYMERS

The antigenic materials discussed above have all been artificial in the sense that they were all made by the chemical linkage of synthetic haptens to polymer molecules which were themselves chemically synthesized, with the exception of dextran, which is a bacterially produced polysaccharide. It is valid to question whether or not natural, biologically important polymeric immunogens, such as bacterial polysaccharides, would behave in a similar fashion with respect to their unaltered internal epitopes. Accordingly, we carried out analogous studies on the immunogenicity of the type specific bacterial polysaccharide produced by the pathogen,

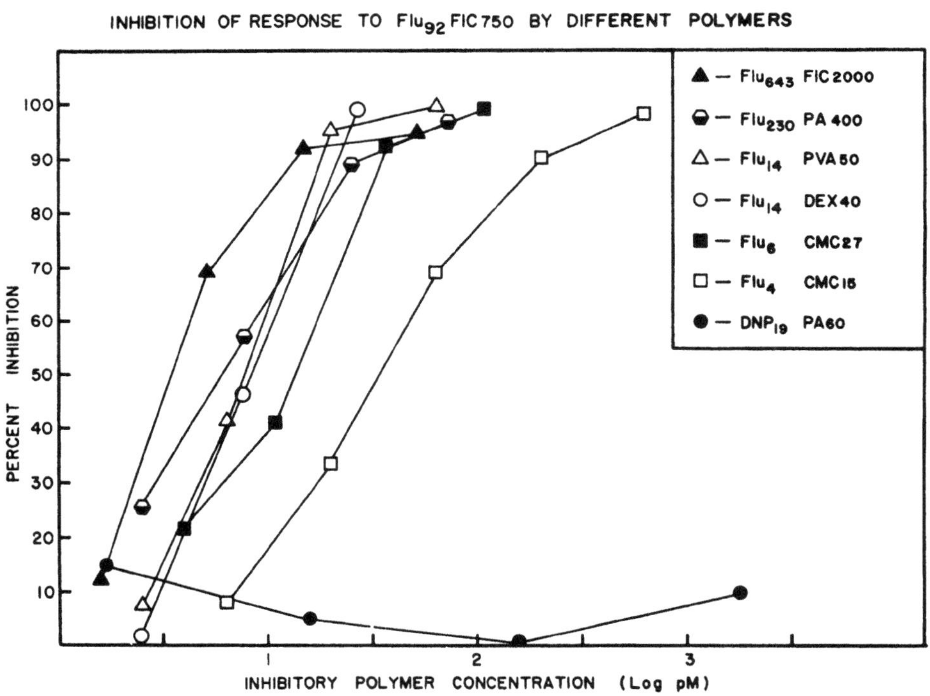

FIGURE 9 *In vitro* dose-inhibition curves showing "cross inhibition" of the response to a constant and near optimal dose of a stimulatory fluoresceinated ficoll polymer by a series of other haptenated polymers.[10] Polymer designations are as in Figure 8.

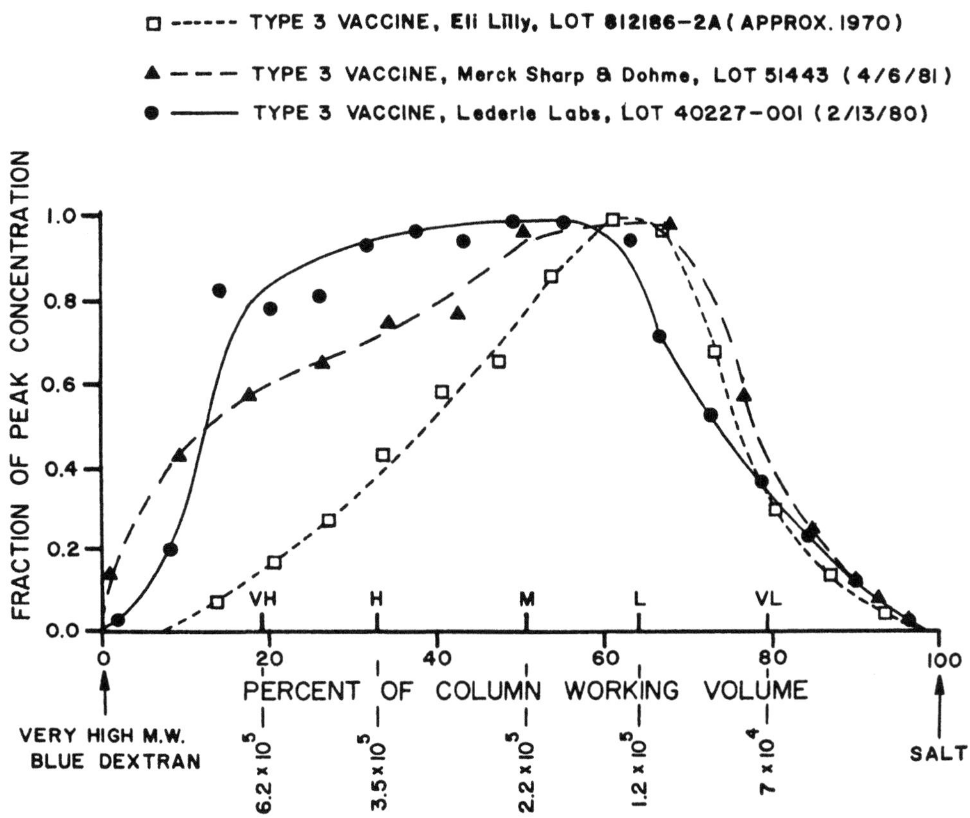

FIGURE 10 Fractionation of three different human vaccine preparations of PP3 by molecular size (gel filtration) on a CL-2B column. Refractionation of the material at the five points indicated gave relatively homogeneous fractions of the weight average molecular weights in Daltons as indicated.[11]

Streptococcus pneumoniae Type 3. This polysaccharide is often called pneumococcal polysaccharide Type 3, or PP3, and is routinely available for use in human vaccines. It is a linear structure made up of alternating glucose and glucuronic acid subunits. It is purified from cultures of the pathogenic bacteria. Several PP3 preparations which had been prepared for use in human vaccines were size-fractionated by us.

It may be seen (Figure 10) that the preparations were quite heterogeneous in their behavior on gel filtration columns (which separate on the basis of molecular size). Fractions obtained from such columns gave widely different molecular weights, as determined by measurement at sedimentation equilibrium in the analytical ultracentrifuge. When twice fractionated materials of five different sizes were tested

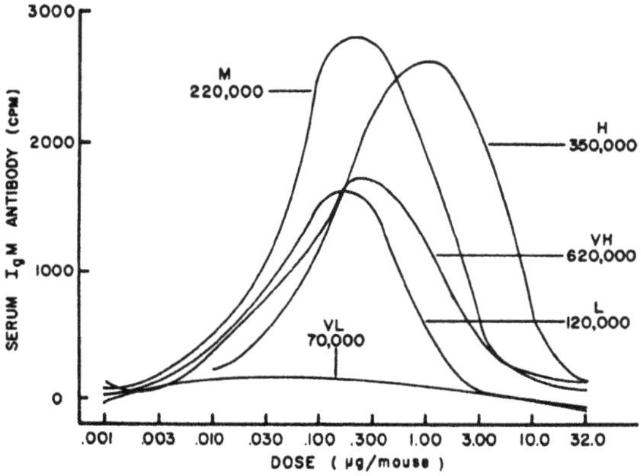

FIGURE 11 Dose-response curves *in vivo* for the five different fractions of PP3 shown in Figure 10. Each curve is the smoothed result of individual measurements on serum taken from each of 50 mice 6 days after injection of PP3 fractions. At each of the 10 indicated doses, 5 mice were used.

in dose-response curves *in vivo*, it again was found that preparations of molecular weight less than 100 kD were not immunogenic, Figure 11.

In order to preclude the possibility that the non-immunogenicity of smaller fractions was due to an unexpected difference in chemical composition between the material smaller than 100 kD and the larger material, the following experiment was carried out. A center fraction of molecular weight 250 kD was prepared from a recent human vaccine preparation and was found to be immunogenic *in vivo* and *in vitro*. It

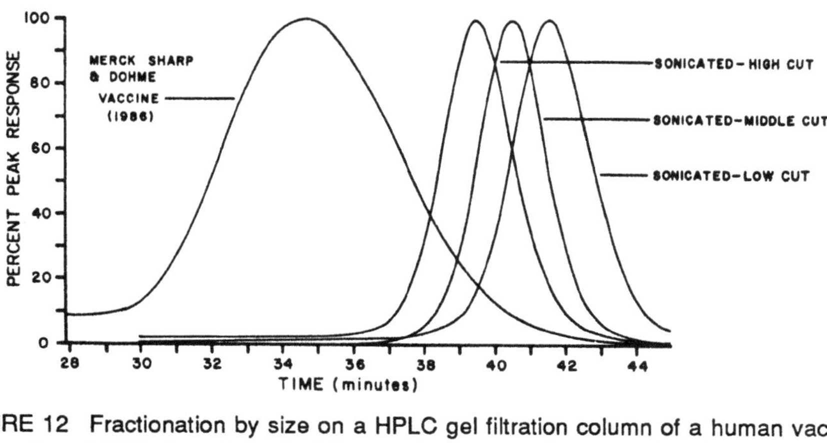

FIGURE 12 Fractionation by size on a HPLC gel filtration column of a human vaccine preparation of PP3. A center cut of the vaccine preparation was taken for immunogenicity studies before and after sonication as shown in Figure 13.

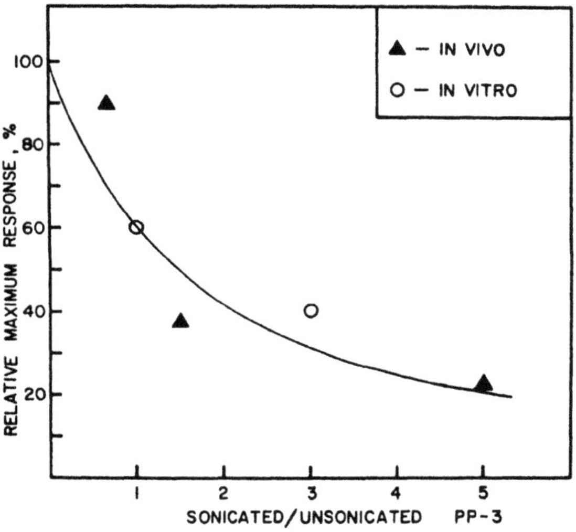

FIGURE 13 Inhibition of *in vivo* and *in vitro* responses to an optimal dose of immunogenic PP3 (250 kD) by sonicated fragments (35 kD, middle cut) derived from it.[11]

was then exposed to intense ultrasonic treatment in solution, which caused the large, linearly extended molecules of polysaccharide to be fragmented into smaller pieces by shear forces, yielding fragments in the size range of 20 to 50 kD. The fragmented preparation was then itself size fractionated on an appropriate gel filtration column, giving the result shown in Figure 12. Not only were the smaller molecules of PP3 completely non-immunogenic, *in vivo* and *in vitro*, but when administered with a comparable (and stimulatory) dose of the larger molecules, they could strongly inhibit the immune response to the larger molecules, from which they had been fragmented by ultrasonication. This is shown in Figure 13.

DISCUSSION

The experimental findings described above, based on the results of studies on a variety of soluble polymeric antigens, are consistent with the expectations of the immunon model. This apparent generality supports the conviction that the immunogenic effects of any polymeric antigen may be predictable provided one knows the molecular characteristics of that antigen. Among the more theoretically useful generalizations is the observation that polymeric antigen material may be either inhibitory (if it contains closely spaced epitopes of low multiplicity), or amphigenic (if it contains closely spaced epitopes of high enough multiplicity). The amphigenic material will stimulate the immune response to its multiple epitopes in a low dose range, but will suppress the immune response to its epitopes at higher doses by competitive self-inhibition.

An essential component for the modeling of regulating systems is a source of negative signal capable of maintaining a steady state condition in the presence of positive signal (see Segel and Perelson, this volume). Antigenic material, in general, has not been considered a source of negative signal since it has been viewed as being either ineffective (and needing the "help" of "adjuvants") or effective in stimulating an immune response. When negative effects have been found, they have often been ascribed to the "special" chemical nature of the carrier, to a special property such as "epitope density," or to the ability of special molecular characteristics of the antigen to induce populations of suppressor cells.

The immunon model proposes that, for soluble polymeric molecules, there can exist a range of sizes which have sufficient valency of epitopes to bind well to B cell receptors specific for such epitopes, but not enough to stimulate the cells. By acting as competitive inhibitors, these oligomeric molecules can act as negative molecular signals in regulating the immune response to their epitopes. A variety of molecular sizes and epitope valences is often present within a population of antigen molecules. The resulting immune response may reflect the sum of the positive or stimulatory effects of the polyvalent (amphigenic) molecules and the negative effect of the oligovalent (intrinsically inhibitory) molecules.

In our *in vitro* studies, where the concentration of soluble inhibitory oligovalent polymer molecules could be experimentally defined, 50% inhibition of stimulation typically occurs at concentrations of inhibitory polymer near 10^{-12} M. Inhibition is essentially complete at 10^{-11} M concentration. The effective binding of inhibitory polymer, which these extremely low concentrations demonstrate, is a consequence of the oligovalence of the epitopes on the polymer. This oligovalence results in a binding avidity which is much greater than the intrinsic binding affinity of monovalent epitopes. It should be noted that monovalent epitopes can also inhibit immune responses against such epitopes on carriers, but only at concentrations 7 orders of magnitude higher, i.e., 10^{-4} M.[2]

To date the immunon model is based on our experimental results derived from studies of the immune response to chemically simple epitopes contained in soluble polymeric molecules of relatively simple chemistry. Although the polymers we have used in these studies have differed from each other in chemical nature, internal geometric connectedness, molecular shape, size, electrical charge, hydrophilic character and synthetic or natural origin, they have shared two important characteristics: 1) the polymer backbones were all highly repetitious polymers of simple chemical moieties, and 2) the haptens or epitopes were multiple and homogeneous. This type of molecule stimulates the predominantly IgM antibody response associated with "T-cell independent" antigens. Such responses are believed to occur in the relative absence of "help" from antigen-specific T cells, with which the antigenic material could interact.

When appropriate experimental evidence is available, it will be interesting to see if, and how well, the simple immunon model holds for antigenic materials of more complex chemical composition, such as proteins. The response to such antigens is generally thought to be "T-cell dependent." Of particular interest with regard to the "molecular logic" of T-cell interaction with antigenic material has been a series of experiments conducted by Siliciano et al. in collaboration with our laboratory.[14–16] The ability of a number of our fluoresceinated polymers to bind to and to activate a variety of murine and human cytotoxic and helper T-cell clones was studied. Importantly, a number of our larger, highly haptenated, soluble polymers were found to bind to and to activate such clones. It is generally believed that T cells "see antigen" not in soluble form, but only when complexed with certain specific cell surface molecules on other cells. This conventional view may not be in accord with the data mentioned above.

Our paradigm may have to be modified to describe situations where the antigen is not in soluble form, but is adsorbed to or incorporated within complex structures such as cell surfaces, extracellular matrices, immune precipitates, etc. To test the applicability of the immunon model to these more complex situations, it will be necessary to obtain quantitative data using as experimental models physically complex antigenic materials. Importantly, although complex, such materials would have to be relatively homogeneous and well defined both chemically and physically. In guiding such future experimental efforts, the theoretical basis for modeling the systems should evolve in parallel with experiment, so that theory and experiment can continue to guide each other.

We have outlined above a hypothetical mechanism whereby the molecular state of the antigen could control the immune response. It is interesting to speculate on the possible biological significance of such a molecular control mechanism. A primary role of the immune response is to aid in the defense against foreign microorganisms and viruses. These organisms have external structures composed of highly repetitious and closely spaced arrays of molecules which are chemically different from those in the animal body. It seems very appropriate, therefore, that large dense arrays of "foreign" molecular structures are responded to by immunological defense mechanisms.

Normally, bodily defense mechanisms function efficiently and only very small numbers of microorganisms or viruses enter the body before they are destroyed and hydrolyzed, and the breakdown products eliminated. It is, therefore, unlikely that a steady supply of oligomeric (inhibitory) forms of foreign antigenic material from such sources would be present to inhibit lymphocytes with receptors specific for such antigens. However, if the foreign organisms produce and secrete large quantities of inhibitory oligomeric material of a chemical composition similar to their potentially immunogenic surface molecules, they may inhibit the immune response against themselves and thus escape immune system attack. Hence, the potential for immune inhibition exists when the sustained formation of oligovalent forms of antigenic material occurs. This may be one of the mechanisms by which such diverse troublemakers as highly pathogenic microorganisms, parasites and cancer cells escape immune surveillance.

To extend speculation still further, we will attempt to explain the long-standing immunological mystery of the difference between "self" and "nonself" in terms of the immunon concept. Applying the immunon concept in its simplest form, one would expect the many large arrays of identical molecules found on cell surfaces and extracellular matrices all over the animal body to be highly immunogenic. Such polymeric forms should, therefore, constitute a highly diverse collection of positive immunological signals. However, these diverse molecular arrays within the body are "self" and must not be immunologically attacked. What and where then, are the counterbalancing negative signals, i. e., the inhibitory oligomers preventing anti-self response, which are implied by the immunon model?

We hypothesize that:

1. Inhibitory oligomers are continuously being produced by degradative turnover of the macromolecules of the body through normal catabolic processes including the activities of macrophages.
2. Normal degradative turnover of body components produces such oligovalent molecular "dust" or "lint" within the body at a steady state concentration of 10^{-11} M or greater for each of the potentially immunogenic molecules of the body. This oligomeric molecular "self-dust" could bind nonproductively to the receptors on potential "anti-self" cells, thus inhibiting immunon formation and subsequent "anti-self" immune response. The "self-dust" could exist in soluble form or be "presented" by cells such as macrophages. "Anti-self" cells could be any B or T lymphocyte with antigen receptors which recognize a "self" molecule.
3. It is the oligomeric (oligovalent) nature of some of the "self-dust" which allows it to inhibit an anti-self immune reaction at such extremely low concentrations as 10^{-11} M. Interference with the steady state production of any particular kind of "self-dust" may lead to immunological response against any tissue which displays arrays of the molecules from which that "dust" was derived. This could lead to an anti-self response, i.e., an autoimmune response. The autoimmune response could be implemented by B cells, T cells, or by the combined activities of both.

It is well documented that the blood contains in soluble form many molecular components derived from cells and tissues, as well as a diverse collection of material in "immune complexes." The physical and chemical state of this material is not known. An appropriate fraction of it might correspond to the "missing molecular dust" required by the immunon model of "self."

This intriguing biological puzzle of the missing "self-dust" has analogies to the puzzle of the "missing mass of the universe" which is currently perplexing cosmologists. In both cases one can have reasonable confidence that a proper combination of theory and experiment will clarify the questions.

ACKNOWLEDGMENTS

We should like to thank Dr. Eaton E. Lattman for suggesting the word amphigen and for critical reading of the manuscript. The work reported here has been supported in part by Contract DAMD 17-86-C-6038 from the U.S. Army Medical Research and Development Command, but these views, opinions, and/or findings of the research do not necessarily reflect the position, or decision of the United States Army and no official endorsement should be inferred. The work was additionally supported by funds from Cortech Inc.

REFERENCES

1. Dintzis, H. M., R. Z. Dintzis, and B. Vogelstein (1976), "Molecular Determinants of Immunogenicity: The Immunon Model of Immune Response," *Proc. Natl. Acad. Sci. U.S.A.* **73**, 3671.
2. Vogelstein, B., R. Z. Dintzis, and H. M. Dintzis (1982), "Specific Cellular Stimulation in the Primary Immune Response: A Quantized Model," *Proc. Natl. Acad. Sci. U.S.A.*. **79**, 395.
3. Dintzis, R. Z., B. Vogelstein, and H. M. Dintzis (1982), "Specific Cellular Stimulation in the Primary Immune Response: Experimental Test of a Quantized Model," *Proc. Natl. Acad. Sci. U.S.A* . **79**, 884.
4. Dintzis, R. Z., M. H. Middleton, and H. M. Dintzis (1983), "Studies on the Immunogenicity and Tolerogenicity of T-Independent Antigens," *J. Immunol.* **131**, 2196.
5. Dintzis, R. Z., M. H. Middleton, and H. M. Dintzis (1985), "Inhibition of Anti-Dnp Antibody Formation by High Doses of Dnp-Polyacrylamide Molecules; Effects of Hapten Density and Hapten Valence," *J. Immunol.* **135**, 423.
6. Dintzis, R. Z., M. H. Middleton, and H. M. Dintzis (1985), "Use of Systematic Structural Variation of Antigenic Molecules to Study Signal Generating Receptor Aggregates in Immunocyte Membrane," *Biophysical J.* **47**, 441a.
7. Dintzis, H. M., M. H. Middleton, and R. Z. Dintzis (1985), "Antigen Molecular Structure is a Primary Controlling Factor in the Regulation of the Immune Response," *Biophysical J.* **47**, 200a.
8. Dintzis, R. Z., M. J. Middleton, and H. M. Dintzis (19850, "Induction of a Cellular Differentiation Signal by Critical Membrane Receptor Aggregation," *J. Cell. Biol.* **101**,428a.
9. Okajima, M., H. M. Dintzis, and H. M. Dintzis (1986), "Molecular Parameters Influencing Immunological Properties of Haptenated Polymers," *6th International Congress of Immunology, Toronto, Canada*, 240

10. Okajima, M., A. Lees, H. M. Dintzis, and R. Dintzis (1986), "B -Lymphocyte Activation via Receptor Cross-Linkage by Macromolecules: Molecular Properties Determine Activation Capabilities," *J. Cell Biol.* **103**, 506a.
11. Dintzis, H. M., M. H. Middleton, A. Wolitzky, and R. Z. Dintzis (1987), "The Murine Antibody Response to Pneumococcal Polysaccharide Type 3 is Stimulated by Molecules of 100 kD or Larger and is Inhibited by Low Molecular Weight Oligomers," *Fed. Proc.* **46**, 777.
12. Macken, C. A., and A. S. Perelson (1986), "Renewal Theory, Geiger Counters, and the Maximum Number of Receptors Bound to a Randomly Haptenated Polymer Chain," *IMA J. Math. Appl. in Med. and Biol.* **3**, 71.
13. Perelson, A. S. (1986), in *Paradoxes in Immunology* , Eds. G. W. Hoffmann, J. G. Levy, and G. T. Nepom (Boca Raton, Florida: CRC Press, Inc.), 199–214.
14. Siliciano, R. F., A. D. Keegan, R. Z. Dintzis, H. M. Dintzis, and H. Shin (1985), "The Interaction of Nominal Antigen with T Cell Antigen Receptors. I. Specific Binding of Multivalent Nominal Antigen to Cytolytic T Cell Clones," *J. Immunol.* **135**, 906.
15. Siliciano, R. F., R. C. Colello, R. Z. Dintzis, H. M. Dintzis, and H. S. Shin (1985), "The Interaction of Nominal Antigen with T Cell Antigen Receptors. II. Antigen Valence Determines the Binding of Nominal Antigens to T Cell Clones," *J. Exp. Med.* **162**, 768.
16. Siliciano, R. F., T. J. Hemesath, J. C. Pratt, R. Z. Dintzis, H. M. Dintzis, O. Acuto, H. S. Shin, and E. I. Reinherz (1986), "Direct Evidence for the Existence of Nominal Antigen Binding Sites on T Cell Surface Ti α-β Heterodimers of MHC-Restricted T Cell Clones," *Cell* **47**, 161.

GEORGE I. BELL
Theoretical Division, Los Alamos National Laboratory, Los Alamos, New Mexico 87545

Cell-Cell Adhesion Resembles Receptor Crosslinking

INTRODUCTION

In the immune system, *transient* and *specific* adhesions between cells are commonplace. The adhesions are transient so that the highly mobile cells involved, including macrophages and lymphocytes, do not become permanently stuck to one another, but can continue their explorations and functions. The adhesions are specific in that specific cell surface molecules, "receptors," on each of the cells mediate the adhesion. For some of the adhesions, the nature of one or both of the receptors is fairly well known, but even in the absence of such information, we can draw some conclusions as to the consequences of an adhesive interaction, given only that the receptors have some mobility on the plane of the membrane. In particular, we will argue that in such cell-cell adhesions there will be competition between specific bonding and nonspecific repulsion such that the particular receptors will become highly concentrated in relatively small areas of cell-cell contact. Thus, the adhesive interaction that involves bonding between receptors on two different cells may achieve many of the same results as crosslinking of receptors on either cell because, in either case, a high local concentration of receptors is achieved in the membrane of the cell. We are less able to explain why the adhesions are transient; one or both cells must do something active to terminate the adhesion, and some possibilities are discussed.

RECEPTORS

In the development and functioning of an immune system, there are probably many important interactions between lymphocytes or their progenitors and other cells, including macrophages, epithelial, dendritic, and endothelial cells. Perhaps the first such interactions occur in bone marrow during the development of pre-B cells and pre-thymocytes from the pluripotent stem cells. In the thymus, the situation seems clearer, and adhesion of the thymocytes to epithelial and perhaps dendritic cells probably plays a role in generating self-tolerance and MHC restriction of T-cell responses (Gefter and Marrack, 1986; Lo and Sprent, 1986). These adhesions would involve interaction between T-cell receptors (Marrack and Kappler, 1986) and molecules of the major histocompatibility complex (MHC) on the other cell. In this connection, Farr et al. (1985) have shown by electron microscopy, cortical thymocytes binding epithelial cells—and with a high concentration of the T-cell receptors on that part of the thymocyte membrane that is in contact with the epithelial cell.

During recirculation of lymphocytes from blood to lymph, a particular receptor on the lymphocyte membrane mediates attachment of the lymphocytes to high endothelial cells lining the post-capillary venules that pass through lymph nodes and other lymphoid tissues. These receptors have been found to be branched-chain glycoproteins carrying the small polypeptide, ubiquitin, on one or more of its branches (Siegelman et al., 1986). Inasmuch as different kinds of lymphocytes show preferences for adhesion to gut-associated high endothelium as opposed to that in peripheral nodes, it is likely that there are corresponding differences in the receptors involved. The complementary receptors on the endothelial cells are uncharacterized as yet.

Antigens, and/or fragments thereof, are evidently involved in the adhesion of lymphocytes to one another or to antigen-pressing cells (APC). In particular, T-lymphocytes bind strongly to APC (Powell et al., 1980), presumably binding by their receptors to a complex of antigen fragments together with MHC molecules. In any case, this adhesion leads to antigen-specific and genetically restricted lymphocyte activation (Rosenthal, 1980; Unanue, 1980). In this connection, it may be noted that the T-cell receptor, a heterodimer of two glycoproteins called α and β (Marrack and Kappler, 1986), is physically associated with several other proteins on the T-cell membrane including three polypeptides comprising, on human cells, the T-3 complex (van den Elsen et al., 1984; Gold et al., 1986) and possibly additional proteins (Weissman et al., 1986; Oettgen et al., 1986). Surface expression of the T-3 complex appears to require the presence of T-cell receptor proteins (Ohashi et al., 1985). Moreover, monoclonal antibodies to T-3 are mitogenic for T-cells and/or block cell adhesion. These findings suggest that the association of the T-3 complex with receptors is functional with the T-3 proteins delivering activation signals to the cell, following binding of receptors to the APC and the consequent aggregation of receptor and T-3 proteins in the region of cell-cell contact.

During the effector phase of an immune response to cell-associated antigens, cell-cell adhesion is required for the elimination of the antigen. Thus, cytotoxic T-cells form tight conjugates with their target cells before killing them (Bongrand et al., 1983), while phagocytes bind to and engulf bacteria to which antibodies have become bound. In the latter case, the phagocytes such as macrophages have on their surfaces specific receptors for the constant regions (F_c) of the antibodies. No doubt additional examples of cell-cell adhesion in the immune system will be clarified in the future. For example, in B-cell activation by helper T-cells, the relative importance of cell-cell contact as contrasted with communication by soluble factors (interleukins) remains to be determined.

SPECIFIC ADHESION VS. NON-SPECIFIC REPULSION

In the immune system, as in other tissues of multicellular creatures, it is important that the right cells stick to one another and the wrong ones do not. Interactions between the right cells is mediated by specific cell surface receptors such as those noted above. That cells do not indiscriminately stick to one another is provided by the nature of the cell-surface coat; namely, the hydrophilic glycocalyx (Alberts et al., 1983) which is composed of the extracellular portions of cell-surface glycoproteins and other associated macromolecules. These molecules have the general property that they are hydrophilic, and in addition, many are negatively charged. When two cells approach each other such that their glycocalyces begin to interpenetrate or compress each other, water must be removed from the space between the cells. The removal of water from the space adjacent to hydrophilic polymers increases the free energy of the system, and thus, gives rise to a repulsive force between the cells—although the repulsion is highly screened by ions in the aqueous medium.

These repulsive forces have been estimated (Bongrand et al., 1982; Bongrand and Bell, 1984) for a representative model of the glycocalyx, resembling that of the red cell. The hydration force was estimated using a theory of polymers in solution that has been successfully applied to analyze the stabilization of colloidal suspensions by solvated polymer attached to their surfaces (Napper, 1977). It was found that this so-called steric stabilization force is about an order of magnitude larger than the electrostatic repulsion, which is, in turn, larger than the van der Waals attraction between the cells for membrane separations of interest. It follows that there is a net repulsion between cells in the absence of specific receptor-mediated interactions between them.

Thus, in considering cell-cell adhesion, it is necessary to evaluate the competition between specific bonding and nonspecific repulsion. Three aspects of this competition may be considered. First of all, we may seek the configuration of adhering cells at *equilibrium*, ignoring any active intervention by the cells. Theoretical results are known for this case and will be considered in the next section. But a second aspect concerns the time required to achieve or approach equilibrium. This involves the spatial extension of the receptors and glycocalyx perpendicular to the cell membrane; if, for example, the glycocalyx extends much farther out than the receptors, it may be very hard to establish any bonding. In that case, a strong force pushing the cells together, and/or a long time waiting for favorable fluctuations and/or cell activity perhaps locally separating receptors and glycocalyx, may be required to initiate adhesion. The latter possibility may be important since first cell contacts are likely to involve microvilli or other extensions from at least one of the cells, and it is likely that such extensions would be specialized for purposes of exploration, contact, and adhesion. In addition, it will be seen that, at equilibrium, a substantial redistribution of receptors into the contact area is to be expected. This redistribution can take place by diffusion of the receptors in the plane of the membrane. If the diffusion constant for such motion is D, then the time, t, for receptors to accumulate in a contact area, A, is $t \simeq A/D$. If, for example, $A \simeq 1\mu m^2 = 10^{-8} cm^2$ and $D \simeq 10^{-10} cm^2/sec$, as is common for many receptors, then $t \simeq 100\, sec$. We shall return to receptor diffusion when later considering *thresholds* for adhesion.

Finally, what is the role of cell activity in establishing or maintaining adhesion? Since, as will be seen, a common result of cell adhesion is cell activity, it should not be surprising if this activity would modify the adhesion; we will argue that a common result for cells of the immune system must be termination of the adhesion. In other cases, cell activity may favor adhesion; a mobile cell, for example, may expand its contact area simply by moving its membrane in a vigorous manner, which we have referred to as the "tar baby" effect (Bell, Dembo, and Bongrand, 1984).

EQUILIBRIUM PREDICTIONS

Consider two cells having receptors capable of forming intercellular bonds and also possessing glycocalyces producing nonspecific repulsion. We assume that the receptors are mobile in the planes of cell membranes. Suppose that the cells are in contact over an area, A, in the sense that intercellular bonds can form only in this area. Assuming the adhesion takes place, it is then clear that receptors will accumulate in the contact area. We wish to estimate the number of bonds per unit area, the contact area, and the separation between cell membranes—all at thermodynamic equilibrium. To this end, we have developed some thermodynamic models of cell adhesion. Independent variables include the area of each cell and number of receptors per cell, together with characterizations of the free energy and deformability of the bonds and of the nonspecific repulsive forces. For the bonds, we may take representative antigen-antibody bonds, while for the repulsion we use values estimated earlier (Bongrand and Bell, 1984).

Three different models have been considered. In the first (Bell et al., 1984), receptor molecules were assumed to be mobile in the plane of each membrane, but the elements of the glycocalyx (repellers) are assumed immobile. In the second (Torney et al., 1986), the repellers are likewise treated as mobile, while a third treats the adhesion of a cell with mobile receptors and fixed repellers to a surface having immobile ligands that can bind the receptors (Dembo and Bell, 1987). These are all reviewed and compared in Bell (1988).

In each case, we begin by writing down an expression for the Gibbs free energy of the system consisting of two cells in contact over an area, A. We then seek to minimize the free energy with respect to independent variations of A; of the number of bonds per unit area, n_b; and of the cell-cell, or really membrane-membrane separation distance, S—subject to constraints on the allowable values of A, n_b, and S. The major results from the first model (Bell et al., 1984) with mobile receptors and immobile repellers may be summarized as follows.

There will be a *threshold* for cell adhesion when

$$n_1 n_2 K(S) = \Gamma(S)/kT, \tag{1}$$

where n_i is the number of free receptors per unit area on cell i, $K(S)$ is the bond equilibrium constant, $\Gamma(S)$ is the energy of repulsion per unit area, k is Boltzmann's constant, and T the temperature. When the left size of Eq. (1) exceeds the right side, adhesion is thermodynamically expected; otherwise, it is not. Note that this is an equilibrium threshold. A more stringent kinetic threshold may be defined by considering that adhesion must be established in a time so short that additional receptors cannot diffuse into a contact area. This case of locally mobile, but globally immobile, receptors is analyzed in Bell (1988) where it is seen that for $n_i K(S) >> 1$, the kinetic threshold is substantially more stringent. The applicability of this threshold to various experiments is also considered in Bell (1988).

For parameters above, the adhesion threshold, and for particular and reasonable forms of $K(S)$ and $\Gamma(S)$, we derived simple expression of S, A, and n_b at equilibrium. Of greatest interest for present purposes is the general expression

$$n_b = \Gamma(S)/kT. \tag{2}$$

This equation may be understood by noting that $n_b kT$ represents a pressure tending to expand the contact area, while $\Gamma(S)$ is a pressure tending to decrease the area. At equilibrium, these two pressures must be equal as stated in Eq. (2). A representative value for the surface energy is $\Gamma(S) = 0.2\, ergs/cm^2$, and when used in Eq. (2), this yields $n_b = 0.05/nm^2$. This means that in any area of 100 Å on a side, we would expect about five bonds in the contact area. For large and bulky receptors such as the T-cell receptor and T3 complex, such high values of n_b may be impossible to achieve at close packing. A more complex model in which the density of receptors is limited has been developed (Bell, unpublished), but in any case, we predict that high receptor concentrations are to be expected in regions of cell-cell adhesion.

To some extent, high concentrations of receptors are to be expected whenever they mediate adhesion over a small portion of the cell surface, independent of nonspecific repulsion. The repulsion serves, then, to limit the contact area and provide a threshold for adhesion; without repulsion, the cells would tend to spread over each other so as to increase bond entropy until limited by mechanical or other constraints. To see this, note that Eqs. (1) and (2) may be combined to give

$$n_b = K(S)n_1n_2, \tag{3}$$

as would be expected on the basis of simple mass action (Bell, 1978). $K(S)$ is an equilibrium constant for reactants on a surface, which is approximately the regular equilibrium constant; K' for reactants in solution divided by some distance, d, within which reactants may be localized relative to a cell surface. (Alternatively, we could replace $K(S)$ by K' in Eq. (3) and replace one of the receptor surface densities, say, n_1, by a volume density, n_1/d.) Suppose, for example, that K' has the modest value of $10^6\, M^{-1} = 1.7 \times 10^{-3} \mu m^3$ and that $d \simeq 5\, nm$, so that $K(S) \simeq 0.3 \mu m^2$. If n_1 and n_2 correspond to 10^5 receptors on a cell of area $200 \mu m^2$, as for Ig receptors on a B cell, then $n_1 = n_2 = 500 \mu m^{-2}$; and from Eq. (3), $n_b = 7.5 \times 10^4 \mu m^{-2}$. For this example, $n_1 d = 10^5 \mu m^{-3} = 1.7 \times 10^{-4} M$. Thus, the very high local concentration of reactants near the cell surface can lead to very high concentrations of bonds in a small contact area even for a modest equilibrium constant. Note, however, that if, for the above numerical values, the contact area were as large as $1 \mu m^2$, the predicted number of bonds would be 7.5×10^4, which would seriously reduce the number of available *free* receptors on either cell. When nonspecific repulsion is included with $\Gamma = 0.2\, ergs/cm^2$, the analysis (Bell et al., 1984) gives a predicted contact area of $0.37\ \mu m^2$ within which, as seen earlier, $n_b = 0.05/nm^2$, so that about 17% of the receptors are engaged in the binding. This example illustrates that the nonspecific repulsion plays an essential role in limiting the contact area to a small value within which the very high local concentration of reactants leads to very high concentrations of receptor bonds.

It is also possible for mechanical forces to constrain contact areas. For smooth cells, the mechanics have been analyzed by Evans (1985a and 1985b), but for the rather fluid cells of the immune system, the constraints are probably not of general importance. Of course, the contact area may also be limited to the top of a microvillus or other specialized protrusion.

Accumulation of receptors in contact areas has been seen for a number of systems, but has been difficult to quantitate (Bell, 1988). Predicted and observed contact areas for thymocytes agglutinated by Con A were found to be in reasonable agreement (Bell et al., 1984; Capo et al., 1982). In general, functional contacts between cells seem to take place at highly local and specialized areas, including tight and gap junctions, desmosomes, and synapses (Alberts et al., 1983). In the case of gap junctions, Abney et al. (1987) have shown that the high concentration of junctional proteins in a gap junction is not maintained by attractive forces between the proteins within the membrane of either cell. Rather, these forces are repulsive, and the junction appears to be maintained by confining pressure of the excluded glycocalyx.

CONSEQUENCES OF RECEPTOR ACCUMULATION IN CONTACT AREAS

The thermodynamic model of the preceding section shows that there is a natural tendency of receptors and associated molecules to become highly concentrated in local regions of cell-cell contact. Hence, specialized structures may become self-organized in contact areas. The predicted lateral spacings of receptors can be quite comparable with spacings between receptors that have been crosslinked by an external ligand. Hence, we may expect that any effects that can be achieved by crosslinking of receptors on a cell can also be obtained in cell-cell adhesions mediated by these same receptors. For example, it was shown (Daëron and Voisin, 1978) that mast cells armed with IgG specific for certain MHC proteins will degranulate on adhesion to cells of the appropriate MHC specificity. Similarly, binding of a mast cell to a lectin-coated particle can lead to receptor clustering and a localized degranulation (Lawson et al., 1978).

It is likely that clustered receptors serve as, or lead to, the formation of specific ion channels or other activation signals to cells. Moreover, receptor clustering appears to play a role in the organization of the cytoskeleton, as may be seen in the process of capping whereby receptors crosslinked by various lectins or antibodies can be moved, probably by cytoskeletal action, to one pole of a cell. It has also been noted that binding of lectin-coated particles to a lymphocyte surface leads to a general decrease in the mobility of cell surface molecules; this was attributed to an overall anchorage modulation of surface proteins caused by mobilization of the cytoskeleton (Edelman, 1976). In addition, crosslinking of IgE molecules on rat basophilic leukemia cells rapidly leads to cytoskeletal changes and to their immobilization (Pfeiffer et al., 1985; Menon et al., 1986; J. Oliver et al., this volume; B. Baird et al., this volume).

HOW IS ADHESION TERMINATED?

Adhesive interactions are sometimes long lived as, for example, between epithelial cells, but for the mobile cells of the immune system, they are generally transient. Unsticking cannot be achieved simply by waiting for the bonds to break, for although individual bonds are breaking and reforming spontaneously all the time, they are quite strong and stable collectively (Bell, 1978). In general, it appears that cells must do something active in order to terminate the adhesion. In principle, this could be done in many ways. Motile cells crawling on adhesive substrates will often break off cytoplasmic tails to escape their tethers and leave behind portions of themselves on the substrate. The force to effect tail rupture presumably arises from contractions in the cytoskeleton. For adhering cells in the immune system, it is unclear how cytoskeletal forces can come to pull the cells apart, although forces tending to peel the cells apart could be much smaller than those required to rupture all bonds at once.

There are other contexts in which cell receptors must release ligands to which they have become bound. During receptor-mediated endocytosis (Albers et al., 1983), receptors plus ligands become internalized within vesicles inside the cell. A mechanism for uncoupling of the ligand-receptor bond is provided by proton pumps in the vesicle wall that can rapidly decrease the pH in the vesicle (Stone et al., 1983). Presumably such proton pumps also exist in cell plasma membranes where they could be used to acidify the microenvironment in the contact area between two cells, thereby dissociating intercellular bridges. A difficulty with this hypothesis is that, while protons may be readily confined within a vesicle, it is not so clear that they can be adequately confined against lateral diffusion in an extracellular microenvironment (Bell, 1988). Nevertheless, osteoclasts do appear to acidify an extracellular microenvironment (Baron et al., 1985), while cells of the digestive system evidently acidify a large volume.

A further clue may be provided by the observation that cells that cap receptor-ligand complexes often either internalize or shed such complexes. The mechanisms governing such disposal of complexes are not known, nor is it clear just how two cells would use them to terminate adhesion. Nevertheless, it seems that mechanisms for cap disposal may be related somehow to those for adhesion termination. Many possibilities could be imagined; for example, a form of cytokinesis removing a cap or adhesive contact from the rest of the cell, perhaps by contraction of a ring of cytoskeletal proteins around the circumference of the contact area. Further experimental evidence is needed.

SUMMARY

We have seen that whenever cell adhesion is mediated by receptors that are otherwise able to move in the plane of the cell membrane, a marked accumulation of receptors in a limited contact area is to be expected. The accumulation results from the high local concentration of membrane-associated reactants in the contact area, while nonspecific repulsive forces between cell glycocalyces cause the contact area to be small, or zero below the adhesion threshold. Since the densities of receptors in a contact area can be quite comparable to those locally achieved in crosslinked receptor-ligand complexes, we expect that cell-cell adhesion can produce all of the effects on cell behavior that are associated with receptor crosslinking. In some cases, the receptors or associated molecules, such as the T3 complex, may serve as selective ion channels.

We have also seen that the cells must actively terminate adhesion. Acidification of the extracellular microenvironment of the contact area is one possibility, as is enzymatic cleavage of receptors. Perhaps there is some endocytosis of material in the contact area. Whatever the process, it is likely for cells of the immune system that the mechanisms terminating adhesion are initiated by the adhesion itself.

ACKNOWLEDGMENTS

Work supported by the U.S. Department of Energy.

REFERENCES

1. Abney, J. R., J. Braun, and J. C. Owicki (1987), "Lateral Interactions among Membrane Proteins: Implications for the Organization of Gap Junctions," *Biophys. J.* **52**, 441.
2. Alberts, B., D. Bray, J. Lewis, M. Raff, K. Roberts, and J. D. Watson (1983), *Molecular Biology of the Cell* (New York: Garland).
3. Baron, R., L. Neff, D. Louvard, and P. J. Courtay (1985), ' 'Cell-Mediated Extra-Cellular Acidification and Bone Resorption: Evidence for Low pH in Resorbing Lacunae and Localization of a 100kD Lysosomal Membrane Protein at the Osteoclast Ruffled Border," *J. Cell Biol.* **101**, 2210.
4. Bell, G. I. (1978), "Models for the Specific Adhesion of Cells to Cells," *Science* **200**, 618–626.
5. Bell, G. I. (1988), "Models of Cell Adhesion Involving Specific Binding," *Cell Adhesion*, Ed. P. Bongrand (Boca Raton, Florida: CRC Press), in press.
6. Bell, G. I., M. Dembo, and P. Bongrand (1984), "Cell Adhesion: Competition between Nonspecific Repulsion and Specific Bonding," *Biophys. J.* **45**, 1051–1064.
7. Bongrand, P., and G. I. Bell (1984), "Cell-Cell Adhesion: Parameters and Possible Models," *Cell Surface Dynamics: Concepts and Models*, Eds. A. S. Perelson, C. DeLisi, and F. W. Wiegel (New York: Marcel Dekker).
8. Bongrand, P., C. Capo, and R. Depieds (1982), "Physics of Cell Adhesion," *Prog. Surf. Membr. Sci.* **12**, 217–285.
9. Bongrand, P., M. Pierres, and P. Golstein (1983), "T Cell-Mediated Cytolysis: On the Strength of the Effector-Target Cell Interaction," *Eur. J. Immunol.* **13**, 424–429.
10. Capo, C., F. Garrouste, A. M. Benoliel, P. Bongrand, A. Ryter, and G. I. Bell (1982), "Concancavalin-A Mediated Thymocyte Agglutination: A Model for a Quantitative Study of Cell Adhesion," *J. Cell Sci.* **56**, 21–48.
11. Daëron, M., and G. A. Voisin (1978), "H-2 Antigens on Mast Cell Membrane as Target Antigens for Anaphylactic Degranulation," *Cell Immunol.* **37**, 467–472.
12. Dembo, M., and G. I. Bell (1987), "The Thermodynamics of Cell Adhesion," *Current Topics in Membranes and Transport* **29**, 71–90.
13. Edelman, G. M. (1976), "Surface Modulation in Cell Recognition and Cell Growth," *Science* **192**, 218.

14. Evans, E. A. (1985a), "Detailed Mechanics of Membrane-Membrane Adhesion and Separation. I. Continuum of Molecular Cross-Bridges," *Biophys. J.* **48**, 175.
15. Evans, E. A. (1985b), "Detailed Mechanics of Membrane-Membrane Adhesion and Separation. II. Discrete Kinetically Trapped Molecular Cross-Bridges," *Biophys. J.* **48**, 185.
16. Farr, A., S. K. Anderson, P. Marrack, and J. Kappler (1985), "Expression of Antigen-Specific Major Histocompatibility Complex-Restricted Receptors by Cortical and Medullary Thymocytes *in Situ*," *Cell* **43**, 543–550.
17. Gefter, M., and P. Marrack (1986), "Development and Modification of the Lymphocyte Repertoire," *Nature* **321**, 116–118.
18. Gold, D. P., J. M. Puck, C. L. Pettey, M. Cho, J. Coligan, J. N. Woody, and C. Terhorst (1986), "Isolation of cDNA Clones Encoding the 20k Nonglycosylated Polypeptide Chain of the Human T-Cell Receptor/T-3 Complex," *Nature* **321**, 431–433.
19. Lawson, D., C. Fewtrell, and M. Raff (1978), "Localized Mast Cell Degranulation Induced by Concanavalin A-Sepharose Beads," *J. Cell Biol.* **79**, 394.
20. Lo, D., and J. Sprent (1986), "Identity of Cells that Imprint H-2-Restricted T-Cell Specificity in the Thymus," *Nature* **319**, 672–675.
21. Marrack, P., and J. Kappler (1986), "The T-Cell and its Receptor," *Sci. Am.* **254(2)**, 36.
22. Menon, A. K., D. Holowka, W. W. Webb, and B. Baird (1986), "Cross-Linking of Receptor-Bound IgE to Aggregates Larger than Dimers Leads to Rapid Immobilization," *J. Cell Biol.* **102**, 541.
23. Napper, D. H. (1977), "Steric Stabilization," *J. Colloid Interface Sci.* **58**, 390–407.
24. Oettgen, H. C., C. L. Pettey, M. L. Malow, and L. Terhorst (1986), "A T3-Like Protein Complex Associated with the Antigen Receptor on Murine T Cells," *Nature* **320**, 272–274.
25. Ohashi, P. S., T. W. Mak, P. van den Elsen, Y. Yanagi, Y. Yoshikai, A. F. Calman, J. D. Stobo, L. Terhorst, and A. Weiss (1985), "Reconstitution of an Active Surface T3/T-Cell Antigen Receptor by DNA Transfer," *Nature* **316**, 606–609.
26. Powell, L. W., P. Hart, M. H. Nielsen, and O. Werdelin (1980), "Antigen-Dependent Physical Interaction between Human Monocytes and T Lymphocytes," *Scand. J. Immunol.* **12**, 467–473.
27. Pfeiffer, J. R., J. C. Seagrave, B. H. Davis, G. G. Deanin, and J. M. Oliver (1985), "Membrane and Cytoskeletal Changes Associated with IgE-Mediated Serotonin Release from Basophilic Leukemia Cells," *J. Cell Biol.* **101**, 2145.
28. Rosenthal, A. S. (1980), "Regulation of the Immune Response-Role of the Macrophage," *N. Engl. J. Med.* **303**, 1153–1156.
29. Siegelman, M., M. W. Bond, W. M. Gallatin, T. St. John, H. T. Smith, V. A. Fried, and I. L. Weissman (1986), "Cell-Surface Molecule Associated with Lymphocyte Homing is a Ubiquitinated Branched-chain Glycoprotein," *Science* **231**, 823–829.

30. Stone, D. K., X.-S. Xie, and E. Racher (1983), "An ATP-Driven Proton Pump in Clathrin-Coated Vesicles," *J. Biol. Chem.* **258**, 4059.
31. Torney, D. C., M. Dembo, and G. I. Bell (1986), "Thermodynamics of Cell Adhesion. II. Freely Mobile Repellers," *Biophys. J.* **49**, 501–507.
32. Unanue, E. R. (1980), "Cooperation between Mononuclear Phagocytes and Lymphocytes in Immunity," *N. Engl. J. Med.* **303**, 977–985.
33. van den Elsen, P., B.-A. Shepley, J. Borst, J. E. Coligan, A. F. Markham, S. Orkin, and C. Terhorst (1984), "Isolation of cDNA Encoding the 20k T3 Glycoprotein of Human T-Cell Receptor Complex," *Nature* **312**, 413–418.
34. Weissman, A. M., L. E. Samelson, and R. D. Klausner (1986), "A New Subunit of the Human T-Cell Antigen Receptor Complex," *Nature* **324**, 480–482.

GAYLE DELMONTE WETZEL
Basel Institute for Immunology, Basel, Switzerland

B Cell Clonogenesis Induced by Anti-Immunoglobulin (Anti-Ig) and Interleukins

INTRODUCTION

Experiments by Mitchison demonstrated that the hapten-carrier effect that is, the collaboration of thymus-derived lymphocytes, was required to induce antibody secretion by B lymphocytes.[1] When high specific-activity thymidine was included in culture, antigen-specific B cell responses were ablated (so-called antigen suicide experiments) showing that B cell proliferation was required for a specific immune response.[2] From these observations, a model was constructed in which antigen induced B cell proliferation and T cells induced B cell differentiation.[3] It was realized that this model was not entirely correct when T cell-derived B cell growth factors (BCGFs) were discovered. There are several of these BCGFs of which the most important for B cell proliferation are probably interleukin (IL) 4, IL5 and IL2.

TABLE 1 Cell Density Dependence of Anti-IgM Induced Entry Into Cell Cycle[1]

Cells/cm^2 x 10^{-4}	% of Culture in Cell Cycle Phase G_0	$G1a$	$G1b$, S, $G2+M$
2	82	16	2
4	70	25	5
8	56	29	15
16	50	31	19

[1] Resting splenic, murine B cells were cultured at various cell densities for 2 days in the presence of goat anti-mouse immunoglobulins and vinblastine. Cell cycle analyses were performed with acridine orange on a FACS.

ROLE OF GROWTH FACTORS IN B CELL ACTIVATION

To investigate the role of growth factors in B cell activation, I chose anti-Ig to model antigen effects and serve as a polyclonal B cell activator. Although polyclonal, anti-Ig is thought to mimic antigen by interacting with B cell surface antigen receptors and inducing crosslinkage and, hence, early activation steps.

Several reports in the literature have shown that high concentrations of anti-Ig (usually about 5–100 μg/ml) alone are sufficient to induce B cell proliferation.[4,5] I investigated this phenomenon using different numbers of purified B cells in the culture vessels. The results are shown in Table 1. I used acridine orange to assess the cell cycle status of the cultured cells after 2 days of stimulation. The results show that an increasing proportion of the cells remain in the resting state (G_0) as the number of initial cells per culture decreases. I should point out that these results are not unique.

Similar results were obtained using several different polyclonal and monoclonal anti-Ig reagents. This was unexpected according to the model that anti-Ig alone is sufficient to induce blastogenesis in all B lymphocytes and DNA synthesis in roughly half of them.[5]

A simple interpretation of these results is that a necessary contaminating cell is present and functioning in high-density cultures, but its effect is diluted out as the number of cells cultured decreases. One interleukin, B Cell Stimulatory Factor-1

TABLE 2 Low Cell Density Restoration of Anti-IgM Induced Cell Cycle Entry by IL4[1]

% of Culture in	Addition to Culture			
Cell Cycle Phase	Anti-IgM	IL4	Anti-IgM + IL4	LPS + DxS
G_0	78	95	42	36
$G1a$	18	5	32	36
$G1b$	1	0	2	8
$S+G2+M$	3	0	25	20

[1] Resting splenic, murine B cells were cultured at low cell density for 2 days with vinblastine and the indicated stimulants. Cell cycle analyses were performed with acridine orange on a FACS.

or Interleukin 4, is known to enhance anti-Ig-mediated B cell proliferation. Cells were cultured at low numbers and IL4 was added to test whether its presence could replace the requirement for high cell numbers. The results are shown in Table 2. LPS and DxS are included as a positive control since they are known to activate clonogenesis in an accessory-cell-independent fashion. At low cell numbers, anti-Ig plus IL4 induced a significant fraction of B cells to enter cycle apparently replacing the requirement for high cell density. In the previous experiments, mitotic inhibitors were present to allow the attempt to quantitate the fractions of cells which responded. From these data, we can say that one B cell growth factor, specifically IL-4, is definitely involved in the transition from the resting state (G_0) into early $G1$ ($G1a$). It may also function in later transitions; however, Table 2 shows that most of the activated cells are in $G1a$.

To analyze transitions later than G_0 to $G1a$, another system was used. BCL1 is an *in vivo* passaged B cell leukemia. Spleens from tumor-bearing mice can be used as an almost pure source of these cells. When freshly excised, most of the BCL1 cells are in early $G1$. The effect of a second B cell growth factor called BCGF2 or IL5 was analyzed and the results are shown in Table 3. Cultures with mitotic inhibitors and IL5 contained significant numbers of cells in later stages of the growth cycle. Hence, these data suggest that IL5 is active in the $G1a$ to $G1b$ transition and the $G1b$ to S phase transition.

A MODEL

A model can now be formulated for the action of antigen (anti-Ig) and interleukins in inducing B cell proliferation. Antigen plus IL4 should drive B cells from the resting state (G_0) to a transitional state in early $G1$ ($G1a$). Together with IL5, B cells should be driven from the transitional state to late $G1$ ($G1b$) and perhaps to S phase. Others have shown that the probability of S phase entry is directly proportional to cell size[6] and to RNA content.[7] The parameters I measure using acridine orange to determine cell cycle distributions are DNA content and RNA content per cell.[8] The RNA content of $G1b$ cells is equal to that of cells in early S phase. Since cell size is proportional to RNA content,[8] $G1b$ cells are similar in size to cells in early S phase. Hence, cells driven to $G1b$ should have a high probability of S phase entry (parenthetically, it should be mentioned that because of these observations, the assignment of an S phase-entry, transition, probability function probably describes the $G1a$ to $G1b$ transition). It is generally accepted that cells entering S phase are committed to mitosis, i.e., first-order growth kinetics are observed in nearly all cell proliferation systems. Hence, a single B lymphocyte stimulated with anti-Ig, IL4 and IL5 should generate a clone. This is the system I have used to test this model. B cells are cultured at limiting numbers with these additions and proliferative responses are observed. Poisson statistics are used to calculate the frequencies of input cells giving rise to clones.[9] The results of this type of experiment are shown in Table 4. Two additional polyclonal activators,

TABLE 3 Interleukin 5 Induced Cell Cycle Progression by BCL1 Cells[1]

	% of Culture in Cell Cycle Phase			
Addition to Culture	$G1a$	$G1b$	S	$G2/M$
None	96	0.9	0.4	2.8
25% IL5 SN	64.6	7.2	7.2	21.1
LPS	62.0	9.0	7.6	21.4

[1] BCL1 B cell leukemia cells were isolated from mouse spleen and cultured 3 days with the indicated additions, in the presence of vinblastine. IL5 SN is a gel filtration fraction of a stimulated T cell thymoma supernatant which contained no IL4, IL2 or IL3.

TABLE 4 B Cell Clonogenesis Induced by Anti-IgM and Interleukins[1]

Addition to Culture	% of Input B Cells Yielding Clones
None	< 0.1%
Anti-IgM	4 ± 3%
IL4	0.5 ± 0.5%
Anti-IgM + IL4	19 ± 6%
Anti-IgM + IL4 + IL5	25 ± 8%
LPS + DxS	72 ± 10%

[1] Resting splenic murine B cells were cultured at limiting numbers and growth was assessed on day 5. Cells which had divided at least twice were scored as clones. IL4 and IL5 were supernatants of a myeloma cell line transfected with cDNA for the respective proteins. They were used at approximately 50 U/ml. Anti-IgM is a monoclonal rat anti-mouse IgM used at 15 μg/ml. LPS and DxS were used at 50 and 5 μg/ml, respectively.

lipopolysaccharide (LPS) and dextran sulfate (DxS), were used as positive controls. Stimulation with anti-Ig + IL4 + IL5 activated a significant proportion of B cells to clonal proliferation. The fraction stimulated, however, was not significantly greater than that with anti-Ig + IL4. Moreover, the fraction stimulated with these reagents was only about one-third of the response with the positive control. Clearly, some B lymphocytes can be stimulated to proliferate with surface antigen-receptor crosslinking reagents and interleukin 4. This mode of stimulation, however, is sufficient only for a fraction of the resting B cell population under these stringent conditions. Furthermore, interleukin 5 appears to be irrelevant to this particular population of B lymphocytes.

A SUBPOPULATION OF B CELLS RESPONSIVE TO IL5

It is possible to divide B cells into many distinct populations according to anatomical location, size, density of surface markers and other characteristics. It is curious, however, that the response of BCL1 cells to IL5 was so remarkable and yet resting splenic B cells could not be made to respond to IL5. BCL1 is an *Ly*1

TABLE 5 DNA Synthetic Responses of Various B Cells to IL5[1]

B Cell Origin and Density	None	LPS	cpm x 10^{-3} DxS	IL5*	DxS + IL5 (U/ml) 200	20	2
Spleen							
1.085 g/cc	0.2	43.3	1.0	0.7	1.9	1.0	1.2
1.08 g/cc	0.5	98.1	2.1	1.5	1.2	1.3	2.4
1.075 g/cc	1.6	126.1	3.5	4.3	6.4	6.5	5.6
Peritoneum							
1.085 g/cc	0.1	7.4	0.5	0.6	1.3	0.7	0.1
1.08 g/cc	0.5	22.2	1.9	4.2	3.9	1.1	0.2
1.075 g/cc	0.5	28.0	1.4	6.4	14.6	11.3	3.6

[1] B cells were separated by density on a Percoll gradient and cultured 2 days with LPS (50 μg/ml), DxS (5 μg/ml) and IL5 containing supernatant from a myeloma cell line transfected with appropriate cDNA.

antigen-bearing cell line. The *Ly*1 antigen was thought to be specific to T cells, but now is recognized to be expressed by part, and I stress PART, of a separate lineage of B cells.[10] *Ly*1 positive ($Ly1^+$) B cells are in a high frequency in the peritoneum and in a very low frequency in the mouse spleen. I examined the IL5-induced proliferative responses of B cells separated by density (size) from these two anatomical sites. The results are given in Table 5. Apparently, the peritoneum is enriched in a population of B cells responsive to IL5, whereas the spleen is poorly represented by such cells. Moreover, the cells responding in the peritoneum are large B cells of which 80% are $Ly1^+$.[10] These results, and a series of others which I will not present, suggest that IL5-responsive cells are $Ly1^+$ and their sister population is $Ly1^-$. These cells responding by proliferation, however, do not appear to overlap with those proliferating in response to IL4. Furthermore, the proliferative response of these cells to IL5 is not induced by anti-Ig, but by other polyclonal activators as shown in Table 6. This implies that anti-Ig does not mimic all antigens.

TABLE 6 Lack of Peritoneal B Cell DNA Synthetic Response to Anti-IgM + IL4

Addition to Culture	cpm x 10^{-3} Splenic B Cells	Peritoneal B Cell
None	0.5	1.7
LPS	89.0	37.9
DxS	2.4	4.2
Anti-IgM	0.5	0.3
IL4	1.3	ND
IL5	1.2	ND
DxS + IL5	3.7	25.5
Anti-IgM + IL4	9.7	1.9

[1] Resting splenic B cells were isolated by density centrifugation. B cells were cultured 2 days with lPS (50 μg/ml), DxS (5 μg/ml), monoclonal anti-IgM (15 μg/ml), recombinant IL4 (10 U/ml) and recombinant IL5 (10 U/ml).

CONCLUSIONS

In summary, there appear to be at least two distinct populations of B lymphocytes with distinct triggering requirements. One major population requires receptor crosslinking and IL4 to transit from G_0 to $G1b$ (and, hence, proliferate). The other apparently does not respond to receptor crosslinking mediated by anti-Ig, but requires a different primary signal and IL5 to transit from G_0 to $G1b$.

ACKNOWLEDGMENTS

The Basel Institute for Immunology was founded and is supported by F. Hoffmann-La Roche & Company, Basel, Switzerland.

REFERENCES

1. Mitchison, N. A. (1971), "The Carrier Effect in the Secondary Response to Hapten-Protein Conjugates. I. Measurement of the Effect with Transferred Cells and Objections to the Local Environment Hypothesis. II. Cellular Cooperation," *Eur. J. Immunol.* **1**, 10 and 18.
2. Dutton, R. W., and R. I. Mishell (1967), "Cell Populations and Cell Proliferation in the *In Vitro* Response of Normal Mouse Spleens to Heterologous Erythrocytes. Analysis by the Hot Pulse Technique," *J. Exp. Med.* **126**, 443–454.
3. Dutton, R. W. (1975), "Separate Signals for the Initiation of Proliferation and Differentiation in the B Cell Response to Antigen," *Transplant. Rev.* **23**, 66.
4. Parker, D. C. (1980), "Induction and Suppression of Polyclonal Antibody Responses by Anti-Ig Reagents and Antigen-Nonspecific Helper Factors: A Comparison of the Effects of Anti-Fab, Anti-IgM and Anti-IgD on Murine B Cells," *Immunol. Rev.* **52**, 115.
5. de Franco, A. L., E. S. Raveche, R. Asofsky, and W. E. Paul (1982), "Frequency of B Lymphocytes Responsive to Anti-Immunoglobulin," *J. Exp. Med.* **155**, 1523.
6. Baserga, R (1984), "Growth in Size and Cell DNA Replication," *Exp. Cell. Res.* **151**, 1.
7. Darzynkiewicz, Z., D. Evenson, L. Stainanco-Coico, T. Sharples, and M. R. Melamed (1979), "Relationship between RNA Content and Progression of Lymphocytes through S Phase of Cell Cycle," *Proc. Natl. Acad. Sci. USA* **76**, 356.
8. Wetzel, G. D., S. L. Swain, R. W. Dutton, and J. R. Kettman (1984), "Evidence for Two Distinct Activation States Available to B Lymphocytes," *J. Immunol.* **133**, 2327.
9. Wetzel, G. D., and J. R. Kettman (1981), "Activation of Murine B Lymphocytes. Stimulation of B Lymphocyte Clonal Growth with Lipopolysaccharide and Dextran Sulfate," *J. Immunol.* **126**, 723.
10. Herzenberg, L. A., A. M. Stall, P. A. Labor, C. Sidman, W. A. Moore, D. R. Parks, and L. A. Herzenberg (1986), "The Ly-1 B Cell Lineage," *Immunol. Rev.* **93**, 81–102.

Antigen Presentation and the Structure of Antigenic Determinants

JACQUES NINIO
Institut Jacques Monod, Tour 43, 2 Place Jussieu, 75251 Paris cedex 05, France

Where Did the Macrophage Get the Antigen From?

It is commonly admitted that an antigen, or a processed antigen, is first presented to a T cell by a macrophage or some other antigen-presenting cell, bearing an MHC molecule compatible with the T-cell receptor. The T cell thus activated, when encountering a B cell, tests for the presence of the same antigen, through a similar mechanism of MHC-dependent antigen presentation. If the test is positive, the B cell is rewarded with a variety of growth factors. While every detail in this picture is backed up by solid experimental evidence, I am not aware of a single work demonstrating the sequential occurrence of the two steps. The scheme is somewhat incongruous, as suggested by the following medical metaphor, where B, T and AP cells are impersonated by patients, doctors and nurses, respectively.

We are in a hospital. There are many patients in a waiting room, and nurses busying around.

"I do not feel in my normal condition," says a patient to a nurse. "It itches here and there, my head aches and I have digestive troubles..."

"Do not worry, we'll make a blood tests."

"I'd rather have a complete check up."

"It's no use. It is our doctors' opinion that blood tests are the only reliable ones."

"And then?"

"If the test is negative, you go home. If it is positive, whatever your problem, you will be administered our special unique home-made drug."

I happened, this particular day, to visit a colleague on the second floor, in the Immunological Laboratory, Department of Obscure Thinking.

"What is going on down there," says I.

"Well, we have a bunch of nurses. They go around and collect all the blood they can, they test the samples all together, and present the results to the doctors."

"I assume they apply a strict procedure for labelling the tubes in which the blood is collected, so as not to confuse one patient with another."

"Oh, this is not necessary. We see no reason to get into that trouble."

"But what happens if one sample gives a positive reaction? How do you locate the patient from which the blood was extracted?"

"Elementary. We call for extra doctors, and we send them all down in the waiting room, to retest all the patients and find out the positive ones."

My colleague pointed out an advantage of this system. Most often, all tests are negative so that, over a long period, the savings on the nurse's labelling work more than compensate the expenses on the doctor's retesting time. Next, labelling the blood samples without misspelling the patient's names is, in any event, a rather complex task requiring high qualifications. I was not convinced. Since the doctors had to test all the patients once, pretesting was a loss of time. So why not dispense with nurses? I, thus, expounded a heretical theory. All right, the nurses do not label the blood tubes. They just get hold of a patient, and do not release it until the analysis has been carried out and the result forwarded to the doctor. No retesting is needed. The doctor would just have to give the medication to the patient brought to his office by the nurse.

But, objected my friend, performing a blood test is a rather elementary task, that most patients can carry out by themselves, if provided with the appropriate equipment. So why should nurses be needed in this scheme? Having no answer at this point, I pushed the debate sideways: "I thought that, in a modern institution, a patient would be entitled to go through a complete check-up, including all those sophisticated tests that have been developed the last ten years..."

Enough metaphors. Let us now return to the world of cells and molecules. There are two sensory systems by which the organism perceives the presence of antigen. One is constituted by the set of antibodies and the B cells that carry them. The other is constituted by the set of T-cell receptors, the MHC molecules, and the corresponding cells. While the first system may, in the abstract, provide a detailed diagnosis, the important decisions are taken on the basis of the crude, but reliable, tests performed by the second system.

Each antigen may, in principle, be uniquely characterized by its set of binding energies to a large collection of antibodies. One might store in the memory files of a digital computer, the various ΔG (binding free energy) spectrums of all antigens encountered by the immune system, and use these data for antigen identification. This, however, cannot be translated into a reasonable molecular or cellular strategy. In most biological processes, decisions are taken on the basis of rather scarce information. There is an encounter between two partners (an enzyme and a substrate, an antigen and an antibody) which leads to a kind of decision: abort, or go one step further in the process. The major question, for the immune system, is not: "What is the antigen?" but "Is it foreign?" which is a more limited question, requiring a yes/no answer. For this task, the immune system simplifies the problem, limiting its analysis to a few selected features of the antigens. The sacrifice made in analyzing—through the MHC and T-cell receptor system—only a restricted part of the antigen, is the price paid for having rapid and highly reliable answers.[1,2] Thus, the use of "blood testing" versus "extensive check-up" is easily rationalized.

Next, since some B cells do have the capacity to present antigen, why is this property not extended to all B cells? Or, assuming as modern orthodoxy admits, that all B cells ultimately present antigen, why must there be, in many cases, a pre-presentation by APCs? I have a very simple theoretical answer to this question. When a B cell engages into a program of extensive somatic diversification of its immunoglobulin genes, there is a potential risk of increased mutation rate in other genes[3] and, in particular, the MHC genes. But mutations in MHC genes would be catastrophic for self-/non-self discrimination. Thus, it would be wise to couple somatic diversification with a shut-off of antigen presentation. Thus, a separation of labour between B cells and APCs makes sense.

The point which is still bizarre, in current orthodoxy, is the lack of coupling between the first APC presentation of antigen, and the second presentation by B cells. We would be in the situation of the nurses collecting the blood samples without labelling the tubes. My answer, which I believe is compatible with experimental evidence,[4,5] is that the nurse grabs the patient and brings him to the doctor, i.e., an APC collects the antigen from a B cell, and presents this antigen to the T cell while the B cell is still around. Then, the T cell stimulates the B cells without antigen representation.[1,2] There would be two classes of B cells. A minority of B cells would be capable of antigen presentation. Then, the immune system would work according to the Lanzavecchia model,[6] and T help or suppression would depend upon the MHC type of the B cell. The majority of B cells, issued from extensive somatic diversification would have extremely limited capacities for antigen presentation. Then, the immune system would work according to my scheme[1,2] and T help or suppression would mainly depend upon the MHC type of APCs.

REFERENCES

1. Ninio, J. (1986), "Adaptation and Evolution in the Immune System," *Evolutionary Processes and Theory,* Eds. S. Karlin and E. Nevo (New York: Academic Press), 143–165.
2. Ninio, J. (1986), "A Ternary Cellular Complex between B and T Lymphocytes and Antigen Presenting Cells?" *Immunology Today* **7**, 354.
3. Wabl, M., H.-M. Jack, J. Meyer, G. Beck-Engesser, R. C. von Borstel, and C. M. Steinberg (1987), "Measurements of Mutation Rates in B Lymphocytes," *Immunological Reviews* **96**, 91–107.
4. DeFranco, A. L., J. D. Ashwell, R. H. Schwartz, and W. E. Paul (1984), "Polyclonal Stimulation of Resting B Lymphocytes by Antigen-Specific T Lymphocytes," *J. Exp. Med.* **159**, 861–880.
5. Rock, K. L., S. I. Haber, D. Liano, B. Benacerraf, and A. K. Abbas (1986), "Antigen Presentation by Hapten-Specific B Lymphocytes. III. Analysis of the Immunoglobulin-Dependent Pathway of Antigen Presentation to Interleukin 1-Dependent T Lymphocytes," *Eur. J. Immunol.* **16**, 1407–1412.
6. Lanzavecchia, A. (1985), "Antigen-Specific Interaction between T and B Cells," *Nature* **314**, 537-539.

JOHN L. SPOUGE† and JAY A. BERZOFSKY‡
† Laboratory of Mathematical Biology and ‡ Metabolism Branch, National Cancer Institute, National Institutes of Health, Bethesda, MD 20892

Statistics in Immunology and Biochemistry: An Application to T Cell Antigenic Sites

ABSTRACT

This paper discusses the application of statistics to biochemistry and, in particular, the application of statistical matching techniques to the problem of T cell antigenic peptides. Statistics are indispensable in the practice of clinical medicine; they have not found similar use in biochemistry, probably because biochemically interesting null hypotheses have not been available for statistical testing.

1. INTRODUCTION

Statistics have been indispensable to the practice of clinical medicine. Pre-statistical clinical medicine had to rely on anecdotal evidence and often drew mistaken conclusions from statistically inadequate samples. The clinical physician of yesteryear often had no objective criteria by which to verify his ideas. The cellular biochemist of today sometimes finds himself in much the same unfortunate position.

Let us review some relevant immunology before considering specifics.

Theoretical Immunology, Part One, SFI Studies in the Sciences of Complexity
Ed. A. S. Perelson, Addison-Wesley Publishing Company, 1988

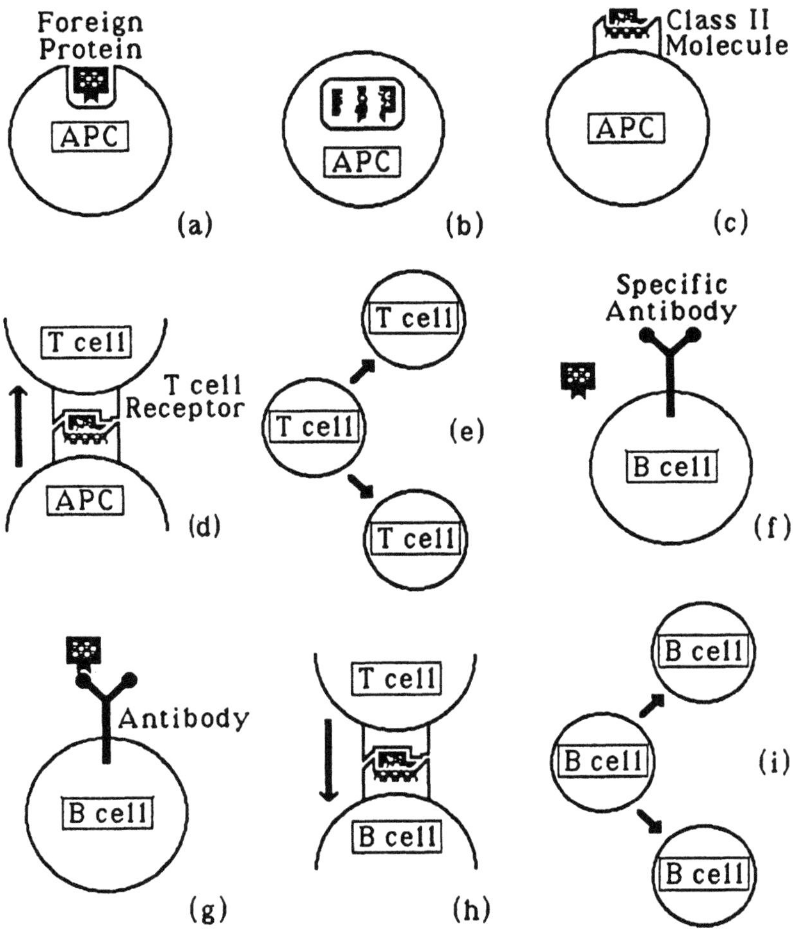

FIGURE 1 (a) Ingestion of a foreign protein by an antigen-presenting cell (APC); (b) Digestion of the foreign protein into smaller peptides; (c) Association of a peptide fragment with a class II major histocompatibility complex protein on the APC surface; (d) Presentation of the peptide-class II complex to a T cell which recognizes it; (e) T cell division into a clone of T cells; (f) B cell production of the antibody specific to the foreign protein; (g) B cell recognition of the foreign protein; (h) B cell presentation of the peptide-class II complex to one of the T cell clones and T cell stimulation of the B cell; (i) B cell division into a clone of B cells.

Immune recognition of foreign proteins as antigens begins with T cell recognition of fragments of the protein on the surface of another cell of the immune system. An antigen-presenting cell, usually a macrophage, dendritic or B cell, ingests the protein (Figure 1a) and digests it into smaller peptides (Figure 1b; Shimonkevitz et al., 1983; Streicher et al., 1984; Unanue, 1984). Either within the cell or on its surface, some of the resulting peptides become associated with a class II major histocompatibility complex protein (Figure 1c). The cell then presents the peptide-class II complexes on its surface to T cells (Figure 1d; Benacerraf, 1978; Rosenthal, 1978; Berzofsky, 1980,1987). If an appropriate helper T cell recognizes the complex, it divides to produce a clone of T cells (Figures 1d and 1e). B cells can produce antibodies specific to foreign proteins (Figure 1f). When a B cell recognizes the antigen via a specific antibody on its surface (Figure 1g), it proceeds to ingest (Figure 1a), digest (Figure 1b) and present the antigen to one of the T cell clones (Figures 1c and 1d). The T cell then stimulates the B cell to divide (Figure 1h and 1i; reviewed in Berzofsky, 1987), thereby increasing the production of the antibody which recognizes the foreign protein.

Within a given foreign protein, some peptides stimulate B cells; others, T cells; yet other peptides appear to be relatively inert. Understanding the biochemistry of these specificities would speed the development of artificial vaccines. Several theories for these specificities have been proposed.

B cell antigenicity is related to the tertiary structure, the global three-dimensional geometry, of a protein. As Figure 2a indicates, when compared to other structures, a turn or coil protruding from a protein surface is probably more accessible to an antibody (Lee and Richards, 1971; Connolly, 1983, Novotny et al., 1986; Thornton et al., 1986). In addition, as Figure 2b indicates, free peptides tend to be very mobile, so protein chains with high crystallographic mobilities are probably better mimics of immunizing peptides than chains with low mobilities (Westhof et al., 1984; Tainer et al., 1984; Tainer et al., 1985). Accessibility and mobility are correlated, however, since protruding structures on protein surfaces tend to be more mobile. If, e.g., protrusion is the causal factor, then mobility, though correlated, might be a biologically irrelevant. Debate over this issue has continued for several years and can be resolved by the statistical matching methods we discuss.

T cell antigenicity, in contrast to B cell antigenicity, appears to be related to secondary structure, the local geometry of a peptide segment (Pincus et al., 1983; Schwartz et al., 1985; Spouge et al., 1987a; Carbone et al., 1987; Vasquez et al., 1987). Various hydrophobicity (DeLisi and Berzofsky, 1985; Corradin, 1987; Margalit et al., 1987) and charge properties (Muckerheide et al., 1987 a,b; Fox et al., 1987) are also implicated. We review some of the hydrophobicity properties before considering their biological relevance.

As the Greek roots imply (hudor = water; phobos = fear; philos = friend), hydrophobic structures avoid water, while hydrophilic structures seek it. A structure is *amphipathic* (amphi = on both sides) when it has both a hydrophobic part and a hydrophilic part (Tanford, 1973). For example, an amphipathic interface is an interface between water and an apolar environment (e.g., a cell surface or hydrophobic cleft in a protein; see Figure 3a).

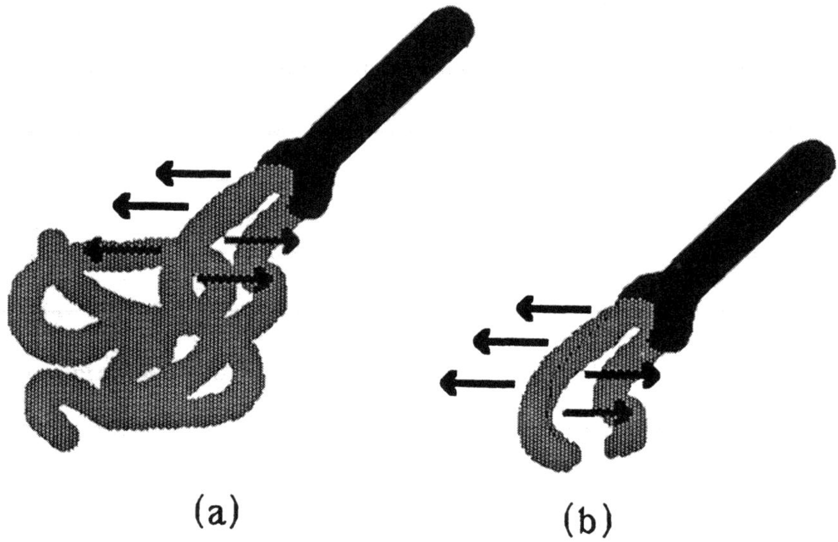

FIGURE 2 (a) Antibody recognition of a protruding, mobile peptide chain in a protein; (b) Antibody recognition of the same peptide, now free.

An α-helix is a secondary structure where the peptide chain has 3.6 amino acid residues per helix turn. The peptide is α-helically amphipathic (Figure 3b) if one face of the α-helix is hydrophobic, the other face hydrophilic (Segrest and Feldman, 1977). When the peptide is at an amphipathic interface, α-helical amphipathicity helps to organize the α-helical conformation (i.e., shape) (Spouge et al., 1987b; Figure 3b). Both the hydrophobic and hydrophilic parts of the peptide can organize into separate faces on the α-helix. With the hydrophilic face in water and the hydrophobic face in an apolar environment, the peptide has optimized its hydrophobic interactions at the amphipathic interface. α-helical amphipathicity is related to, but not equivalent to, α-helical *interphilicity*, the tendency of an α-helical peptide to partition from water to an amphipathic interface (Spouge et al., 1987b).

A peptide is segmentally amphipathic (Figure 3c) when the peptide contains at least two disjoint subpeptides, one hydrophobic, the other hydrophilic (DeLisi and Berzofsky, 1985). Although promoting favorable interactions at an amphipathic interface, segmental amphipathicity, unlike α-helical amphipathicity, does not help to organize a particular peptide conformation (Figure 3c).

Several biochemical properties might influence T cell antigenicity. The class II molecule probably has at least one cleft for binding peptides (Bjorkman et al., 1987). Like a lock and key, the class II cleft might accept only peptides of an α-helical or other fixed shape (Buus et al., 1987). The cleft might also be hydrophobic (Allen et al., 1984; Streicher et al., 1984; Berzofsky, 1985; Berkower et al., 1986), in which case α-helical amphipathicity would help to organize the α-helical conformation at the amphipathic interface between the cleft and the surrounding water. A correlation between T cell antigenicity and either α-helical folding free energy or α-helical amphipathicity would support these hypotheses.

A cell surface, when in contact with water, is an amphipathic interface. α-helical interphilicity, therefore, aids the retention of α-helical peptides at the cell

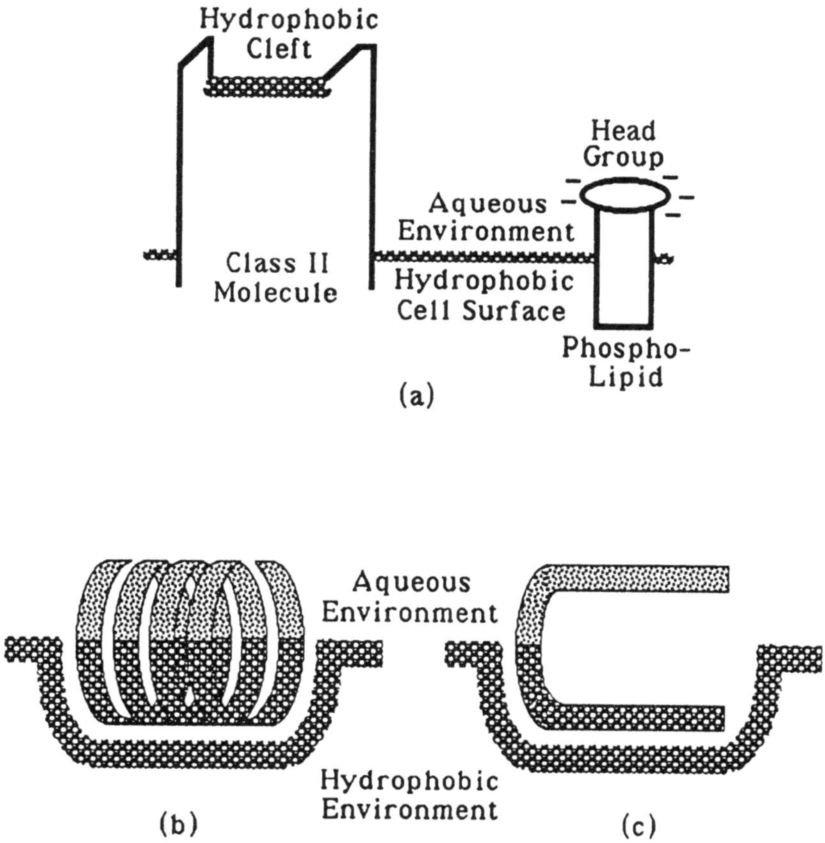

FIGURE 3 (a) The figure shows the amphipathic environments between an aqueous environment and: i) the hydrophobic cleft in a protein and ii) a hydrophobic cell surface; (b) an α-helically amphipathic peptide at an amphipathic interface; (c) a segmentally amphipathic peptide at an amphipathic interface.

surface. Likewise, because of negatively-charged phospholipid head groups on the cell surface (Figure 3a), positive peptide charges should also aid retention. Hence, if class II molecules associate with peptides on the cell surface, then α-helical interphilicity or positive peptide charge should correlate with T cell antigenicity. Absence of this correlation would suggest that the association of peptides and class II molecules might not require these non-specific retentive mechanisms and might well occur intracellularly.

The α-helical conformation (Pincus et al., 1983), α-helical amphipathicity (DeLisi and Berzofsky, 1985; Margalit et al., 1987), and segmental amphipathicity (DeLisi and Berzofsky, 1985; Corradin, 1987) are suggested correlates of T cell antigenicity. Formal statistical tests (Spouge et al., 1987a,b) indicate the independent correlation of T cell antigenicity with both α-helical folding free energy and α-helical amphipathicity. By contrast, α-helical interphilicity and peptide charge do not correlate with T cell antigenicity. As explained above, these statistical results support: a) the importance of secondary structure in peptides which are T cell antigenic sites, possibly through a lock-and-key mechanism between the T cell peptide and class II molecule; b) the existence of a hydrophobic cleft in the class II molecule; and c) the association of peptides and class II molecules does not require non-specific retentive mechanisms.

Formal statistical testing has usually vindicated most proposed correlates of B and T cell antigenicity. Despite this, one of the proposed T cell correlates, segmental amphipathicity (DeLisi and Berzofsky, 1985; Corradin, 1987), is, in fact, the most consistent *anti*correlate yet found (Spouge, 1987a,b). In other words, had theoretical selections of T cell peptides been based on segmental amphipathicity, they probably would have been worse than random selection. This case, perhaps more than any other, underlines the need for formal statistical testing in cellular biochemistry.

2. STATISTICAL METHODS IN T CELL ANTIGENICITY

Section 2 summarizes the essential statistical methods that Spouge et al. (1987a,b) applied to T cell antigenicity. In what follows, these papers will be abbreviated individually as *Sa* and *Sb* and together as *Sa,b*. Section 3 then addresses some statistical problems in biochemistry exemplified by T cell antigenicity. *Sa,b* dealt only with the specifics of the T cell antigenicity problem; Section 3 addresses some important general issues which *Sa,b* had to omit for specificity's sake. The lettered subsections (A., B., etc.) preceding specific points in Section 2 will be used to designate the corresponding generalities in Section 3.

A. DATABASE

Experimental immunologists had determined that, within a given protein, only certain peptides stimulated large numbers of T cells to divide. A database of these peptides (T cell sites) was collected. The sequences of the T cell sites and the proteins from which they came (parent proteins) were entered into a computer.

B. SITE STATISTICS

Each biochemical property under consideration (e.g., segmental amphipathicity, tendency to an α-helical conformation, etc.) corresponded to a "site statistic." The site statistic was a number A which quantified how much a peptide possessed the corresponding property. *Sa,b* used only peptide sequences to calculate site statistics, although in principle non-sequential information (e.g., crystallographic data) could have been used as well.

C. STATISTICAL TEST

For any site statistic A, sum A for the T cell sites to produce an "overall statistic" $S_0 = \sum A_0$ for the T cell sites. This procedure can also produce an overall statistic $S = \sum A$ for any "random" group of peptides. For any given distribution of S, e.g., a normal distribution, the p-value of S_0 is the probability that S exceeds S_0. (*Sa,b* used the Monte Carlo computer method to sample the distribution of the statistic S (Hammersley and Handscombe, 1964). *Sa* probably overemphasized this purely technical use of the Monte Carlo method.) When the p-value is small, S_0 is large compared to random S's, and so on average, when compared to a random peptide, a T cell site must have a larger A. The p-value depends crucially, of course, on the definition of the word "random."

D. NULL HYPOTHESIS

Sa,b defined "random" in the following way. Consider the parent protein of a particular T cell site and, within that protein, all peptides (including the T cell site) which have the same length as the T cell site. These peptides form a theoretical control population for the T cell site. The null hypothesis (H_0) used by *Sa,b* was: the experiments were equally likely to choose any peptide within the theoretical control population. Any deviation from this null hypothesis was supposed to reflect non-random peptide selection.

E. MATCHING TO REMOVE BIASES

The above procedure does not always produce a reasonable control population. For example, certain experiments used trypsin to digest the parent protein and then examined the digested peptides for T cell antigenicity. Trypsin cleaves protein sequences after arginine and lysine residues. Therefore (with the exception of the carboxy-terminus of the whole protein) these experiments could examine only those peptides having either arginine or lysine at their carboxy-terminus for T cell antigenicity. For tryptic experiments, any peptide not ending in arginine or lysine was eliminated from the theoretical control population of the corresponding T cell site. Similar elimination was used for cyanogen bromide, which cleaves after methionine residues.

Trypsin and cyanogen bromide bias the carboxy-terminus of T cell sites. The above elimination of peptides from the control population *matches* the peptides remaining within the control population, so that their carboxy-terminus falls into the same class (arginine/lysine or methionine) as the corresponding T cell site, if that site was produced by trypsin or cyanogen bromide. This matching compensates for carboxy-terminal biases.

Sb also found that T cell sites have a systematic bias in their hydrophobicity, presumably because experiments require soluble peptides (Lerner et al., 1981). *Sb*, therefore, eliminated additional peptides from the control populations, so that the average hydrophobicity of the remaining control peptides was about the same as the hydrophobicity of the corresponding T cell site. This *matched* hydrophobicities within the control populations to the hydrophobicity of the corresponding T cell site.

Matching ensures that the T cell sites are unexceptional with respect to the matched property (Cochran, 1983). Within their matched control populations, the T cell sites can not show a bias with respect to the matched property. Matching can, therefore, remove biases.

F. MATCHING TO ASSESS INDEPENDENCE

Matching can also assess the independence of correlated significances. A classic example of matching involves smoking and cholesterol levels as factors in heart disease. Both are statistically significant: is the significance of smoking due to causality, while that of cholesterol is due to correlation? Matching subpopulations by their smoking habits and re-examining the significance of cholesterol within each subpopulation answers the question. A significant p-value after matching strengthens arguments that cholesterol is an independent factor in heart disease.

There is an obvious analogous use for matching in peptide biochemistry. Let P_1 and P_2 be two biochemical properties (e.g., α-helical amphipathicity and the tendency to α-helical conformation) which are both significant and correlated. Matching P_1 makes the T cell sites 'average' members of their control populations with respect to P_1. If, after matching P_1, the T cell sites still significantly tend to possess P_2, they possess P_2 independent of their tendency to possess P_1.

Matching to assess independence can be used to explain statistical significances as well. For example, certain amino acids, e.g., glycine, can destabilize the α-helical conformation (Chou and Fasman, 1974). *Sa* found that, within unmatched control populations, T cell sites tend both to lack glycine and to have an α-helical conformation. Since the two properties are correlated, is the infrequency of glycine due to glycine's ability to destabilize α-helices? *Sb* found that, when compared to a control population matched by glycine destabilization (P_1), T cell sites did not particularly lack glycine (P_2). T cell sites do not, therefore, lack glycine relative to other peptides with comparable amounts of glycine destabilization. Hence, glycine's ability to destabilize α-helices is sufficient to explain the infrequency of glycine within T cell sites.

This summarizes the relevant methods in *Sa* and *Sb*. We now discuss some of the general issues underlying their use in biochemical problems.

3. SOME GENERAL STATISTICAL METHODS IN BIOCHEMISTRY

The experienced statistician should forgive some statements which are now of almost historical interest (e.g., Subsection 3.B. below parallels Cournot's objection to social statistics; see Cournot, 1843; from Stigler, 1986). In defence of inclusion of these statements, they often indicate errors that we, coming from a biochemical perspective, very nearly made.

The general problem we discuss begins with experimenters who have investigated a particular biological property (in Section 2, this was T cell antigenicity). Each experimenter, having his favorite protein, has determined that certain peptides in his protein have the biological property, certain others do not. As in T cell antigenicity, let us call the peptides with the property "sites"; for contrast, let us call those without it "non-sites."

The correspondence between experimental studies of peptides and retrospective studies of human disease is striking: there are two groups of peptides, those possessing the biological property in question, the sites (patients with the disease), and those without, the non-sites (healthy individuals); the biological property is often dependent on several biochemical properties (human disease is often multifactorial); and correlations between the biochemical properties complicate attempts at causal interpretation. In general, techniques for testing hypotheses developed for retrospective studies of human disease can be transferred without change to experimental studies of biological activity of peptides.

A. DATABASE

Given a list of proteins containing some experimentally determined sites and non-sites, the general question this section addresses is: how can statistics be used to determine the biochemical properties which cause a peptide to be a site? As in T cell antigenicity, the answer to this question might give important clues identifying relevant cellular pathways.

B. SITE STATISTICS

An essential step in the investigation is to decide which biochemical properties are likely to be relevant. Investigating all imaginable parameters will inevitably produce small p-values; by chance alone, about one result in every twenty will have a p-value of less than 0.05. Using available biochemical theory to restrict the investigation will produce more trustworthy statistical significances.

The next step is to produce site statistics corresponding to the relevant properties. There are, for this purpose, two types of site statistics: theoretical and empirical. Assume, e.g., that equilibrated folding into an α-helix might be a relevant peptide property. Because the free energy is the physical agent of the process, any approximation to the free energy is a "theoretical" site statistic; an "empirical" site statistic is any correlate of this free energy, e.g., the infrequency of helix-breakers like glycine. Theoretical site statistics are preferable for causality arguments.

A causal agent must always be significant on its own, independent of any associations. Since the free energy of α-helical folding is the physical agent of the process in question, its statistical significance implicates the process more directly. If, for example, the infrequency of glycine was used as a statistic, any significance could be ascribed to other glycine correlates, e.g., the fact that glycine is the smallest of all amino acids.

Restricting the investigation to theoretical statistics also has the same advantage as avoiding irrelevant statistics. It reduces the number of statistics investigated, thereby minimizing the random risk of low p-values.

C. STATISTICAL TEST

A non-parametric statistical test (Siegel, 1956) could have substituted for the test actually used. Non-parametric tests have the property that occasional outliers (i.e., anomalous extreme values) do not change their significances much. For the study of T cell antigenicity this was undesirable, since the T cell sites might take conformations other than the α-helical one, e.g., the β-strand conformation (Sette et al., 1987). If this were the case, for α-helical statistics, the β sites would be noise and vice versa. Because testing sensitivity was important, non-parametric tests were not used.

D. NULL HYPOTHESIS

As mentioned above, techniques for testing hypotheses from studies of human disease can be transferred to biochemical studies. The usual impediment to this transfer is the absence of a biologically interesting null hypothesis (H_0). Scientific interest usually lies in showing that the experimental results are improbable under a null hypothesis.

The ideal null hypothesis in the case of T cell antigenicity would be: all experimental peptides from the protein had an equal chance of being sites. Any deviation from this hypothesis would presumably represent biological selection. The null hypothesis actually used in *Sa,b* subtly differs: there, the theoretical control population was not the experimental peptides, but peptides of the same length from the parent protein. As a general rule, the theoretical parent population for each site should resemble the experimental parent population as much as possible, but sometimes this is inconvenient.

In the case of the T cell sites, the experimental parent populations varied considerably in peptide length. If the theoretical control population had duplicated this variation, then peptide length would have complicated the biochemical interpretation of the statistical results. (To avoid this complication, experiments should examine only peptides of equal length.) Because each T cell site was compared to peptides of the same length, length did not enter into theoretical interpretations.

In the problem of T cell antigens, the theoretical parent population models the experiment sufficiently well to make the null hypothesis interesting to an analysis of causality. Matters are remarkably different in several other biological situations.

Take, for example, the usual null hypothesis used in statistical analyses of DNA. The analyses usually scramble the sequences in question and then assume that the actual DNA was selected at random from the population of scrambled sequences. This could be thought of as a model of evolution, the process for selecting real DNA. This model would be poor, however, because most of the scrambled sequences never had any chance to be selected. Rejecting the scrambling model as a null hypothesis is, therefore, irrelevant to any analysis of causality in evolution.

Probably nobody has ever (consciously) intended sequence scrambling to be an evolutionary model. In the absence of a viable evolutionary model, DNA statistics are in fact used observationally. As an example of the observational use of statistics, consider the following: many objects in nature are green, far more than would occur if their colors had been selected uniformly from the set red, orange, yellow, green, blue, violet. The observation that the color green is statistically significant, though far removed from a causal statement, does in fact reflect the importance of chlorophyll in nature.

Advanced statistical techniques like matching are very helpful in resolving questions of causality. In biochemistry, the lack of causally interesting null hypotheses has probably prevented the use and adoption of advanced statistical techniques.

E. MATCHING TO REMOVE BIASES

Despite theoretical control populations being subject to certain restrictions (such as avoiding the interpretative complications of length in T cell antigenicity), they should otherwise be as close as possible to the experimental control populations. The theoretical control populations should, however, account for experimental biases produced by, e.g., tryptic digests or hydrophobicity preferences. If no other means are available, matching can be used to remove biases. Its use, however, forces the sites to become 'average' with respect to the matched property. This prevents the statistical study from making any statement about the biochemical relevance of the matched property.

F. MATCHING TO ASSESS CAUSALITY

As any tobacco company will happily confirm, statistics demonstrate correlation, not causality. The statistical technique of matching does, however, strengthen causality arguments.

If a biochemical property P *causes* the biological property in question, and if sufficient data is available, no amount of matching by other properties can prevent an accurate site statistic representing P from being significant. This, in fact, is analogous to establishing causality in ordinary experience: one attempts to dissociate the probable cause from its effect; in statistics, one attempts to dissociate the causal property from its significance. This process also highlights the importance of using theoretical site statistics whenever possible, since these can not be dissociated from the physical process that they represent.

4. SUMMARY

This paper has used the problem of T cell sites to discuss the general application of statistics to biochemistry. The lack of biochemically interesting statistical null hypotheses, particularly those relating to biochemical causality, has delayed the widespread introduction of statistical methodology in biochemistry.

Use of statistical methodology in the case of T cell sites has led to the discovery of several correlates of T cell antigenicity, some of which were previously unsuspected (see Spouge et al., 1987a, for, e.g., the confirmation of an unusual number of carboxy-terminal lysines amongst T cell sites, even after accounting for tryptic digests). More importantly, statistical methodology is available to examine the causality (as opposed to the mere correlation) of various biochemical properties with T cell antigenicity or other biological properties.

We expect that the methodical application of statistics to biochemistry will become increasingly important, particularly because most biochemical parameters, for example, those used in biomolecular energy calculations or secondary structure prediction, are (for the near future at least) likely to remain subject to considerable random noise.

REFERENCES

1. Allen, P. M., D. J. Strydom, and E. R. Unanue (1984), "Processing of Lysozyme by Macrophages; Identification of the Determinant Recognized by Two T Cell Hybridomas," *Proc. Natl. Acad. Sci. USA* **81**, 2489.
2. Benacerraf, B. (1978), "A Hypothesis to Relate the Specificity of T Lymphocytes and the Activity of I Region-Specific Ir Genes in Macrophages and B Lymphocytes," *J. Immunol.* **120**, 1809.
3. Berkower, I. J., G. K. Buckenmeyer, and J. A. Berzofsky (1986), "Molecular Mapping of a Histocompatibility-Restricted Immunodominant T Cell Epitope with Synthetic and Natural Peptides: Implications for T Cell Antigenic Structure," *J. Immunol.* **136**, 2498.
4. Berzofsky, J. A. (1980), "Immune Response Genes in the Regulation of Mammalian Immunity," *Biological Regulation and Development*, Ed. R. F. Goldberger (New York, Plenum), vol. 2, 467–594.
5. Berzofsky, J. A. (1985), "The Nature and Role of Antigen Processing in T Cell Activation," *The Year in Immunology 1984–1985*, Eds. J. M. Cruse and R. E. Lewis (Basel: Karger), 18–24.
6. Berzofsky, J. A. (1987), "Ir Genes: Antigen-Specific Genetic Regulation of the Immune Response," *The Antigens*, Ed. M. Sela (New York: Academic Press), in press.
7. Bjorkman et al. (1987), "The Foreign Antigen Binding Site and T Cell Recognition Regions of Class I Histocompatibility Antigens," *Nature* **329**, 512.
8. Buus, S., A. Sette, S. M. Colon, D. M. Jenis and H. M. Grey (1986), "Isolation and Characterization of Antigen-Ia Complexes Involved in T Cell Recognition," *Cell* **47**, 1071.
9. Carbone, F. R., B. S. Fox, R. H. Schwartz and Y. Paterson (1987), "The Use of Hydrophobic, α-Helix-Defined Peptides in Delineating the T Cell Determinant for Pigeon Cytochrome C," *J. Immunol.* **138**, 1838.
10. Chou, P. Y., and G. Fasman (1974), "Prediction of Protein Conformation," *Biochemistry* **13**, 222.
11. Cochran, W. G. (1983), *Planning and Analysis of Observational Studies* (New York: Wiley & Sons).
12. Connolly, M. L. (1983), "Solvent-Accessible Surfaces of Proteins and Nucleic Acids," *Science* **221**, 709.
13. Corradin, G. P., C. J. A. Wallace, A. E. I. Proudfoot, and S. Baumhuter (1987), "Murine T Cell Response Specific for Cytochrome C," *The Immunogenicity of Protein Antigens: Repertoire and Regulation*, Eds. E. E. Sercarz and J. A. Berzofsky (Boca Raton, Florida: CRC Press), in press.

14. Cournot, A. A. (1843), *Exposition de la Theorie des Chances et des Probabilités* (Paris: Hachette); reprinted as Cournot's *Oeuvres Complètes*, Ed. B. J. Bru (Paris: Vrin).
15. DeLisi, C., and J. A. Berzofsky (1985), "T Cell Antigenic Sites Tend to be Amphipathic Structures," *Proc. Natl. Acad. Sci. USA* **82**, 7048.
16. Fox, B. S., C. Chen, E. Fraga, C. A. French, B. Singh, and R. A. Schwartz (1987), "Analysis of Functionally Distinct Agretopic and Epitopic Sites on the Dominant T Cell Determinant of Moth and Pigeon Cytochromes c Using Synthetic Peptide Antigens," *J. Immunol.* **139**, 1578.
17. Hammersley, J. M., and D. C. Handscombe (1964), *Monte Carlo Methods* (London: Methuen).
18. Lee, B., and F. M. Richards (1971), "The Interpretation of Protein Structures: Estimation of Static Accessibility," *J. Mol. Biol.* **55**, 379.
19. Lerner, R. A., N. Green, H. Alexander, F.-T. Liu, J. G. Sutcliffe, and T. M. Shinnick (1981), "Chemically Synthesized Peptides Predicted from the Nucleotide Sequence of the Hepatitis B Virus Genome Elicit Antibodies Reactive with the Native Envelope Protein of Dane Particles," *Proc. Natl. Acad. Sci. USA* **78**, 3403.
20. Margalit, H., J. L. Spouge, J. L. Cornette, K. B. Cease, C. DeLisi, and J. A. Berzofsky (1987), "Prediction of Immunodominant Helper T Cell Antigenic Sites from the Primary Sequence," *J. Immunol.* **138**, 2213.
21. Muckerheide, A., R. A. Apple, A. J. Pesce, and J. G. Michael (1987a), "Cationization of Protein Antigens. I. Alteration of Immunogenic Properties," *J. Immunol.* **138**, 833.
22. Muckerheide, A., P. L. Domen, and J. G. Michael (1987b), "Cationization of Protein Antigens. II. Alteration of Regulatory Properties," *J. Immunol.* **138**, 2800.
23. Novotny, J., M. Handschumacher, E. Haber, R. E. Bruccoleri, W. B. Carlson, D. W. Fanning, J. A. Smith, and G. D. Rose (1986), "Antigenic Determinants in Proteins Coincide with Surface Regions Accessible to Large Probes (Antibody Domains)," *Proc. Natl. Acad. Sci. USA* **83**, 226.
24. Pincus, M., F. Gerewitz, R. H. Schwartz, and H. A. Scheraga (1983), "Correlation between the Conformation of Cytochrome c Peptides and Their Stimulatory Activity in a T-Lymphocyte Proliferation Assay," *Proc. Natl. Acad. Sci. USA* **80**, 3297.
25. Rosenthal, A. S. (1978), "Determinant Selection and Macrophage Function in Genetic Control of the Immune Response," *Immunol. Rev.* **40**, 136.
26. Schwartz, R. H., B. S. Fox, E. Fraga, C. Chen, and B. Singh (1985), "The T Lymphocyte Response to Cytochrome c. V. Determination of the Minimal Peptide Size Required for Stimulation of T Cell Clones and Assessment of the Contribution of Each Residue Beyond This Size to Antigenic Potency," *J. Immunol.* **135**, 2598.
27. Segrest, J. P., and R. J. Feldman (1977), "Amphipathic Helixes and Plasma Lipoproteins: A Computer Study," *Biopolymers* **16**, 2053.

28. Sette, A., S. Buus, S. Colon, J. A. Smith, C. Miles, and H. M. Grey (1987), "Structural Characteristics of an Antigen Required for Its Interaction with Ia and Recognition by T Cells," *Nature* **328**, 395.
29. Shimonkevitz, R., J. W. Kappler, P. Marrack, and H. Grey (1983), "Antigen Recognition by H-2 Restricted T Cells. I. Cell Free Antigen Processing," *J. Exp. Med.* **158**, 303.
30. Siegel, S. (1956), *Nonparametric Statistics for the Behavioral Sciences* (New York: McGraw Hill).
31. Spouge, J. L., H. R. Guy, J. L. Cornette, H. Margalit, K. Cease, J. A. Berzofsky, and C. DeLisi (1987a), "Strong Conformational Propensities Enhance T-Cell Antigenicity," *J. Immunol.* **138**, 204.
32. Spouge, J. L., H. R. Guy, K. Cease, and J. A. Berzofsky (1987b), "A Matched Statistical Analysis of the Conformational and Biochemical Properties of Helper T Cell Antigenic Sites," *J. Mol. Biol.*, submitted.
33. Stigler, S. M. (1986), *The History of Statistics* (Cambridge, MA: Belknap Press of Harvard University Press).
34. Streicher, H. Z., I. J. Berkower, M. Busch, F. R. N. Gurd, and J. A. Berzofsky (1984), "Antigen Conformation Determines Processing Requirements for T-Cell Activation," *Proc. Natl. Acad. Sci. USA* **81**, 6831.
35. Tainer, J. A., E. D. Getzoff, H. Alexander, R. Houghten, A. J. Olson, R. A. Lerner, and W. A. Henderson (1984), "The Reactivity of Anti-Peptide Antibodies is a Function of the Atomic Mobility of Sites in a Protein," *Nature* **312**, 127.
36. Tainer, J. A., E. D. Getzoff, Y. Paterson, A. J. Olson, and R. A. Lerner (1985), "The Atomic Mobility Component of Protein Antigenicity," *Ann. Rev. Immun.* **3**, 501.
37. Tanford, C. (1973), *The Hydrophobic Effect: Formation of Micelles and Biological Membranes* (New York: Wiley Interscience).
38. Thornton, J. M., M. S. Edwards, W. R. Taylor, and D. J. Barlow (1986), "Location of 'Continuous' Antigenic Determinants in the Protruding Regions of Proteins," *EMBO J.* **5**, 409.
39. Unanue, E. R. (1984), "Antigen-Presenting Function of the Macrophage," *Ann. Rev. Immunol.* **2**, 395.
40. Vasquez, M., M. R. Pincus, and H. Scheraga (1987), "Correlation between Computed Conformational Properties of Cytochrome c Peptides and Their Antigenicity in a T-Lymphocyte Proliferation Assay," *Biopolymers* **26**, 373.
41. Westhof, E., D. Altschuh, D. Moras, A. C. Bloomer, A. Mondragon, A. Klug, and M. H. V. Van Regenmortel (1984), "Correlation between Segmental Mobility and the Location of Antigenic Determinants in Proteins," *Nature* **311**, 123.

PETR KLEIN and R. L. SOMORJAI
Division of Biological Sciences, National Research Council of Canada, Ottawa, Ontario, Canada K1A OR6

Hydrophilicity, Solvent Accessibility and Location of Antigenic Determinants in Proteins

ABSTRACT

Solvent accessibility (a property derived from protein tertiary structure) and hydrophilicity of protein segments (dependent on the amino-acid sequence only) are among the properties thought to be correlated with the location of protein antigenic determinants. We discuss these properties in the context of protein structure prediction, and present some quantitative results on the prediction of a region in the folded protein (defined by accessibility values) from sequence properties such as hydrophilicity and charge.

1. INTRODUCTION: IMPORTANCE OF PROTEIN STRUCTURE PREDICTION FOR PROTEIN IDENTIFICATION AND ENGINEERING

It is now relatively easy to obtain the sequence of nucleotides in a gene, and this sequence can be translated into the sequence of amino acids in the corresponding protein. In fact, practically all new protein sequences are now obtained by DNA sequencing. The number of known protein sequences is increasing quite rapidly and already exceeds 4000 (Barker et al., 1987). With increasing numbers of translated proteins there is an increasing demand for computational methods to identify and analyze them, that is to find their function and cellular location and to characterize their structure. Protein structure prediction is at the centre of this effort because structure determines function and often also location (e.g., the structures of membrane and soluble proteins are different).

Methods of genetic engineering can be used to make changes in the DNA sequence resulting in single amino-acid substitutions. Such changes can be used to study protein function and properties, but a model of protein structure is needed to make the changes purposefully (Hartley, 1986).

Ultimately, the knowledge of protein structure, together with some understanding of the interaction between proteins and smaller molecules, can be used to design molecules, such as peptide vaccines, receptors for drugs, or new enzymes.

It is very likely that the amino-acid sequence contains all the information about the protein's structure, but protein structure prediction remains a major unsolved problem in biochemistry despite intensive effort. Protein structure can be determined by X-ray crystallography and 3-dimensional (3-d) structures of more than 100 different proteins are now known and collected in the Protein Data Bank (Bernstein et al., 1977); however, protein crystallography is difficult and laborious. Understanding how these structures arose is connected with understanding the process of protein folding. This is known to be very fast and reversible but details of this process are still unknown, partly because it is difficult to obtain intermediate structures. The factors known to be involved include hydrophobicity, hydrogen bonds and electrostatic interactions (Creighton, 1983).

Although a general solution to the problem of protein folding is not known, there has been some success in special cases. Conformational energy minimization methods can be used to predict structure of short peptides (Pincus et al., 1982) and of those proteins which are homologous to a protein of known structure (Blundell et al., 1987). In conjunction with computer graphics, such methods are now used for designing experiments in protein engineering (White, 1986). Other methods concentrate on some aspects of protein structure, such as folding type (Klein and DeLisi, 1986). A hierarchical approach to protein folding has been advanced by Somorjai and Narang (1987).

Among the most widely used methods are secondary structure prediction and hydrophobicity profile. They accompany practically every new published protein sequence. We will now briefly describe these two complementary methods which could, if they were sufficiently reliable, provide a crude picture of the geometry of a protein backbone.

2. SOME SIMPLE METHODS FOR STRUCTURE ANALYSIS

Secondary structure refers to regular features of the polypeptide backbone. These include alpha helices, beta sheets and turns. Alpha helices are formed by hydrogen bonds between residues i, i+4. Beta sheets are formed by hydrogen bonding between beta strands which are extended segments of a protein. Turns usually involve four residues and occur where the backbone reverses its direction, which is usually at the surface.

Methods for secondary structure prediction (Chou and Fasman, 1978; Garnier et al., 1978) try to classify each residue as part of an alpha helix, beta sheet, turn or none. These methods are usually based on statistical analysis of the occurrence of the 20 amino acids in these structures, and they also take into account the information about the neighbours in the sequence. The best of these methods have less than 60% success in the prediction of 3 states: alpha, beta, everything else (Kabsch and Sander, 1983). It is difficult to improve these results because the local structure is influenced not only by sequentially near residues but also by residues which happen to be close in space. These so-called long-range interactions are especially important for the formation of beta sheets (Kolaskar and Rambrahman, 1984).

As mentioned earlier, hydrophobicity is an important factor in protein folding. There is a tendency for residues with hydrophilic side chains to be near the protein surface where they can interact with water, and for hydrophobic residues to be inside where they are shielded from water (Chothia, 1976). Efforts to quantify this tendency led to the development of scales of hydrophobicity which assign numerical values to amino-acid residues. These scales were derived either experimentally, by measuring free energies of transfer from water to non-polar solvents, or by statistical analysis of known structures (Rose et al., 1985).

Hydrophobicity plots are constructed by sliding a window along the sequence and plotting average values. Peaks in these plots are interpreted as showing possible membrane-spanning regions and sometimes also as showing inside and outside regions of water-soluble proteins (Kyte and Doolittle, 1982).

The extent to which amino-acid residues are buried or exposed in the protein can be expressed in several ways from 3-d coordinates; for example, by the distance from the center or by counting the number of residues (or atoms) in a fixed-size sphere around each residue (packing profiles). The most widely used method derives solvent accessibility from solved 3-d structures by rolling a sphere of the size of a water molecule over the surface of the protein and calculating the surface area of amino-acid residues accessible to this probe (Lee and Richards, 1971). It would be extremely useful to have an approximation of accessibility that could be derived from the linear sequence of amino acids. Hydrophilicity is an obvious candidate for such an approximation.

There is much interest in these concepts also from the immunological point of view, because hydrophilicity and accessibility of protein segments have been found to be correlated with the location of antigenic determinants for B cells. This has implications for the prediction of antigenic determinants as well as for peptide vaccine design.

3. PREDICTION OF ANTIGENIC DETERMINANTS FROM THE TERTIARY AND PRIMARY STRUCTURES

Immunogenicity (antigenicity) for B cells is the ability to stimulate (bind) antibodies; antigenic determinants are those segments of the protein (or other antigen) which interact with antibodies. (We shall be discussing only antigenicity for B cells; it seems that T cell antigenicity does not depend on the 3-d structure—Berzofsky et al., 1986.)

Several articles have been published relating both tertiary and primary structure properties to the location of antigenic determinants. Solvent accessibility or accessibility to a probe larger than a water molecule (Novotny et al., 1986) and high segmental mobility derived from crystallographic temperature factors (Westhof et al., 1984; Tainer et al., 1984) are among properties of tertiary structure found to be indicative of antigenic sites or peptides. Antigenic sites were also found to correlate with regions protruding from the globular protein (Thornton et al., 1986). Properties derived from the primary structure used for the prediction of antigenic sites include hydrophilicity (Hopp and Woods, 1981), an approximation of mobility (Karplus and Schulz, 1985) or both (Parker et al., 1986).

The nature of antigenicity makes it difficult to evaluate its correlation with various properties. The main problem is that antigenicity is not an absolute property. In addition to what is called intrinsic antigenicity and which is well correlated with properties such as accessibility, segmental mobility and, to some extent hydrophilicity, antigenicity is also determined by host-dependent extrinsic factors, such as genetic control and similarity between the antigen and corresponding proteins of the host (Berzofsky, 1985).

Another problem is that there are a variety of ways of identifying antigenic determinants and each assay has a different bias (Van Regenmortel, 1986). Both polyclonal and monoclonal antibodies are used and these are directed against whole proteins or just peptide segments. Often the response to peptides is evaluated by reactivity with the whole protein and vice versa; in this case what is measured is really cross-reactivity (this is especially true about studies relating the location of antigenic determinants to mobility).

Another difficulty, which makes the prediction of antigenic sites from the sequence information particularly troublesome lies in the finding that most such sites are assembled from short segments from different parts of the chain which are close in 3-d structure (Barlow et al., 1986).

Because of the dependence of antigenicity on extrinsic factors (i.e., host) it is practically impossible to prove that a certain segment is not antigenic, except for segments which are buried in the protein interior and thus inaccessible to antibodies.

Accessibility seems to be a necessary condition for interaction with antibodies, although this also should not be taken absolutely—it has been shown recently that both antibody and antigen can undergo conformational changes upon the formation of the complex (Colman et al., 1987) and it has also been suggested that antibody binding can perhaps expose some buried residues for interaction (Getzoff et al., 1987).

4. RELATION BETWEEN ACCESSIBILITY IN FOLDED PROTEINS AND HYDROPHOBICITY AND OTHER SEQUENCE PROPERTIES

Rose and Roy (1980) noted the resemblance between hydrophobicity profiles and atom packing profiles and Kyte and Doolittle (1982) also showed correspondence between peaks in hydrophobicity profiles and interior and exterior segments. Lipman et al. (1987) noted an important difference between solvent accessibility and hydrophobicity. They found that while there is significant clustering of exposed and buried residues neighbouring in the sequence, clustering in hydrophobicity values

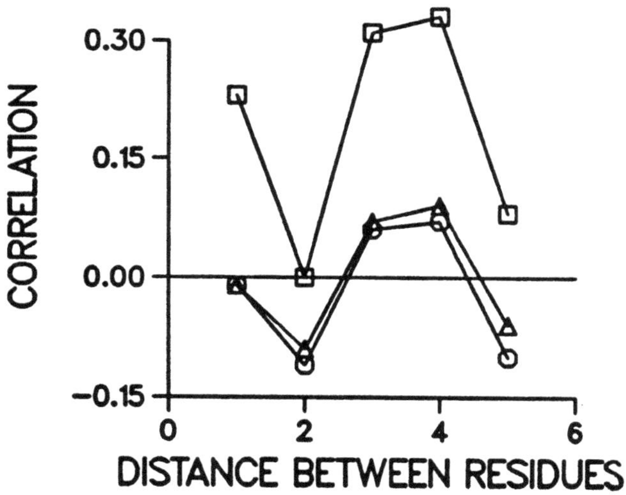

FIGURE 1 Correlation in accessibility values (□), Nozaki-Tanford (△), and Parker et al. (o) hydrophobicities between residues $1, 1+j$ $(j = 1, 2, \ldots, 5)$ in alpha helical proteins.

is not better than random. This is important especially in view of the probable role of hydrophobic clusters in the initiation of protein folding.

We decided to evaluate quantitatively the relation between solvent accessibility and sequence properties such as hydrophobicity. The methods used were correlation analysis—primarily to probe the relation between near neighbours, cluster analysis (Bezdek, 1981) to define inside and outside regions, and discriminant analysis (Mardia et al., 1979) to assess the power of sequence properties or their combinations to predict inside and outside positions in proteins.

Correlation analysis revealed marked differences between two extreme types of protein architecture—alpha rich and beta rich [this classification is that of Levitt and Chochia (1976); the 29 alpha-rich and 27 beta-rich proteins we used are listed in Klein (1986)]. Serial correlation patterns of accessibilities in alpha proteins follow helicity: residues $i, i+3$ and $i, i+4$ which are on the same side of a helix have correlated accessibilities. The pattern in hydrophobicity is much weaker, but follows surprisingly closely that of accessibility (Figure 1). In beta proteins, correlation in accessibility decreases with the lag between residues. The pattern in hydrophobicity is very weak and different (Figure 2). It is interesting to note that it corresponds, albeit very weakly, to amphipathic beta sheets (see Eisenberg et al., 1984). Also, correlations between hydrophobicity and accessibility smoothed over different windows were different in these two types of proteins (Figure 3). These correlations do not depend very much on the choice of hydrophobicity

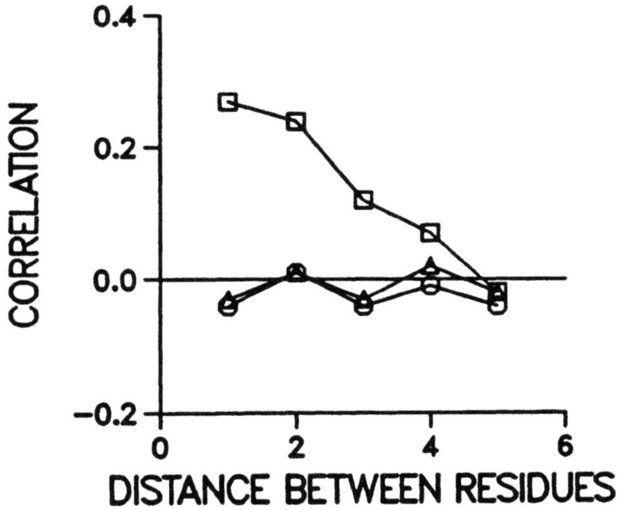

FIGURE 2 Correlation in accessibility values (□), Nozaki-Tanford (△), and Parker et al. (○) hydrophobicities between residues $i, i+j$ ($j = 1, 2, \ldots, 5$) in beta-rich proteins.

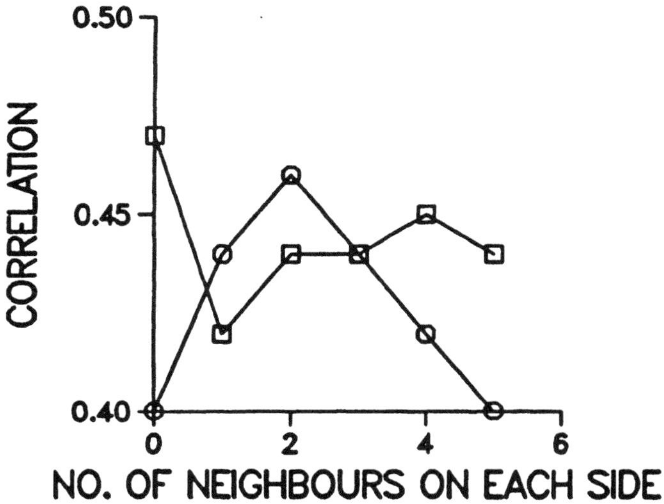

FIGURE 3 Correlation between accessibility values and hydrophobicity values (Parker et al.), smoothed over different windows, in alpha-(□) and beta-(○) rich proteins.

scale, although the scale of Parker et al. (1986) was marginally preferable and was used in further analyses.

Cluster analysis of accessibility values showed that *inside* is much better defined that *outside*. The center of the innermost cluster was much less dependent on the number of clusters than the center of the most exposed cluster (Figure 4). This reflects much tighter packing inside.

The results of discriminant analysis are expressed as a matrix showing, e.g., for two clusters, numbers of inside residues allocated inside and outside on the basis of, say, hydrophobicity, and the same for outside residues (the allocation is based on distributions of sequence properties such as hydrophobicity in residues which are inside or outside according to their accessibility). From this matrix one can easily derive overall probability of correct allocation, *sensitivity* (probability that an inside or outside residue will be correctly allocated) and *specificity* (probability that a residue allocated inside or outside really belongs there). (In the terminology of medical diagnosis, which is probably the most common application of discriminant analysis, a highly sensitive test will detect most cases of a disease, but might have many false positives while a highly specific test will have few false positives.) For our application, high sensitivity in detecting outside residues might be useful in the search for potential antigenic peptides while high specificity for either inside or outside might be useful in searching for constraints for a protein folding algorithm.

Residue accessibilities were first clustered into two groups, inside (about 60% of residues) and outside. In discriminant analysis using hydrophobicity or charge as attributes, overall probability of correct allocation was around 0.7 and the prediction was a little better in alpha than in beta proteins (see Table 1). The most interesting result from this analysis was that hydrophobicity and charge differed in sensitivity and specificity for inside or outside prediction. So, for example charge (or rather lack of it) was very sensitive (0.9) for inside while hydrophobicity was highly specific, especially in alpha proteins (0.9). Hydrophobicity was highly sensitive for outside prediction (0.94)—see Table 1. This suggests that different amino-acid properties will be useful for different predictive purposes.

It is not realistic to force the residues with intermediate accessibilities into the inside or outside cluster and we, therefore, also performed discriminant analysis based on three clusters either of equal size or defined by optimal clustering. Clearly, the amino-acid properties that we tried were not sensitive enough to allow prediction of the intermediate class and we, therefore, tried to distinguish the innermost cluster from the rest and the outside cluster from the rest. With equal-sized clusters, the inside was easier to predict than the outside (in alpha proteins hydrophobicity was the best attribute for inside, giving 76% correct prediction; charge gave the best prediction for outside vs. rest—71%). Inside and outside regions in beta proteins were more difficult to predict (by about 5%) but, in contrast to the situation in alpha proteins, it was possible to improve the prediction somewhat by combining attributes (hydrophobicity, charge, turn propensity) and by smoothing over a small window, which takes into account information from nearest neighbours. Optimal clustering assigned about 45% residues inside and about 20% outside. Overall reliability of prediction of inside based on amino-acid properties was about the same as

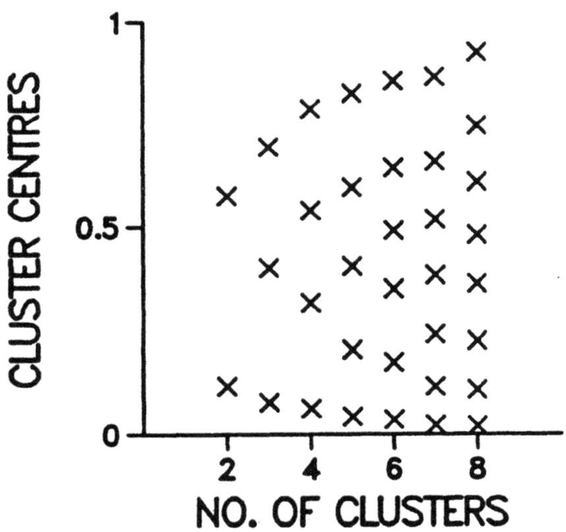

FIGURE 4 The use of accessibility values (accessible surface are divided by standard state accessibility) to define up to 8 regions in protein structures by cluster analysis. Centers of the regions are shown.

TABLE 1 Prediction of Position in Inside and Outside Regions of Alpha-rich and Beta-rich Proteins using Amino-acid Properties

Property	Probability of correct allocation	Sensitivity[1] Inside	Sensitivity[1] Outside	Specificity[2] Inside	Specificity[2] Outside
Hydrophobicity (Parker et al.)	0.71/0.65[3]	0.51/0.58	0.94/0.76	0.90/0.81	0.62/0.51
Charge	0.68/0.66	0.90/0.90	0.43/0.24	0.65/0.67	0.79/0.58

[1] See text for definition.
[2] See text for definition.
[3] The first number refers to alpha-rich, the second to beta-rich proteins.

with equal-sized clusters. Hydrophobicity was again the best attribute; but while its sensitivity for inside did not change, its specificity increased by 0.17 to 0.81. We also found that the inside prediction in alpha proteins could be slightly improved by taking hydrophobicities of residues $i, i+3$ and $i+4$ as attributes. In beta proteins, hydrophobicities of neighbours did not improve discrimination.

The water-exposed cluster was very difficult to predict. This reflects the distribution of amino acids. While there are 8 amino acids which appear in alpha proteins more often inside than not and 10 such amino acids in beta proteins, there is only a single amino acid (Asp) which appears more often outside than not in alpha proteins and there is none in beta proteins. This again illustrates that especially outside residues are difficult to predict reliably (see also Lipman et al., 1987). It is interesting to compare the residues which prefer the innermost of the 3 optimal clusters with the published hydrophobicity scales. The scale of Parker et al. (1986) agrees with the preferred positions of residues very well, better than other commonly used scales (Nozaki and Tanford, 1971; Kyte and Doolittle, 1982).

In summary, prediction of the position of amino-acid residues in the protein globule from sequence information would provide a useful crude picture of the molecule. There have been attempts at such predictions using specific influence of amino acids on their neighbours (Prabhakaran and Ponnuswamy, 1980; Nishikawa and Ooi, 1980, 1986). We explored the relation of the amino-acid properties to the position in two or three regions in the folded proteins.

We found that different folding types as well as different regions require different treatment, and that sensitivity and specificity of the prediction can be changed by using different properties. Therefore, the property chosen to predict accessibility should depend on the purpose of the analysis.

ACKNOWLEDGMENTS

We are grateful to Dr. David Lipman for sending us his file of proteins with accessibility values.

REFERENCES

1. Barker, W. C., L. T. Hunt, D. G. George, L. S. Yeh, H. R. Chen, M. C. Blomquist, E. I. Seibel-Ross, A. Elzanowski, J. K. Bair, D. A. Ferrick, M. K. Hong, and R. S. Ledley (1987), *Protein Sequence Database of the Protein Identification Resource* (Washington, D.C.: Natl. Biomed. Res. Foundation).
2. Barlow, D.J., M. S. Edwards, and J. M. Thornton (1986), *Nature* **322**, 747–748.
3. Bernstein, F.C., T. F. Koetzle, G. J. B. Williams, E. F. Meyer, M. D. Brice, J. R. Rodgers, O. Kennard, T. Shimanouchi, and M. Tasumi (1977), *J. Mol. Biol.* **112**, 535–542.
4. Berzofsky, J.A. (1985), *Science* **229**, 932–940.
5. Berzofsky, J. A., J. Cornette, H. Margalit, I. Berkower, K. Cease, and C. DeLisi (1986), *Curr. Top. Microbiol. Immunol.* **130**, 13–24.
6. Bezdek, J. C. (1981), *Pattern Recognition with Fuzzy Objective Function Algorithms* (New York: Plenum Press).
7. Blundell, T.L., B. L. Sibanda, M. J. E. Sternberg, and J. M. Thornton, (1987), *Nature* **326**, 347–352.
8. Chothia, C. (1976), *J. Mol. Biol.* **105**, 1–14.
9. Chou, P. Y., and G. D. Fasman (1978), *Annu. Rev. Biochem.* **47**, 251–276.
10. Colman, P.M., W. G. Laver, J. N. Varghese, A. T. Baker, P. A. Tulloch, G. M. Air, and R. G. Webster (1987), *Nature* **326**, 358–363.
11. Creighton, T.E. (1983), *Proteins. Structures and Molecular Properties* (New York: Freeman).
12. Eisenberg, D., R. M. Weiss, and T. C. Terwilliger (1984), *Proc. Natl. Acad. Sci. USA* **81**, 140–144.
13. Garnier, J., D. J. Osguthorpe, and B. Robson (1978), *J. Mol. Biol.* **120**, 97–120.
14. Getzoff, E. D., H. M. Geysen, S. J. Rodda, H. Alexander, J. A. Tainer, and R. A. Lerner (1987), *Science* **235**, 1191–1196.
15. Hartley, B. S. (1986), *Phil. Trans. R. Soc. Lond. A* **317**, 321–331.
16. Hopp, T. P. and K. R. Woods (1981), *Proc. Natl. Acad. Sci. USA* **78**, 3824–3828.
17. Kabsch, W., and C. Sander (1983), *FEBS Lett.* **155**, 179–182.
18. Karplus, P. A., and G. E. Schulz (1985), *Naturwissenschaften* **72**, 212–213.
19. Klein, P. (1986), *Biochim. Biophys. Acta.* **184**, 205–215.
20. Klein, P., and C. DeLisi (1986), *Biopolymers* **25**, 1659–1672.
21. Kolaskar, A. S., and V. Rambrahmam (1984), *Int. J. Peptide Protein Res.* **24**, 392–401.
22. Kyte, J., and R. F. Doolittle (1982), *J. Mol. Biol.* **157**, 105–132.
23. Lee, B., and F. M. Richards (1971), *J. Mol. Biol.* **55**, 379–400.
24. Levitt, N., and C. Chochia (1976), *Nature* **261**, 552–558.
25. Lipman, D. J., R. W. Pastor and B. Lee (1987), *Biopolymers* **26**, 17–26.

26. Mardia, K. V., J. T. Kent, and J. M. Bibby (1979), *Multivariate Analysis* (London: Academic Press).
27. Nishikawa, K., and T. Ooi (1980) *Int. J. Peptide Protein Res.* **16**, 19–32.
28. Nishikawa, K., and T. Ooi (1986), *J. Biochem.* **100**, 1043–1047.
29. Novotny, J., M. Handschumacher, E. Haber, R. E. Bruccoleri, W. B. Carlson, D. W. Fanning, J. A. Smith, and G. D. Rose (1986), *Proc. Natl. Acad. Sci. USA* **83**, 226–230.
30. Nozaki, Y., and C. Tanford (1971), *J. Biol. Chem.* **246**, 2211–2217.
31. Parker, J. M. R., D. Guo, and R. S. Hodges (1986), *Biochemistry* **25**, 5425–5432.
32. Pincus, M. R., R. D. Klausner, and H. A. Scheraga (1982), *Proc. Natl. Acad. Sci. USA* **79**, 5107–5110.
33. Prabhakaran, M., and P. K. Ponnuswamy (1980), *J. Theor. Biol.* **87**, 623–637.
34. Rose, G. D., and S. Roy (1980), *Proc. Natl. Acad. Sci. USA* **77**, 4643–4647.
35. Rose, G. D., L. M. Gierasch, and J. A. Smith (1985), *Adv. Prot. Chem.* **37**, 1–109.
36. Somorjai, R. L., and S. A. Narang (1987), submitted to *Protein*.
37. Tainer, J. A., E. D. Getzoff, H. Alexander, R. A. Houghten, A. J. Olson, R. A. Lerner, and W. A. Hendrickson (1984), *Nature* **312**, 127–134.
38. Thornton, J. M., M. S. Edwards, W. R. Taylor, and D. J. Barlow (1986), *EMBO J.* **5**, 409–413.
39. Van Regenmortel, M. H. V. (1986), *Trends Biochem. Sci.* **11**, 36–39.
40. Westhof, E., D. Altschuh, D. Moras, A. C. Bloōmer, A. Mondragon, A. Klug, M. H. V. Van Regenmortel, (1984), *Nature* **311**, 123–126.
41. White, D. N. J. (1986), *Phil. Trans. R. Soc. Lond. A* **317**, 359–369.

Dynamics of the Immune Response

Theoretical Immunology, Part One, SFI Studies in the Sciences of Complexity
Ed. A. S. Perelson, Addison-Wesley Publishing Company, 1988

WULF DRÖGE
Institute of Immunology and Genetics, The German Cancer Research Center, D-6900 Heidelberg, F.R.G.

The Regulation of Lymphocyte Proliferation by Proliferating Lymphocytes: Is Lactate a Central Regulatory Element in the Immune System?

Environmental pathogens constantly trigger proliferative processes in the lymphoid system as part of the normal immune response, but the total number of lymphocytes in the periphery remains, nevertheless, essentially constant. The number of newly formed lymphocytes is obviously balanced by the decay rate. This important link between renewal rate and decay rate suggests the possibility that the proliferating lymphocyte itself may provide the signal for the decay. In this report, I will summarize experiments suggesting that lactate may be the physiological mediator which provides the hypothetical signal for the decay and which also downregulates the lympho-proliferative activity. Most interestingly, lactate inhibits DNA synthesis, but augments the net production of the T-cell growth factor interleukin 2 in activated T-cell populations.[1] I believe that this seemingly counterproductive principle will inspire mathematical minds and may form the basis for a mathematical model of some practical value.

It is quite likely that the regulation by lactate is of general importance, since both, the production of lactate and also the sensitivity against the regulatory effect of lactate, are properties of proliferating T-cells irrespective of their specificity. Moreover, there is evidence from one experimental system that lactate can also be applied *in vivo* as a powerful immunopotentiating agent.[1] In view of the complex and counterproductive effects of lactate, we expect, however, that quantitative mathematical models may be useful if one wants to take advantage of the full immuno-modulating potential of lactate in a broader range of experimental or clinical systems.

THE PRODUCTION OF LACTATE BY IMMUNOLOGICALLY RELEVANT CELLS

Tumor cells and certain types of nonmalignant cells including activated lymphocytes[2] and certain types of macrophages[3–9] exert strong glycolytic activity and release lactate, even under aerobic conditions. In order to determine the maximum lactate concentration that can be generated in the vicinity of glycolytically active macrophages, we have cultured mouse peritoneal macrophages in Dulbecco's MEM (7.5×10^6 cells/0.4 ml) for several days and determined the concentration of lactate in the culture supernatant.

It was found that the cells produced lactate at a linear rate up to a concentration of almost 60mM L-lactate, indicating that lactate-producing macrophages can release lactate even against a relatively high extracellular concentration of lactate.[1] It was calculated that each macrophage produced under this condition 1.2×10^{-13}M/hr of lactate. The production of lactate by activated lymphocytes has been studied by Wang et al.[2] These authors determined the conversion of glucose into lactate in cultures of concanavalin A-activated murine lymphocytes and compared the time course of the lactate production with the time course of RNA and DNA synthesis. The results of these studies revealed that the production of lactate started to increase at 20 hours and reached a peak at about 50 hours after the initiation of the culture. The production of lactate was found to coincide almost exactly with the proliferative activity (DNA synthesis) and was preceded by the peak of RNA synthetic activity. (The relative rate of DNA synthesis was determined by ^{3}H-thymidine incorporation.)

INHIBITION OF LYMPHOCYTE FUNCTIONS BY EXTRACELLULAR CONCENTRATIONS OF LACTATE

The experiments of Wang et al.[2] showed that the increase in extracellular lactate concentration was followed by a decrease in DNA synthesis, but the data were not sufficient to suggest a causal relationship between these two events. In order to test the effect of lactate on lymphocyte proliferation, we incubated concanavalin A-activated splenic T-cell preparations from C3H mice with graded concentrations of lactate and determined DNA synthesis (incorporation of ^{3}H-thymidine) after various time intervals. The results revealed a dose-dependent inhibition of DNA synthesis already within 20 hours after the addition of lactate. The inhibition was

20%, 50%, and 96% for lactate concentrations of 5mM, 20mM, and 80mM, respectively. When DNA synthesis was tested after 40 hours, a 100% inhibition was observed with 40mM of lactate. A concentration of 40mM lactate was also found to cause an almost immediate and total inhibition of the glycolytic activity of the lymphocytes;[1] and a 50% reduction of the number of surviving cells in the culture was detectable after 2 days.

LACTATE AUGMENTS THE PRODUCTION OF T-CELL GROWTH FACTOR INTERLEUKIN 2

In contrast to the inhibitory effects of lactate on the glycolytic activity, DNA synthesis and survival rate, we found a strong augmentation of the production of interleukin 2 (IL-2) in mitogenically-stimulated murine spleen cell cultures.[1] Moreover, accessory cell-depleted splenic T-cell preparations were found to produce no detectable IL-2 in the absence of lactate, but produced quite substantial amounts of IL-2 if lactate was added to the cultures. This effect was dose dependent: a more than 10-fold increase of IL-2 production was already observed with 1mM lactate and plateau levels of IL-2 production were obtained with 50mM lactate in the culture.[1] It has been known for some time that T-cells of the helper variety require macrophage-like accessory cells for optimal stimulation and IL-2 production.[10–13] They were shown to form intimate contact with these cells.[14] Our experiments indicate that the delivery of lactate may be an important component of the accessory cell function.

HYPOTHETICAL MODEL OF THE GROSS IL-2 PRODUCTION, DNA SYNTHESIS AND LACTATE PRODUCTION IN A STIMULATED LYMPHOID ORGAN

Given the regulatory effects of lactate on DNA synthesis and IL-2 production, one can imagine the following scenario in a stimulated lymph node (see Figure 1): Antigenically stimulated helper T-cells which by chance have made contact with a lactate-producing accessory cell will start to produce increasing amounts of IL-2. This growth factor, in turn, is expected to support an increase in DNA synthesis in other cells in the area; and this is again accompanied by increased lactate production. The regional increase in lactate concentration will downregulate the first wave of DNA synthesis and will simultaneously raise the available concentration of IL-2. This, in turn, is expected to support a second wave of DNA synthesis and so forth (Figure 1). The production of lactate, of course, will always prevent the DNA-synthesis from exceeding a certain maximum threshold.

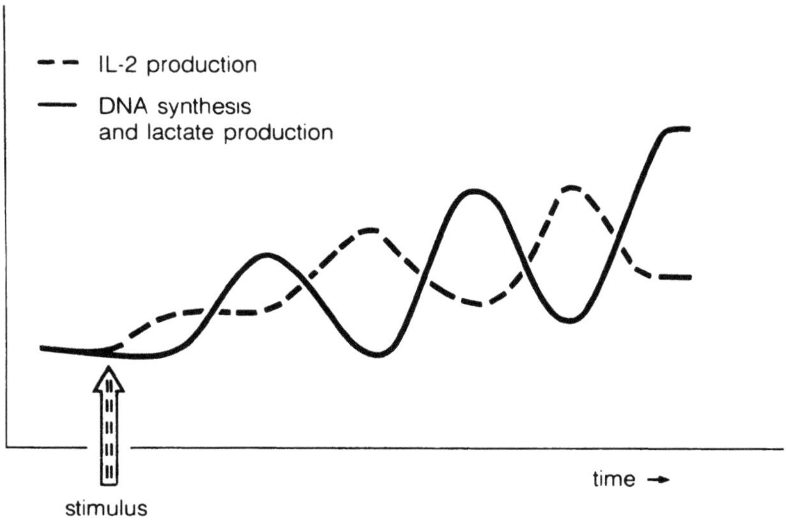

FIGURE 1 Hypothetical model of the gross IL-2 production, DNA synthesis and lactate production in a stimulated lymphoid organ.

Future experiments will show:

- whether the expected oscillation can be demonstrated experimentally;
- how long the periods of this oscillation may be; and
- whether the immune response is augmented by periodical administration of lactate?

THE IMMUNOPOTENTIATING ACTIVITY OF L-LACTATE

A temporary increase of the lactate concentration in a lymphoid organ simultaneous with stimulation by a new antigen is expected to increase the availability of IL-2, and to reduce the background proliferation of cells with unrelated specificities which otherwise would compete for interleukin 2 and nutritional components in the environment. It is expected that these are favorable conditions for an immune response against this new antigenic stimulus. This expectation was, indeed, confirmed in at least one experimental system. The injection of lactate *in vivo* was shown to augment the simultaneous *in vivo* immunization (priming) of DBA/2 mice for a subsequent secondary cytotoxic response *in vitro* against syngeneic tumor cells.[1] This type of experiment suggests that the regulatory effect of lactate may find useful applications.

ACKNOWLEDGMENTS

The assistance of Mrs. Ingrid Fryson in the preparation of this manuscript is gratefully acknowledged.

REFERENCES

1. Dröge, W., S. Roth, A. Altmann, and S. Mihm (1987), *Cell Immunol.* **108**, in press.
2. Wang, T., C. Marquardt, and J. Foker (1976), *Nature* **261**, 702.
3. Roos, D., and A. J. M. Balm (1980), "Oxidative Metabolism of Monocytes," *The Reticuloendothelial System. 2.) Biochemistry and Metabolism*, Eds. A. J. Sbarra and R. R. Strauss (New York and London: Plenum Press), 189.
4. Axline, S. G. (1970), *Seminars in Hematology* **7**, 142.
5. Oren, R. A. E. Farnham, K. Saito, E. Milofsky, and M. L. Karnovsky (1963), *J. Cell. Biol.* **17**, 487.
6. Simon, L. M., E. D. Robin, J. R. Phillips, J. Acevedo, S. G. Axline, and J. Theodore (1977), *J. Clin. Invest.* **59**, 443.
7. Para, M., A. Sagone, S. Balcerzak, and A. LoBuglio (1972), *Clin. Res.* **20**, 742A.
8. Roos, D., M. Reiss, A. J. M. Balm, A. M. Palache, P. H. Cambier, and J. S. Van der Stijl-Neijenhuis (1979), "A Metabolic Comparison between Human Blood Monocytes and Neutrophils," *Macrophages and Lymphocytes: Nature, Functions, and Interaction*, Eds. M. R. Escobar and H. Friedman (New York: Plenum Press), 29–36.
9. Hard, G. C. (1970), *Br. J. Exp. Pathol.* **51**, 97.
10. Schwab, R., M. K. Crow, C. Russo, and M. E. Weksler (1985), *J. Immunol.* **135**, 1714.
11. Rosentreich, D. L., J. J. Farrar, and S. Dougherty (1976), *J. Immunol.* **116**, 131.
12. Devries, J. E., A. P. Caviles, W. S. Bont, and J. Medelsohn (1979), *J. Immunol.* **122**, 1099.
13. Thiele, D. L., and P. E. Lipsky (1982), *J. Immunol.* **129**, 1033.
14. Rosenthal, A. S. (1978), *Immunol. Rev.* **40**, 136.

I. G. KEVREKIDIS,† A. D. ZECHA† and ALAN S. PERELSON‡
† Department of Chemical Engineering, Princeton University, Princeton, NJ 08544 and
‡ Theoretical Division, Los Alamos National Laboratory, Los Alamos, NM 87545

Modeling Dynamical Aspects of the Immune Response: I. T Cell Proliferation and the Effect of IL-2

ABSTRACT

We apply methods of numerical bifurcation theory in the construction and evaluation of mathematical models of the immune response. We focus on the dynamics of the interaction of T cells with IL-1, IL-2 and antigen at several levels of modeling complexity. We examine the dependence of the qualitative dynamics on key parameters, and attempt to draw analogies between numerically predicted and experimentally observed behavior. The predicted behavior is interesting from a mathematical point of view (we describe a double zero singularity) even for these simple models. The most significant qualitative effect observed is the critical dependence of the final system state on small variations of IL-2 during the response. This makes us believe that coupling of the immune responses to antigenically unrelated epitopes through IL-2 and other lymphokines may be a critical factor in determining the ultimate effectiveness of the immune response to pathogenic organisms.

1. INTRODUCTION

During the last two decades, there has been a continuing research effort aimed towards systematically modeling the immune response (cf. Bell, Perelson and Pimbley, 1978; Bruni, et al., 1978; Marchuk, 1983; DeLisi, 1983; Hoffmann and Hraba, 1985). One of the main difficulties in generating realistic models has been our incomplete understanding of the dynamics of T cell and B cell activation. New molecules that are involved in the regulation of lymphocyte proliferation and differentiation (e.g., the lymphokines IL-1, IL-2, IL-4, IL-5, and interferon-γ) and new regulatory mechanisms (e.g., antigen and IL-2 regulation of IL-2 receptor expression, antigen processing and presentation) have been and are continuing to be uncovered and established. As new regulatory pathways are discovered, they need to be included in models and quantitatively evaluated. Here we begin this program by developing and analyzing a kinetic model of T cell activation that incorporates mechanistic rate expressions consistent with our current level of experimental understanding of the effects of antigen, IL-1 and IL-2.

One can distinguish two computational approaches to analyzing the dynamics of immune system models. The first relies on the solution of an initial value problem: parameters are specified, initial conditions are given, and the fate of trajectories is followed by integration forward in time. It is evident that unless a number of initial conditions are used, certain features of the response may be overlooked (multiple attractors, unstable solutions, separatrices, etc.). Such a method is extremely unwieldy when one approaches a transition boundary in parameter space, since the states involved become marginally stable and increasingly large integration times are required for a transient to approach them. Direct integration becomes almost hopeless in a systematic search of a high-dimensional parameter space due to the enormous number of individual numerical experiments required to analyze even the simplest of models. The alternative is to use numerical implementations of *bifurcation theory* procedures (cf. Rabinowitz, 1977; Kubicek and Marek, 1983) that concentrate on transition boundaries in parameter space. Such an approach often can give complete characterizations of the responses of a given system, along with possible phase planes and information about the quantitative stability of the states involved, unstable solutions, etc., without resorting to integration. Here we use both approaches; we use bifurcation theory to find critical parameter values at which the behavior of the system changes. We then integrate the differential equations describing T cell activation for choices of parameters on either side of the critical ones in order to confirm the predictions of bifurcation theory and to get insight into the detailed kinetics of the process.

We proceed by describing the time-dependent changes in the concentration of T cells, macrophages, antigen, IL-1 and IL-2 by a set of autonomous ordinary differential equations. In constructing the differential equations, choices have to be made about the functional form of various terms that model the interactions between the components in the system. For the T cell activation system, there are not yet sufficient experiments to guide us in all of the choices we have to make. We, thus, rely on a combination of modeling experience and mathematical knowledge about the behavior of various terms that we include in the equations. Experiments will ultimately have to be done to confirm our modeling choices. Here we develop preliminary models which we hope, for reasonable parameter values, predict behaviors that are immunologically sensible. Using our bifurcation theory approach we locally search parameter space, systematically varying the values of important parameters. This procedure reveals a range of behaviors consistent with the model. In general, such a procedure may also reveal deficiencies of the model associated with particular regimes of parameter space that might have been overlooked if only one set of parameter values were studied.

Some elegantly simple dynamic models of immune phenomena have been proposed (cf. Bell, 1973; Perelson, Mirmirani and Oster, 1976, 1978; Bell, Perelson and Pimbley, 1978; Marchuk, 1983; DeLisi, 1983; Perelson and Macken, 1984). However, when realistic network models are considered, the large number of kinetic constants, the vastness of possible network topologies and, when allowed, the changes in topology generated by mutations, can make the systematic tracking of the predicted behavior a complicated task (cf. Farmer, Packard and Perelson, 1986; Perelson, this volume). Another approach to constructing network models, is to use elements that are higher-order building blocks than the individual cell populations or idiotype positive or negative populations that are commonly used in such models (cf. Hoffmann, 1975, this volume; Herzenberg, Black, and Herzenberg, 1980; DeBoer, this volume for examples of the traditional approach). The elements coupled in the network could be the cells and molecules thought to be important in generating a response to a given antigenic determinant. The network would then reflect the properties of the system when responses to individual epitopes were coupled, as would occur when a complex protein or cellular antigen were encountered. Here we consider the possibility of the coupling of immune responses by the soluble mediator IL-2. (Dröge, this volume, also discusses this possibility). In our opinion coupling through soluble mediators should be an important ingredient in any systematic model of the immune system, whether it be an idiotypic network model, helper-suppressor T cell circuit model or a higher order model of the type mentioned above.

This paper is structured as follows. In the first section, we present a brief discussion of the biology of the response of a T cell to antigen, and the requirements for IL-1 and IL-2. We then concentrate on developing a model of T cell activation and proliferation in relation to the lymphokines involved in controlling the response (IL-1 and IL-2). Our modeling effort proceeds in stages, with a simple model preceding a more realistic and more comprehensive one. The simple model is used to illustrate how numerical bifurcation theory can be used to explore the possible qualitative behaviors of the system. For each model, specific results are presented and suggestions of interpretations or verification experiments are discussed. We conclude by discussing directions for future research, both in terms of modification of the existing models, and in terms of coupling several models of this type in an attempt to model a response to an antigen, containing multiple epitopes, that stimulates a variety of T cell subpopulations differing in their antigen specificity.

2. IL-2 AND T CELL PROLIFERATION

In the last decade, since Morgan et al. (1976) first reported the presence of a T cell growth promoting activity in supernatants of activated T cell cultures, knowledge about the control of T cell proliferation has expanded tremendously. A substance, first called T cell growth factor (TCGF), later renamed interleukin-2 (IL-2), is now known to be responsible for this growth promoting activity. IL-2 is a 15,000 dalton glycoprotein. The IL-2 gene has been cloned (Taniguchi et al., 1983), sequenced (Fujita, et al., 1983; Holbrook, et al., 1984), and recombinant forms of IL-2 are commercially available. The IL-2 receptor has been characterized and the gene coding for one of its two chains has been sequenced (Greene and Leonard, 1986; Smith, 1987; Teshigawara et al., 1987). There are both high- and low-affinity receptors for IL-2. The high-affinity receptor is the one relevant for growth control. It has an affinity $K = 10^{11}$ M and is saturated when the concentration of IL-2 is 5×10^{-10} M (Smith, 1987). The actual details of how IL-2 promotes T cell growth at the level of intracellular events are still being worked out (cf. Stern and Smith, 1986). However, it is recognized that, after activation by antigen, some critical level of interaction of IL-2 with its receptor is needed to drive the T cell into DNA synthesis and mitosis (Smith, 1984a,b; Cantrell and Smith, 1984).

One of the more interesting features of IL-2 regulation of T cell growth is the fact that IL-2 is produced by activated T cells. Thus, IL-2 acts as an autocrine hormone, i.e., cells that respond to IL-2 also produce it (cf. Meuer et al., 1984). This autocrine feature of IL-2 regulation of T cell growth lead us to believe that IL-2 might provide the molecular basis for a switch that can convert resting T cells into activated T cells. (In biochemical systems, autocatalytic reactions provide the nonlinearity needed for switching behavior; cf. Kath and Murray, 1985). Here we shall develop a model of T cell activation based on this principle.

Our model of T cell activation relies upon the following facts which are summarized in Figure 1:

1. Activated macrophages produce interleukin-1 (cf. Mizel, 1982).
2. The antigen-specific activation of T cells requires two signals: (i) antigen—processed and presented in the appropriate form on the surface of a histocompatible macrophage or other antigen-presenting cell; (ii) a second signal provided by the antigen-presenting cell. For some subpopulation of T cells this signal is IL-1, whereas other T cells may require different accessory cell molecules. T cell clones given signal one (antigen) but not signal two will go into an unresponsive state (Jenkins, et al, 1987).
3. T cells when activated by both antigen and IL-1, respond by synthesizing and secreting IL-2 (Smith, 1980; Greene and Leonard, 1986) and expressing IL-2 receptors (Cantrell and Smith, 1983).
4. If IL-2 is excluded from an antigen-activated T cell population, IL-2 receptor expression still occurs, but DNA synthesis does not proceed (Smith, 1984a,b).
5. IL-2 receptor expression is antigen dependent. Upon removal of antigen, IL-2 receptor levels decline. The proliferative rate of the cell population slows

in parallel, and all of the cells eventually accumulate in the resting (G_0/G_1) phase of the cell cycle, even in the presence of saturating concentrations of IL-2.

Molecular details about the IL-2 receptor will not play a role in the modeling presented here. However, recent evidence about the receptor clarifies some of the biology and could provide fertile ground for future modeling. Smith and co-workers (Smith, 1987; Teshigawara et al., 1987) have discovered that the IL-2 receptor contains two chains, an α chain and a β chain. Both chains need to be expressed

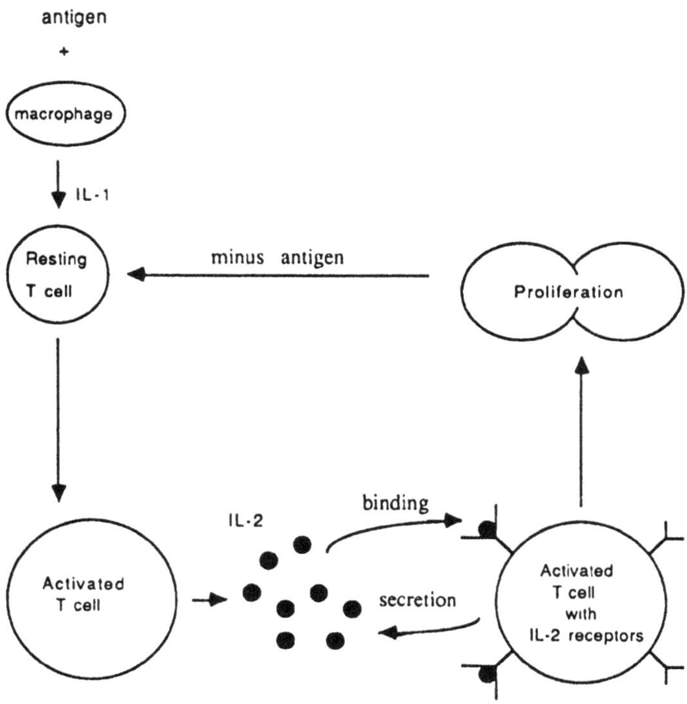

FIGURE 1 Schematic diagram of the interactions involved in T cell activation. Antigen interacts with a macrophage, stimulating the cell and causing it to secrete interleukin 1 (IL-1). When antigen on the surface of the macrophage is presented to the T cell in the presence of IL-1, the T cell becomes activated; it secretes IL-2 and then expresses IL-2 receptors on its surface. The binding of IL-2 to its receptor induces the T cell to divide. As antigen is depleted, the T cell loses its IL-2 receptors and returns to an unactivated state.

in order for the receptor to have high affinity. Either chain by itself may bind IL-2, but only at low affinity. IL-2 causes a marked induction of β-chain expression (Smith and Cantrell, 1985). One may speculate that immunostimulatory ligands that interact with the T cell antigen receptor cause α-chain gene expression. Thus, both immunostimulatory ligands and IL-2 would be necessary for high-affinity IL-2 receptor expression. Further, in the presence of IL-2, high-affinity receptors are lost from the cell surface by a ligand-receptor internalization process (Fujii et al., 1986; Smith and Cantrell, 1985). Thus, repeated stimulation by both antigen and IL-2 would be needed to maintain the high-affinity receptor population and T cell growth.

3. MODELS

Experiments elucidating the control of T cell proliferation by IL-2 are usually done in cell culture. Antigen and cloned T cells specific for the antigen could be used, but more frequently a mitogen replaces the antigen and a heterogeneous population of T cells, all of which are stimulated by the mitogen, is employed. Macrophages or monocytes are added to the culture to supply IL-1 or its equivalent signal. If culture conditions are well maintained and the cells recultured when their density reaches 10^6 cells per ml, then cell death is not frequently observed.

In vivo conditions are very different. Antigen-specific T cells are constantly being supplied via the normal pathway of bone marrow production and thymic differentiation. After thymic involution, T cell production declines, but the thymus may still provide a significant number of new T cells. Cell death is an important feature, particularly among cells that are not antigen stimulated. Spatial compartmentalization occurs, with cells attaining a high density in the lymphoid organs, such as lymph nodes and spleen, and having a much lower density while in the circulation. The rate of encounter of a T cell with an antigen-presenting cell will differ depending on its location. In the lymphoid organs, dendritic cells will collect and present antigen, but T cells will move by crawling and probably encounter antigen-presenting cells infrequently. In the circulation, antigen presentation occurs on the surface of other blood cells (e.g., macrophages and B cells), and the encounter rates depend on the flow conditions in the circulation (cf. Bell, 1981). The ability of a lymphokine such as IL-2 to affect nearby cells will very much depend on the cell's environment. In the circulation, secreted lymphokines will be rapidly diluted. Thus, secretion into local environments created by the apposition of cell membranes may become very important. In lymphoid organs, secreted molecules will diffuse to nearby cells, but might not be significantly diluted by lymphatic flow. Consideration of these factors will be of great importance in designing realistic models of the immune response. Models developed to analyze cell stimulation in culture will provide insights into several, but not all, of the important parameters.

When embarking on a modeling project, it is prudent to isolate the major effects of interest, and to first center one's attention on these aspects of the problem. Here we shall do this by developing two models that are caricatures of the *in vitro* and *in vivo* situations. In the first model, we shall ignore the details of cell supply, cell death, and the effects of the immune response in changing the antigen concentration. The model may, thus, be looked on as a model of an *in vitro* system, or as one of an *in vivo* system in which the time scale of interest is sufficiently short that changes in the T cell and antigen populations can be ignored. Thus, in this first model, we focus on the production of IL-2 by T cells. In the second model, we explicitly include the time-dependent changes in the cell and antigen populations. While this model is more reminiscent of the *in vivo* situation, it still does not address the effects of spatial compartmentalization and transport of cells and molecules. Although macrophages and IL-1 are included in our models, they are not the central focus of our modeling effort. For this reason we have chosen to use rather simplistic terms to model the interaction of macrophages with antigen and their subsequent production of IL-1. There is evidence that factors other than IL-1, that are also produced by macrophages, may be important in T cell activation (cf. Dröge, this volume). For the purposes of this model, we do not distinguish between these factors. Thus, the concentration of IL-1 that we use in our models should be thought of as representing the set of lymphokines produced by macrophages and monocytes that are required for T cell stimulation.

MODEL 1

During the initial stages of an immune response, lymphokine production precedes any increase in the T cell population. We model the dynamics of lymphokine production by the following set of ordinary differential equations:

$$\frac{du_1}{dt} = k_0 G m - k_1 u_1, \tag{1}$$

$$\frac{du_2}{dt} = k_2 G T u_1 + \frac{k_3 T u_2^2}{1 + k_4 u_2^2} - k_5 u_2, \tag{2}$$

where u_1 and u_2 are the concentrations of IL-1 and IL-2, respectively, G is the concentration of antigen or other immunostimulatory molecule such as a mitogen, m is the concentration of macrophages and T is the concentration of T cells that recognize this specific antigen. The k_i are kinetic rate constants.

We have adopted simple, mass action kinetics for the algebraic form of all but one of the terms in Eqs. (1) and (2). The $k_0 Gm$ term represents the rate of IL-1 secretion from macrophages in the presence of antigen G. The antigen is required to stimulate the macrophages and induce IL-1 secretion. Although the process of macrophage stimulation is not considered in detail in this paper, the value of the parameter k_0 must reflect the rate of macrophage activation. The $k_2 GT u_1$ term represents the rate of secretion of IL-2 from T cells stimulated by both IL-1 and antigen. Imbedded in the parameter k_2 are the details of antigen presentation and processing by macrophages or other antigen-presenting cells. At high enough antigen concentrations, the rates of IL-1 and IL-2 secretion should saturate, and thus, these terms would need to be modified. For the present we shall assume that the antigen levels are such that saturation effects need not be considered. The $k_3Tu_2^2$ term represents the rate of secretion of IL-2 from already activated T cells. The fact that we use the square of the IL-2 concentration in this term reflects the autocrine nature of the secretion. Experiments by Cantrell and Smith (1984) clearly show that the IL-2 dose response curve rises faster than linear. More quantitative estimates by Hooten et al. (1985) indicate that IL-2 dose response curves have a slope between one and two. In this simple model, we approximate these findings by using a quadratic dependence on the IL-2 concentration. Because the central focus of this model is IL-2 production, we do incorporate a term $1 + k_4u_2^2$, in the denominator to reflect the saturation of the IL-2 secretion rate in the presence of excess IL-2. The final form of the IL-2 production term, including the denominator, is precisely the same as terms used in developmental biology to model the production of an "activator" molecule, a molecule which stimulates its own production and is postulated to play a role in pattern formation (cf. Meinhardt, 1982). Finally, the k_1u_1 and k_5u_2 terms represent the rate of disappearance of the lymphokines from the system—through spontaneous breakdown, through cellular internalization, or *in vivo* through hepatic removal. In situations where cellular internalization is the major route of interleukin loss, the disappearance terms would need to be modified to include a cell population density effect. In this model, the concentrations of T cells, macrophages and antigen are considered parameters whose values do not vary in time, and hence, this is not necessary.

PREDICTED BEHAVIOR

We examine the steady states of Eqs. (1) and (2) as some of the model parameters vary. Setting the right hand sides of Eqs. (1) and (2) to zero, we obtain the following two algebraic equations:

$$u_1 = \frac{k_0Gm}{k_1}, \tag{3}$$

$$k_4k_5u_2^3 - (k_3 + k_2k_4Gu_1)Tu_2^2 + k_5u_2 - k_2GTu_1 = 0. \tag{4}$$

First consider the situation when antigen is absent. Then $G = 0$ and Eq. (3) implies that $u_1 = 0$. Eq. (4) simplifies to

$$u_2(k_4k_5u_2^2 - k_3Tu_2 + k_5) = 0, \tag{5}$$

which has as possible solutions

$$u_2 = 0 \quad \text{and} \quad u_2 = \frac{k_3 T \pm \sqrt{(k_3 T)^2 - 4k_4 k_5^2}}{2k_4 k_5}. \tag{6}$$

Let $T_c = 2k_5\sqrt{k_4}/k_3$. Depending upon the relationship between T and T_c one, two or three steady state values of u_2 obtain.

(i) If $T < T_c$, then $u_2 = 0$ is the only steady state solution.
(ii) If $T = T_c$, then $u_2 = 0$ and $u_2 = (k_3T)/(2k_4k_5)$ are both steady state solutions.
(iii) If $T > T_c$, then $u_2 = 0$ and

$$u_2 = k_3T/2k_4k_5\left[1 \pm \sqrt{1 - \frac{4k_4k_5^2}{k_3^2T^2}}\right]$$

are all steady state solutions (the middle one being unstable and the other two being stable).

Cases (i) and (iii) are the most relevant, case (ii) being a transition between them. Case (i) can be thought of as corresponding to a virgin state, in which there is no antigen, a low level of T cells and no IL-2. After a successful immune response, the antigen will be eliminated, but the T cell population should be elevated. Case (iii) corresponds to this situation. Note, however, that in this state, high IL-2 is required to maintain the T cell population. This is not realistic since after antigen is removed the IL-2 response eventually also decays. Model 2 below will rectify this shortcoming.

When $G \neq 0$, u_1 is given by Eq. (3) and u_2 is the solution to the cubic, Eq. (4). To simplify the analysis in this case, we arbitrarily set $T = m = k_0 = k_1 = k_2 = k_3 = k_4 = 1$ and vary only k_5, the IL-2 removal rate, and G, the antigen concentration. Figure 2a shows the dependence of the steady state IL-2 concentration on the antigen concentration for k_5=0.65. For each value of G, one single stable steady state exists—the larger G, the larger the amount of IL-2 at steady state. If the rate constant for IL-2 removal, k_5, is decreased (Figure 2b, $k_5 = 0.55$), we observe that for certain values of G, three steady states (two stable and one, the middle, unstable) coexist. This S-shaped hysteresis curve is capable of causing "ignition-extinction" type phenomena in the transient system response. Recall that in this model the antigen concentration is fixed. However, if on a time-scale that is slow compared with the time involved in lymphokine production, we allow the antigen concentration to vary, the following effects may be seen: (1) If the antigen concentration is low, say, below 0.2, then the system will be in the unignited state. As the antigen concentration increases, the IL-2 concentration will slowly increase until the turning point at G=0.28 is reached. At this point, any further increase in the antigen concentration will cause the system to ignite. The IL-2 concentration will increase rapidly and the system will move along the upper, or ignited, branch. (2) If, after the upper steady state is achieved, we allow the antigen concentration

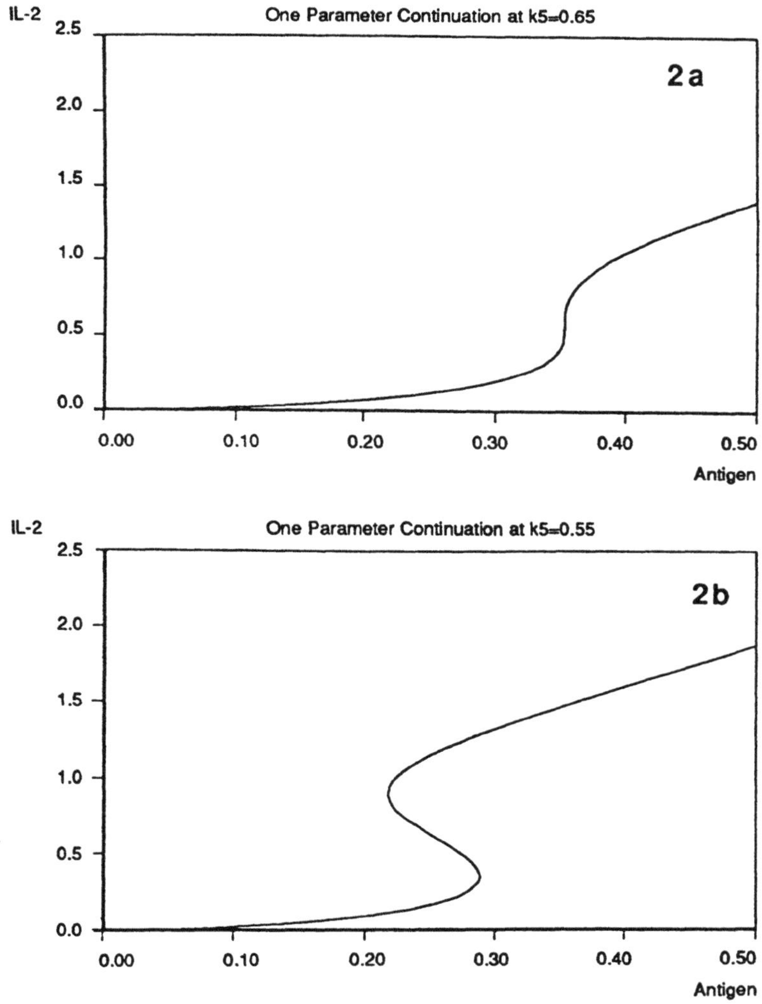

FIGURE 2a-b As the IL-2 removal rate constant k_5 and the antigen concentration are varied, the model's steady state behavior changes. Here we plot the steady state concentration of IL-2, u_2, versus the antigen concentration G. The steady state concentration of IL-1, u_1, varies linearly with G and is not shown. Note that as k_5 is decreased from 0.65, (a) a monotonic curve which exhibits a single steady state value of IL-2 at each antigen concentration is converted (b) into an S-shaped curve for which there are three possible steady state IL-2 concentrations for antigen concentrations $0.2 < G < 0.28$.

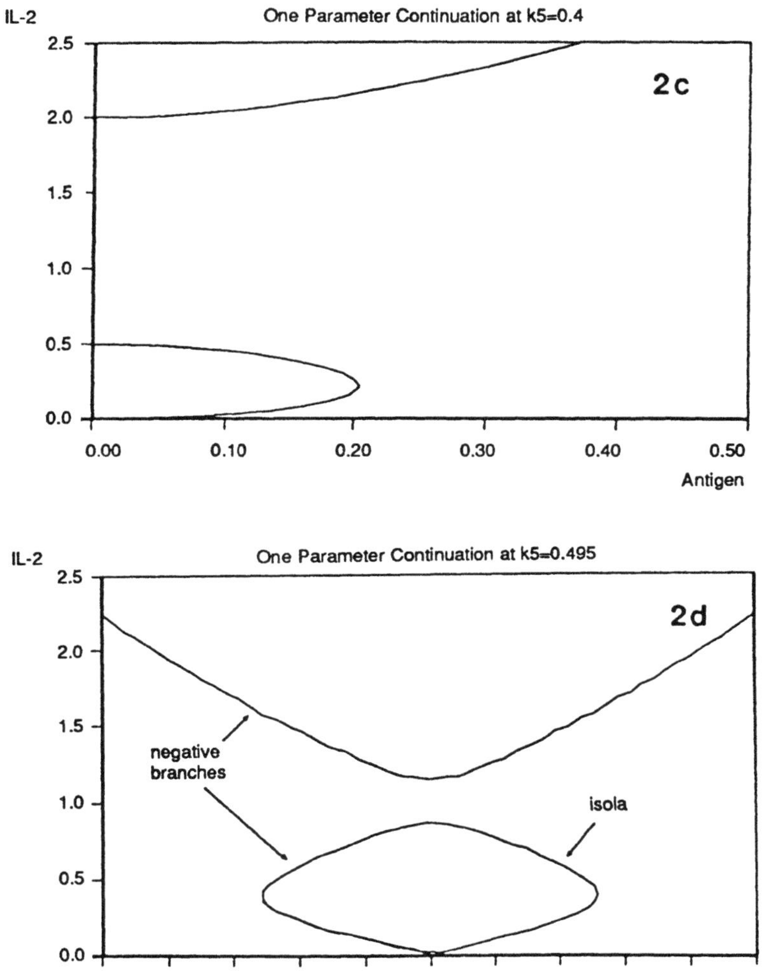

FIGURE 2c-d (c) For values of $k_5 < 0.5$ the upper "activated" branch separates from the lower branch. (d) The separation occurs at an isola point, where the left turning point collides with the rightmost turning point of a negative, physically meaningless solution branch which is the mirror-image of the branch shown in Figure 2b.

to decay, the system will remain close to the upper "ignited" branch of solutions until the left turning point bifurcation is reached. At that point it will abruptly fall back to the "unignited" branch. It is evident that the existence of a secondary response will depend on whether, at the time antigen is re-encountered, the system is near the higher or lower stable steady states.

If the rate constant k_5 is further decreased to $k_5 = 0.4$, an interesting transformation is observed (Figure 2c). The upper ignited branch is now completely disconnected from the lower branch. This lower branch still possesses a turning point, and three solutions—two stable and one unstable—will still coexist for an interval of G values. It is important to observe, however, that after the system lands on the upper branch (for $G > 0.2$), it will never return to the lower steady state, even if the antigen disappears completely.

This simple model is capable of predicting, as its parameters vary, three completely different system responses to stimulation by antigen. In a "normal" situation (Figure 2a), larger stimulation results in a larger response. It is possible, however, as some parameter varies, e.g., the antigen concentration, that ignition-extinction and hysteretic behavior may be observed in the response. Although we have not presented any explicit results of dynamic simulations, the S-shaped steady state curve in Figure 2b is consistent with the long incubation period observed before a primary response takes off. Lastly, in situations corresponding to Figure 2b, if the cause of the stimulation (the antigen) disappears, the system will always return to a resting state ($u_1 = u_2 = 0$). In the situation corresponding to Figure 2c, this is not the case any more. Once ignited, the system will retain a memory of the stimulation even when its cause, the antigen, disappears completely.

These different kinds of steady state structures, along with the associated dynamics, can all be interpreted directly in physical terms. They can all be observed within the parameter space of one simple model. It is even more significant, however, that all of this structure is to be organized around a special point in parameter space, termed an *isola point*, at which a new isolated family of solutions arises. (In this particular model, the isola occurs for $k_5 = 0.5$ at $G = 0$). It is precisely at the parameter values corresponding to the isola that the upper steady state branch disconnects from the lower one. The left turning point (Figure 2b) hits the axis $G = 0$ simultaneously with the right turning point of a negative, physically meaningless solution branch (Figure 2d). The mere existence of such an "organizing center" (Golubitsky and Schaeffer, 1985) is enough to predict that a number of qualitatively different transition diagrams (like 2b and 2c) will be observed for nearby parameter values.

This caricature provides an illustration of our methodology and objectives. We build a model, say, of T cell activation and IL-2 secretion. We then search its parameter space in detail using methods of numerical bifurcation theory in an attempt to classify all (or at least several) dynamical patterns predicted within the framework of this model. Such information may then be used in conjunction with experimental data to validate the model and the mechanisms underlying its construction.

MODEL 2

A more realistic model of T cell activation distinguishes between activated and unactivated T cells, and explicitly accounts for dynamic changes in the antigen population. Here, by an activated T cell we shall mean a cell that has been stimulated by antigen and IL-1 into IL-2 secretion and the expression of IL-2 receptors. The number of IL-2 receptors that a T cell expresses is important in determining its behavior. Because we have chosen to disregard the details of IL-2 receptor expression and receptor occupancy, there is an ambiguity in our model. When an activated T cell divides, it may remain activated or it may return to the unactivated state. The actual outcome may depend on the level of IL-2 receptor expression on the daughter cells. Kendall Smith (1984a, personal communication) believes that under most circumstances the daughter cells are unactivated and need to be re-exposed to antigen and IL-1 to continue their proliferation. The experience of many experimentalists growing T cells in culture is that if IL-2 is present in the medium T cells will continue to grow for a few weeks without antigen restimulation. Whether the culture continues to contain enough antigen to restimulate newly generated daughter cells during these few weeks is not clear. Here we will present two versions of our model, the first (model 2a) assumes that the daughters of activated T cells are unactivated; the second (model 2b) assumes that activated T cells upon division remain activated. The difference in the two models is only in Eqs. (8) and (9) which refer to the T cell populations. Although we present both models, in this paper we shall only analyze the consequences of model 2a.

The differential equations that define these two models are as follows:

$$\frac{du_1}{dt} = k_0 G m - k_1 u_1, \tag{7}$$

$$\frac{dT}{dt} = k_8 - k_9 G T u_1 - k_{10} T + \frac{k_{12} u_2 T^*}{(G+\epsilon)(1+\tilde{k}_{12} u_2)} + \frac{2k_{11} u_2^2 T^*}{1+\tilde{k}_{11} u_2^2}, \tag{8a}$$

$$\frac{dT^*}{dt} = k_9 G T u_1 - k_{10} T^* - \frac{k_{12} u_2 T^*}{(G+\epsilon)(1+\tilde{k}_{12} u_2)} - \frac{k_{11} u_2^2 T^*}{1+\tilde{k}_{11} u_2^2}, \tag{9a}$$

$$\frac{du_2}{dt} = k_{13} T^* - k_5 u_2, \tag{10}$$

$$\frac{dG}{dt} = k_{19} + k_{18} G - k_6 G m - k_7 T^* G. \tag{11}$$

For model 2b, Eqs. (8a) and (9a) are replaced by the following:

$$\frac{dT}{dt} = k_8 - k_9 G T u_1 - k_{10} T + \frac{k_{12} u_2 T^*}{(G+\epsilon)(1+\tilde{k}_{12} u_2)}, \tag{8b}$$

$$\frac{dT^*}{dt} = k_9 G T u_1 - k_{10} T^* - \frac{k_{12} u_2 T^*}{(G+\epsilon)(1+\tilde{k}_{12} u_2)} + \frac{k_{11} u_2^2 T^*}{1+\tilde{k}_{11} u_2^2}. \tag{9b}$$

As before, u_1 and u_2 are the concentrations of IL-1 and IL-2, respectively. T, is now the concentration of antigen-specific, but unactivated T cells, T^*, is the concentration of activated antigen-specific T cells, and G, is the antigen concentration. The macrophage concentration, m, is still a parameter. Eq. (7) is identical to Eq. (1) and thus, as before, k_0Gm represents the rate of IL-1 secretion from macrophages in the presence of antigen, and k_1u_1 represents the rate of loss of IL-1 from the system. k_8 is the (constant) rate of production of new T cells (e.g., from the thymus and the bone marrow). The k_9GTu_1 term represents the rate of differentiation of T cells into activated T cells (T^*), and therefore, appears again, with opposite sign, in the T^* equation. The k_{10}T term, along with the $k_{10}T^*$ term in Eq. (9), represent the normal death rate of T cells in the system. The remaining two terms again appear in both the T and T^* equations of change. The $k_{11}u_2^2T^*$ term represents the division of T^* cells stimulated through IL-2. The term $1+\tilde{k}_{11}u_2^2$ provides a saturation of this division rate in the presence of excess IL-2. Model 2a reflects Smith's (1984a) suggestion that after division in response to IL-2 the daughter cells of activated T^* cells have to be activated once again by antigen. We model this effect by including twice this k_{11} term in Eq. (8a) so that two unactivated T cells are produced in every division of an activated T^* cell. In model 2b, we assume that the daughters of activated T cells are activated and, thus, Eq. (8b) lacks this k_{11} term. In Eq. (9b), the k_{11} term occurs with a positive sign reflecting the growth of the T^* population, whereas in Eq. (9a) this term occurs with a negative sign reflecting the loss of activated cells upon division. The $k_{12}u_2T/(G+\epsilon)(1+\tilde{k}_{12}u_2)$ term provides a mechanism for shutting down the response once the antigen has been contained (G small). For large antigen concentrations, this term is negligible compared to the other terms in Eqs. (8) and (9), but once the antigen concentration has been reduced to a level comparable to ϵ, this term switches activated T^* cells back into unactivated T cells. The rate of IL-2 production has been taken to be proportional to the concentration of activated T^* cells. The term k_5u_2 accounts for the loss of IL-2 from the system. This constant rate of loss is only an approximation since the binding of IL-2 to T cells also removes IL-2 from solution.

The equation for the antigen G represents the most general situation in which the antigen can replicate, while at the same time it is removed by both macrophages and activated T^* cells. Although we do not specify a removal mechanism, removal can be direct, say, by receptor mediated endocytosis, phagocytosis or cytotoxic activity, or indirect, say, through the influences of helper T cells in stimulating B cells and the subsequent effector mechanisms brought into play by antibody recognition of the antigen. The terms k_6mG and k_7T^*G account for these mechanisms. The rate of antigen reproduction once it enters the system, $k_{18}G$ is not resource limited. A more refined model could use a logistic or Gompertzian antigen growth law. An extra constant source term, k_{19}, has been included. If k_{19} is positive this term can be interpreted as continuous exposure to the antigen. Negative values of k_{19} can be interpreted as having a nonspecific mechanism of antigen loss (e.g., hepatic removal) in the system which dominates any constant antigen influx.

In order to study model (2a) via bifurcation methods, the concentrations of all species were normalized using the following concentrations: IL-1 $= 10^{-10}$ M, IL-2

$= 10^{-10}$ M, unactivated T cells $= 2 \times 10^{-15}$ M, activated T^* cells $= 2 \times 10^{-15}$ M, macrophages $= 2 \times 10^{-17}$ M, and antigen $= 2 \times 10^{-6}$ M. The rationale for these choices are given in Appendix A. We have also nondimensionalized all the rate constants and have chosen the unit of nondimensional time to be 50h. We chose as our bifurcation parameters the nondimensional rate constant for antigen replication k_{18} and the nondimensional antigen influx (or nonspecific loss) rate k_{19}. It is in the space of these two parameters that we will perform our bifurcation analysis, trying to locate parameter values at which the antigen escapes from immune system destruction. The other nondimensional parameters in the problem were fixed at values discussed in Appendix A.

A NUMERICAL EXPERIMENT

In an attempt to characterize and classify the possible patterns of dynamic behavior that the above model can predict, we performed a simple numerical experiment: we chose $k_{18} = 2$ (making the nondimensional rate of antigen replication comparable to the nondimensional rates of T cell proliferation in a normal response). We also chose a very small constant antigen source ($k_{19} = 0.0096$), to represent a continuous exposure to antigen. Integration of the nondimensional differential equations with initial conditions $u_1(0) = 0.1, u_2(0) = 0.1, T(0) = 0.4, T^*(0) = .05$ and $G(0) = 0.85$ (a large dose) gives the results shown in Figures 3a–3d. The antigen first grows, its concentration rising sharply (Figure 3a); this is followed by the generation of an immune response as measured by a rise in IL-2 and T cell population levels (Figures 3b–3d). This response causes the antigen concentration to begin declining near $t = 5$. The T cell and IL-2 concentrations then also decline, as would be expected at the termination of an immune response. This apparent success in handling the antigen, however, is deceptive: after a few more time units, the antigen starts to rise again while the immune response having shut down is unable to check the antigen growth. For these parameter values, the antigen overwhelms the system defenses.

We then tried the following modification: starting with the same initial conditions and same parameter values, we let the system run forward in time until t=18. At this point the antigen had already started to run away, $G = 184.4$, but the IL-2 and T cell concentrations were negligible ($u_2 = 0.034, T = 10^{-6}$ and $T^* = 0.018$). We then simulated giving a very small pulse of IL-2 by increasing the IL-2 concentration to 0.3, i.e., we set $u_2(t = 18^+) = 0.3$ leaving all other concentrations unchanged, and continued the integration. This numerical experiment has a physical meaning: IL-2 can be added to an *in vitro* system; whereas *in vivo* it can correspond to IL-2 production by T cells responding to another antigen. The results are presented in Figures 3e–3h. While the antigen concentration rises again at t=20 (Figure 3e), the system response follows it, and now is capable of bringing the antigen down to a low level. It is interesting to observe that the immune response decays after this event to a low steady state, through damped oscillations.

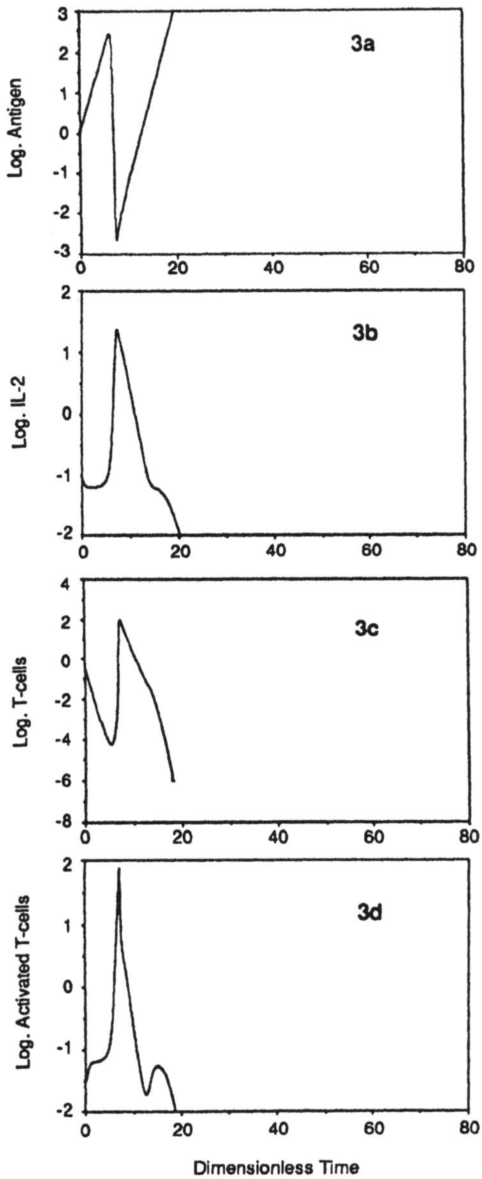

FIGURE 3 a-d. The effect of IL-2 on the outcome of an immune response. Time evolution of the key nondimensional concentrations in two numerical experiments. After an initial rise followed by a decline, the replicating antigen explodes "beating the system" (panel a), while the immune response (panels b-d), shuts down, unable to contain it.

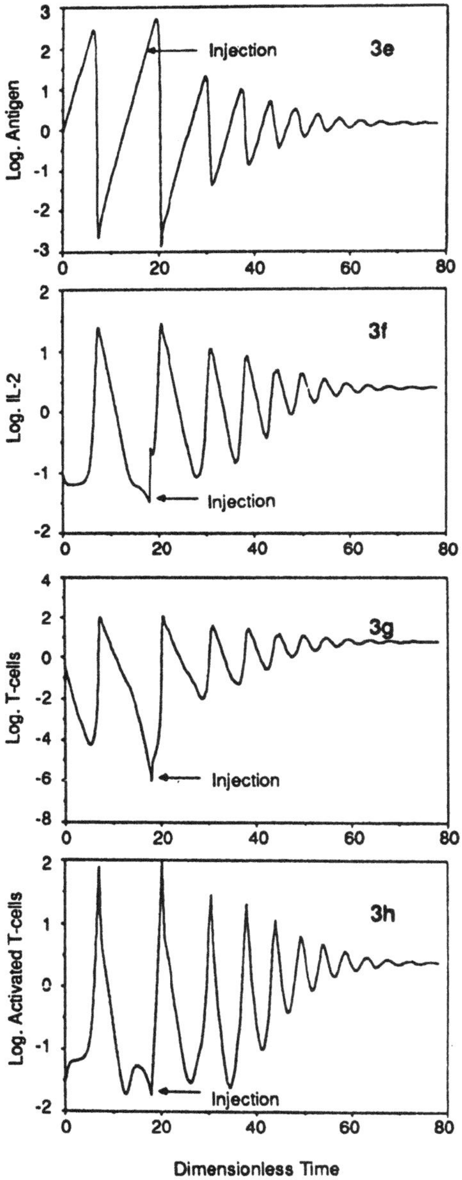

FIGURE 3 e-h. If a small amount of IL-2 is injected at $t = 18$ (panel f), the immune response rises quickly and approaches an "immune" stable steady state in an oscillatory fashion (panel f-h), keeping the antigen at a finite value (panel e). The unit of dimensionless time time is 50 hrs.

It is now evident that, for the same parameter values, two radically different fates are possible for the system: it may conquer the secondary rise of the antigen and return to a normal steady state, or it may be overwhelmed by the growing antigen. There must, therefore, exist a surface in the five-dimensional space that we study, separating the initial conditions that would end on the normal stable steady state—let us call it the "immune steady state"—and those for which the antigen would grow indefinitely—let us say that the system "explodes." Knowledge of the location of this surface may be significant, as the above numerical experiment illustrates: a small dose of IL-2 at the appropriate moment may prove capable of checking the antigen explosion and take the system to the immune steady state. The element of correct timing must also be stressed: if the same IL-2 injection is made in the system slightly later (we tried it at t=22), it is not sufficient to check the antigen explosion.

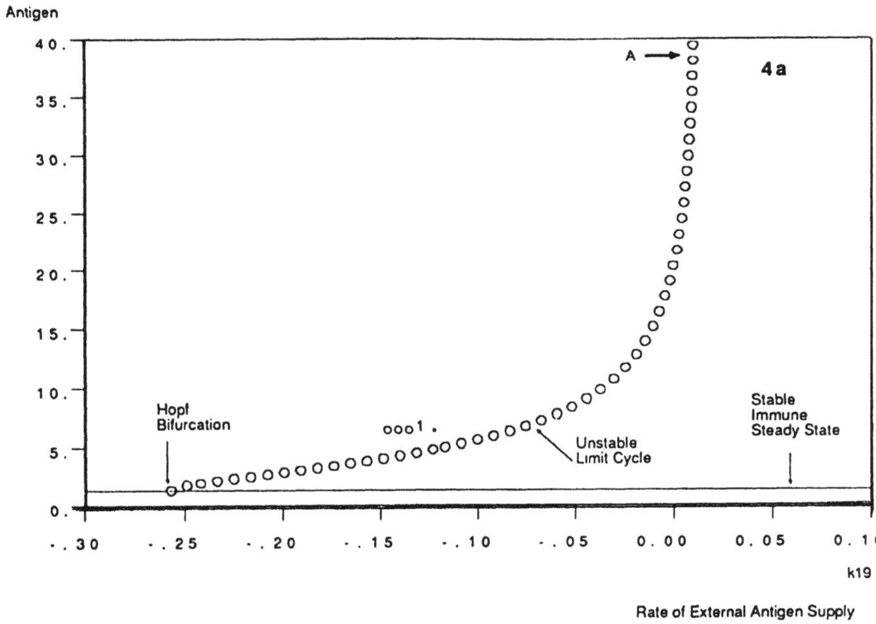

FIGURE 4a Bifurcation diagram of the model equations as the rate of external antigen supply, k_{19} is varied. A stable "immune" steady state branch (5 eigenvalues with negative real part) loses stability at a Hopf bifurcation, at $k_{19} \approx -0.26$. The Hopf bifurcation gives birth to an unstable oscillatory branch (three Floquet multipliers < 1, denoted by o o o , one at 1, and one > 1 denoted by •). These limit cycles play a crucial role in determining for which initial conditions the system will effectively contain the antigen.

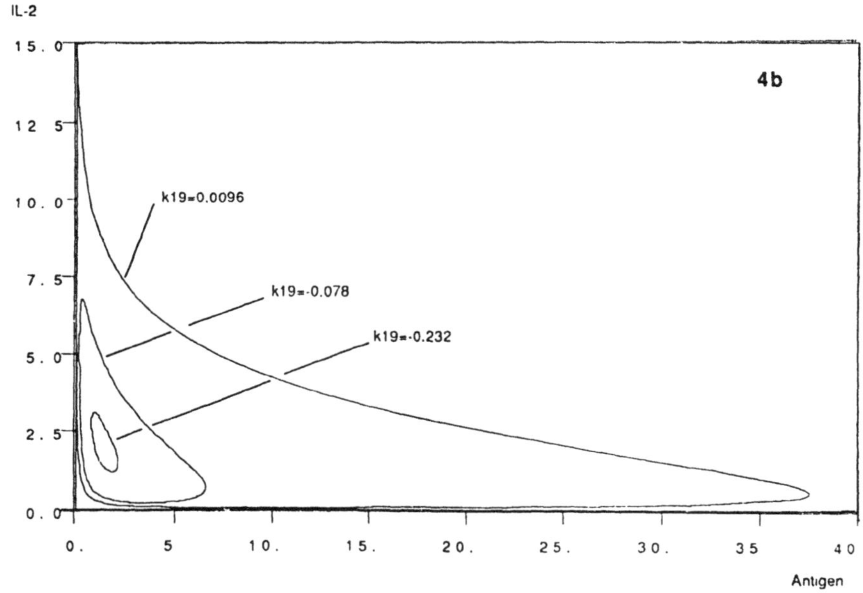

FIGURE 4b Representative shapes of these limit cycles of Figure 4a projected in the (antigen) × (IL-2) plane. They are initially small and surround the steady state ($k_{19} = -0.232$), but as k_{19} grows, they also grow in size, and their shape deforms ($k_{19} = -0.078$). At the conditions of our numerical experiment ($k_{19} = 0.0096$), the limit cycle has approached the two axes. Figures 4a and 4b have been obtained with the antigen replication rate constant $k_{18} = 2$.

THE BIFURCATION BEHAVIOR

Figure 4a shows a one-parameter bifurcation diagram taken at $k_{18} = 2$, as k_{19} varies. The solid line indicates a set of steady states in which the antigen concentration is low, but is not zero. We call these steady states "immune" states, since the system has successfully contained the antigen. Linearization of the model equations around the immune steady state branch at $k_{19} = 0.0096$ gives a 5x5 matrix whose eigenvalues have negative real parts. Three of them are real while the other two are complex conjugates with negative real parts. This is consistent with the oscillatory approach to a stable steady state that we observed in the transient analysis given above (Figure 3e–3h). As k_{19} becomes smaller, the two complex eigenvalues approach the imaginary axis of the complex plane. At $k_{19} \approx -0.26$ they cross this axis becoming positive, rendering the immune steady state unstable. At this point, a Hopf bifurcation occurs and a branch of limit cycles bifurcates from the immune steady state solution. We used AUTO (Doedel, 1981), a FORTRAN code for the automatic bifurcation analysis of ordinary differential equations, to follow this branch. As Figure 4a shows, this branch bifurcates towards higher values of k_{19}. The oscillatory solutions lying on this branch are unstable. The stability of a limit cycle is quantified by its characteristic (or Floquet) multipliers; a limit cycle for a system of n autonomous ordinary differential equations has n characteristic multipliers, one of which is always one. The limit cycle is stable if all of the remaining $n - 1$ multiplers have norms less than one, and is unstable if any of them has a norm greater than one. The branch of limit cycles shown in Figure 4a is unstable, having three multipliers less than one and one greater than one. These then are saddle-type limit cycles (they are attracting in three directions and repelling in one; in the same fashion that saddle-type steady states are attracting in some directions and repelling in others). It is well known, and best illustrated in two-dimensions, that saddle-type steady states lie on *separatrices*, surfaces that separate initial conditions for trajectories that eventually land in different basins of attraction. The same concept applies for this saddle-type limit cycle, even though it is difficult to visualize in 5-dimensional space. The four-dimensional stable manifold of this saddle-type limit cycle constitutes the separatrix that separates initial conditions for which the antigen is contained from those for which the antigen explodes. Figure 4b shows projections of the 5-dimensional limit cycle in two dimensions for three different values of k_{19}. Note that as k_{19} increases, the maximal antigen concentration attained during the limit cycle also increases.

Using AUTO we also performed a two-parameter search of the system behavior as both the external source of antigen (k_{19}) and the rate constant for antigen replication (k_{18}) are varied. The results of this search are shown in Figure 5a. A curve of Hopf bifurcation points runs into a curve of turning points in a double zero singularity. We examined two typical one-parameter cuts across this diagram taken at constant k_{19} values (one below and one above the double zero singularity) as k_{18} varies. These diagrams are shown in Figures 5b and 5c. In Figure 5b, the immune steady state (lower part of the curve) is stable at large values of k_{18}; its linearization has five eigenvalues with negative real part, two of them complex conjugates. As we move towards smaller k_{18}, around $k_{18} = 0.8085$, a subcritical Hopf bifurcation occurs as the two complex conjugate eigenvalues cross the imaginary axis and pass to the right-hand plane. This corresponds to crossing the Hopf bifurcation line in Figure 5a. The eigenvalues then quickly become real and positive, and one of them moves towards zero. At $k_{18} = 0.8079$ it crosses zero and passes again into the left-hand plane. This corresponds to crossing the lower saddle-node bifurcation curve in Figure 5a. This is a turning point bifurcation as can be seen from the steady state branch which turns backward towards higher k_{18} and continues being unstable (saddle type, four real negative and one real positive eigenvalues).

This bifurcation structure is typically observed in the neighborhood of a double-zero point in parameter space. A treatment of this can be found in standard texts on bifurcation theory (e.g., Guckenheimer and Holmes, 1983). As can be seen from their diagram, and is consistent with our numerical observations, the saddle-type limit cycle branch is going to collide with the saddle steady state branch in a so-called infinite-period bifurcation. The physical significance of this for our system is that very long transients with long "almost steady state" intervals may be observed far away from the immune steady state and close to the saddle-type steady state.

A closer inspection of Figure 5c gives rise to the following question: what happens to transients for values of k_{18} lower than the turning point value (to the left of $k_{18} \approx 0.74$). The answer is shown in Figures 6a and 6b. In addition to the steady state branch that we have been examining so far (labeled branch 1), one more branch of steady states exists to the left of the turning point. We have labeled this branch 2. This new branch of stable steady states is characterized by finite antigen concentrations, but by very small immune activity (concentrations of T cells near zero). We call this a "non-responsive" state. In this case the antigen is mainly consumed by the macrophages, without the machinery of the specific immune response. For some set of parameter values (immediately to the right of the turning point on branch 1), three steady states coexist, two of which are stable. One is the immune steady state, that we discussed before, in which the T cell population is elevated, while the other is the "non-responsive" state. As shown in Figure 6a, near $k_{18} = 0.8$ the antigen concentration grows very large. We expect that this nonresponsive branch will disappear for higher values values of k_{18}, although we have not confirmed this yet.

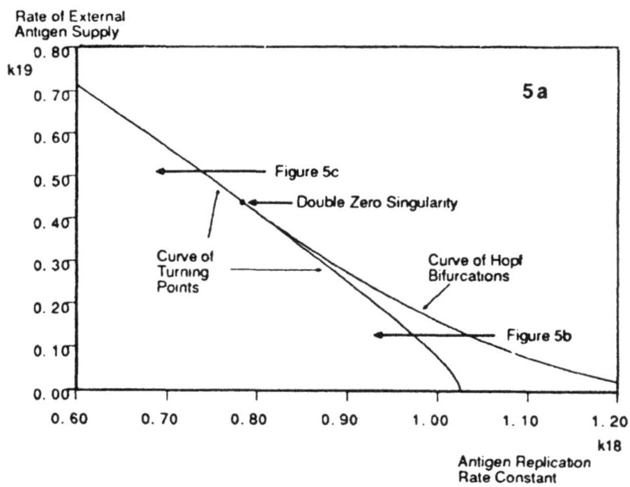

FIGURE 5a Two-parameter bifurcation diagram in k_{18}–k_{19} space, showing the position of a double zero singularity (a point where two eigenvalues are zero) along with the positions of two representative one-parameter cuts, for values of k_{19} below and above the double zero singularity.

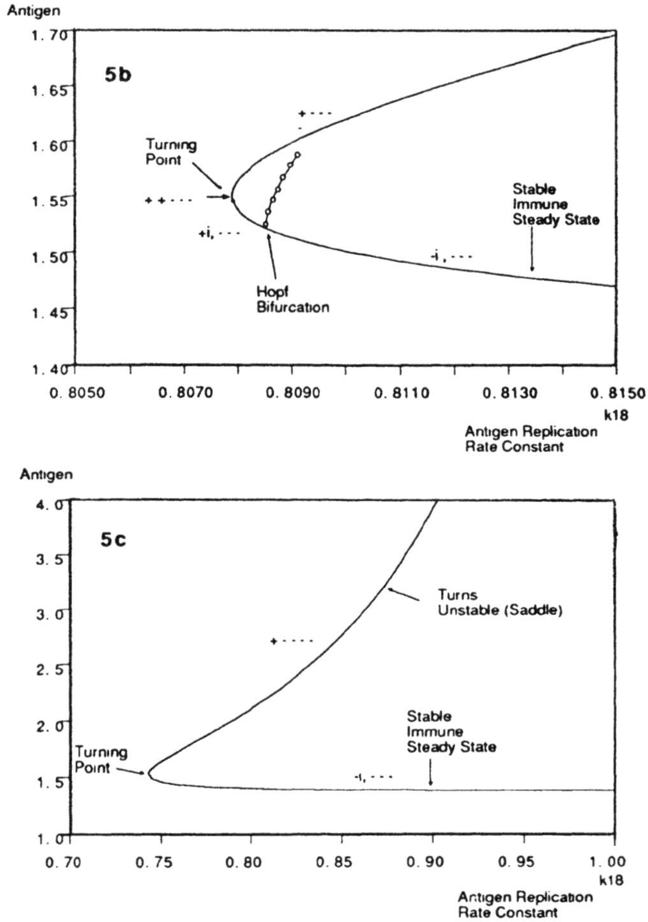

FIGURE 5b-c One-parameter bifurcation diagrams which show the steady state antigen concentration versus the antigen replication rate constant k_{18}. In (b) $k_{19} = 0.4$ and both a turning point and a Hopf bifurcation point are encountered, whereas in (c) $k_{19} = 0.5$ and only a turning point is encountered. Representative spectra of the linearization around these steady states are also shown $-(+)$: real negative (positive) eigenvalue, $-(+)i$: complex pair with negative (positive) real part. The unstable limit cycle branch in (b) is denoted by -o-o-o-o-. It probably collides with the saddle-type steady state branch at $k_{18} = 0.8097$ in an infinite period bifurcation.

DISCUSSION

The model we have discussed above is far from complete; we are currently trying to make it more realistic, as well as to extend it to encompass an entire immune response, including B-cells and the humoral immune response. It serves, however, as an illustration of a modeling methodology. Searching the parameter space of models—through methods of numerical bifurcation theory—and discovering parameter values for which dramatic qualitative changes occur in the system behavior is crucial in model validation or in discrimination between rival models. It is more convenient to look for qualitatively different behaviors rather than attempting to quantitatively fit a response.

Application of the methodology of numerical bifurcation theory to the study of model systems relies heavily on the observation of long-term recurrent dynamic behavior (steady states, sustained oscillations etc.) as opposed to the observation of transient responses. Long-term experiments, such as can be carried out in a chemostat or "flow reactor," are needed in order to test predictions obtained through this approach. Conversely, this approach will help design chemostat experiments at appropriate parameter values in order to probe the validity of a given model.

Another aspect of our approach is the attempt to combine many different dynamical features of an immune response in one single model, each feature corresponding to different areas of the model's parameter space. This might prove helpful in understanding when deviations in some parameter values are responsible for causing certain pathological situations, and thus, pinpoint the underlying mechanisms of deficiencies in the immune response, as well as methods for their modification—like the timely IL-2 injection we described in the numerical experiment above.

Besides these general considerations of modeling methodology, we believe that our simple model has raised two issues. One is the potential explanation of nonresponsiveness to a replicating antigen consistent with the steady states of branch 1 in Figure 6. The second, and more important issue, involves the separatrix between antigen explosion and containment that the IL-2 injection, described in Figure 3, allowed the system to cross. Since the presence of IL-2 is important for the response to any antigen occurring in the system at any given time, it is clear that the timing of a second infection, whether related or unrelated to the given antigen, may dramatically affect the outcome of the first response. The second response by generating additional IL-2 may move the system across a separatrix and lead to the containment of an antigen that might otherwise have escaped. This means that the coupling of immune responses through lymphokines, as well as issues of timing, may be crucial in determining the effectiveness of a specific response.

This study is only a simple, and we hope interesting, initial probe of both the particular model and the methodology. Several research possibilities present themselves. The first involves a more detailed parameter search for the current model—we only presented here what occurs in a small corner of the meaningful

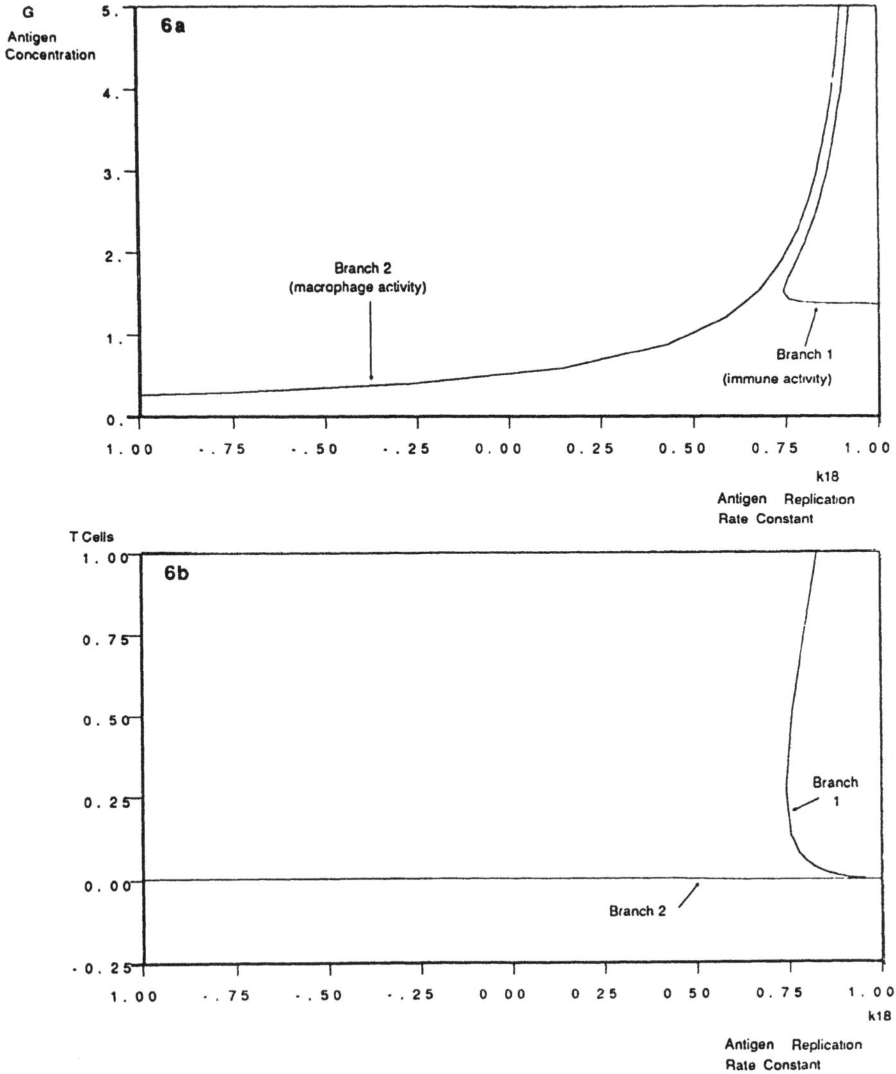

FIGURE 6 For small values of the antigen replication rate constant k_{18}, a new interesting steady state branch is observed (branch 2, Figure 6a). It is characterized by finite antigen concentration but almost no visible immune response (e.g., almost no T-cells, Figure 6b). This branch represents the containment of the antigen via stimulated macrophages as opposed to containment via an immune response. These diagrams have been obtained for $k_{19} = 0.5$.

parameter space. The model can be extended in an attempt to accurately study T cell activation, or to further incorporate B-cells and antibody production. The most challenging possibility, however, lies in the coupling of the responses of many different T cell subpopulations, each responding to a different epitope on one antigen or to totally different antigens. While initially pairs of different responses will be coupled, it is conceivable that large networks of such responses could also be coupled in the spirit of the work by Farmer, Packard and Perelson (1986) in an attempt towards more realistic simulations of the complex nature of the immune system. In that case, detailed searches of the (immense) parameter space are impossible, and efficient ways need to be devised for determining the statistics of the relation between the network configuration and the predicted response. Another challenging aspect of this research is the design of chemostat-type experiments to probe the validity of a model, improve it, estimate its parameters for different antigens, and examine in detail the physical mechanisms underlying its algebraic structure.

APPENDIX: NONDIMENSIONALIZATION OF THE EQUATIONS FOR MODEL 2a

Eqs. (7), (8a-9a), (10-11) in the text have been nondimensionalized by dividing each variable by a "typical" value. Using the subscript 0 to denote a typical value (e.g., $u_{1,0}$ is the typical value of u_1) and a caret to denote a nondimensional variable (e.g., $\hat{u}_1 \equiv u_1/u_{1,0}$), we find

$$\frac{d\hat{u}_1}{d\hat{t}} = \left(\frac{k_0 t_0 G_0 m_0}{u_{1,0}}\right)\hat{G}\hat{m} - (k_1 t_0)\hat{u}_1, \tag{A1}$$

$$\begin{aligned}\frac{d\hat{T}}{d\hat{t}} &= \left(\frac{k_8 t_0}{T_0}\right) - (k_9 t_0 G_0 u_{1,0})\hat{T}\hat{G}\hat{u}_1 - (k_{10} t_0)\hat{T} \\ &\quad + \frac{2(k_{11} t_0 u_{2,0}^2)\hat{u}_2^2\hat{T}}{1+(\tilde{k}_{11}u_{2,0}^2)\hat{u}_2^2} + \left(\frac{1}{\hat{G}+\hat{\epsilon}}\right)\frac{(\frac{k_{12}t_0 u_{2,0}}{G_0})\hat{u}_2\hat{T}^*}{1+(\tilde{k}_{12}u_{2,0})\hat{u}_2},\end{aligned} \tag{A2}$$

$$\begin{aligned}\frac{d\hat{T}^*}{d\hat{t}} &= (k_9 t_0 G_0 u_{1,0})\hat{T}\hat{G}\hat{u}_1 - (k_{10} t_0)\hat{T} \\ &\quad - \frac{(k_{11} t_0 u_{2,0}^2)\hat{u}_2^2\hat{T}}{1+(\tilde{k}_{11}u_{2,0}^2)\hat{u}_2^2} - \left(\frac{1}{\hat{G}+\hat{\epsilon}}\right)\frac{(\frac{k_{12}t_0 u_{2,0}}{G_0})\hat{u}_2\hat{T}^*}{1+(\tilde{k}_{12}u_{2,0})\hat{u}_2},\end{aligned} \tag{A3}$$

$$\frac{d\hat{u}_2}{d\hat{t}} = \left(\frac{k_{13} t_0 T_0}{u_{2,0}}\right) - (k_5 t_0)\hat{u}_2, \tag{A4}$$

$$\frac{d\hat{G}}{d\hat{t}} = \left(\frac{k_{19} t_0}{G_0}\right) + (k_{18} t_0)\hat{G}_0 - (k_6 t_0 m_0)\hat{m}\hat{G} - (k_7 t_0 T_0)\hat{T}^*\hat{G}_0, \tag{A5}$$

where $\hat{\epsilon} \equiv \epsilon/G_0$ and the terms in parentheses involving rate constants define the nondimensional rate constants. Thus, for example, $\hat{k}_0 \equiv k_0 t_0 G_0 m_0/u_{1,0}$. In the text when we deal with the rate constants for the nondimensional equations, we have for convenience dropped the caret.

The choice of accurate "typical" values is not crucial to the analysis. Bifurcation calculations focus on the systematic search of multidimensional parameter spaces. Our purpose in choosing typical parameter values is to select a base state around which to perform the search in parameter space. Here we have made the following choices: $u_{1,0} = u_{2,0} = 10^{-10}$ M, $T_0 = T_0^* = 2 \times 10^{-15}$ M, $m_0 = 2 \times 10^{-17}$ M, $G_0 = 2 \times 10^{-6}$ M and $t_0 = 50$ h. The concentration $u_{2,0}$ was chosen as the value required to saturate the IL-2 receptor (Smith, 1984a). Less is known about the interaction of IL-1 with T cells. Some IL-1 may be bound to the membrane of the antigen-presenting cell and not secreted. Other molecules produced by antigen-presenting cells, such as lactate (Dröge, this volume), may also be involved in stimulating the T cell and, here, are modeled as part of the IL-1 concentration. For simplicity of analysis, the typical concentration of IL-1 was chosen the same as the IL-2 concentration, 10^{-10}

M. *In vitro* T cell concentrations typically range between 10^5 and 10^6 cells/ml. Here we have chosen the higher concentration as our typical value so that the nondimensional variable would be of order one in an immune steady state. In a mouse, similar concentrations could be envisioned. The circulatory system has a volume of approximately one ml and the total number of T cells in the blood is of order 10^8. Estimating that at most one percent of the T cells respond to a cellular antigen, then we again arrive at 10^6 cells/ml as a good choice for the scaling value in an immune steady state. *In vitro* macrophages are added at concentrations of one per 10 or per 100 T cells. Our typical concentration corresponds to the lower value. The antigen concentration we have chosen is appropriate for a protein antigen that would be processed and presented by a macrophage. If the antigen was cellular and growing ($k_{18} > 0$), then a substantially lower value would be typical. However, when the cellular antigen is processed proteins and other molecules would be generated. Here we have used the corresponding molecular concentration as as our scaling value.

The rate constants used in our numerical studies were the nondimensional rate constants defined above. The antigen replication rate $\hat{k}_{18}$ and the antigen influx rate $\hat{k}_{19}$ were used as bifurcation parameters and varied within the range $0 < \hat{k}_{18} \leq 2$, $-1 \leq \hat{k}_{19} \leq 1$. The remaining nondimensional parameter values were fixed with values chosen according to the following considerations. Because of our choice of typical T cell and interleukin concentrations we expect that, if an immune steady state is reached, all the normalized concentrations will be of order one. Similarly, we expect that if our concentration scalings were chosen correctly, then the rates of change of the various nondimensional concentrations will also be of order one. To this end we have chosen all the nondimensional rate constants to be of order one with the following exceptions. For a mouse, the rate of T cell production from the bone marrow is approximately 10^6 cells h^{-1}. Assuming a circulatory system volume of 1 ml and that one out of every 5×10^3 T cells are antigen specific, we chose $\hat{k}_8$ equal to 0.01. In a chemostat experiment, this would correspond to an influx of 200 antigen-specific cells h^{-1} cm^{-3}. If the lifetime of a T cell in culture in the absence of antigen stimulation and IL-2 were of order one day, then the natural death rate $\hat{k}_{10} \approx 2$. When stimulated into proliferation by antigen and IL-2, T cells double their population approximately every 20 h (Cantrell and Smith, 1983). Thus, at saturating concentrations of IL-2, the k_{11} term should be 1/20 h. For both the k_{11} and k_{12} terms at saturation, the denominators should be greater than one. Since the IL-2 concentration has been scaled such that $\hat{u}_2 \leq 1$, both $\hat{\hat{k}}_{11}$ and $\hat{\hat{k}}_{12}$ have been chosen to be 100. This is a somewhat arbitrary choice, but is acceptable since it is the ratios $\hat{k}_{11}/\hat{\hat{k}}_{11}$ and $\hat{k}_{12}/\hat{\hat{k}}_{12}$ that are important at saturating values of IL-2. Using the values given above and the scaling time of 50 h results in $\hat{k}_{11} = 650$. To model the return of activated T cells to the unactivated state in the absence of antigen we have chosen $\hat{\epsilon} = 0.01$ so that $\hat{G}$ must be 1 % of its initial value before this effect comes in to play. Also, we have chosen $\hat{k}_{12} = 1$ so that this decay occurs at a rate that is small (1 %) compared to the rates of the other T cell growth and death processes. The parameter k_7 was chosen somewhat arbitrarily. To summarize, our simulations were done with the following parameter values: $\hat{m} = \hat{k}_0 = \hat{k}_1 = \hat{k}_{12} = \hat{k}_{13} = \hat{k}_5 = \hat{k}_6 = 1$, $\hat{k}_7 = 0.4$, $\hat{k}_8 = \hat{\epsilon} = 0.01$, $\hat{k}_9 = \hat{k}_{10} = 2$, $\hat{k}_{11} = 650$, and $\hat{\hat{k}}_{11} = \hat{\hat{k}}_{12} = 100$.

ACKNOWLEDGMENTS

We thank J. D. Murray for discussions that helped initiate this research while A.S.P. was visiting the Centre for Mathematical Biology, University of Oxford, England. This work was supported by the United States Department of Energy and SERC grant #GR/D/13573 to J. D. Murray.

REFERENCES

1. Bell, G. I. (1973), "Predator-Prey Equations Simulating an Immune Response," *Math. Biosci.* **16**, 291–314.
2. Bell, G. I. (1981), "Estimate of the Sticking Probability for Cells in Uniform Shear Flow with Adhesion Caused by Specific Bonds," *Cell Biophys.* **3**, 289-304.
3. Bell, G. I., A. S. Perelson, and G. H. Pimbley, Jr. (1978), *Theoretical Immunology* (New York: Marcel Dekker).
4. Bruni, C., G. Doria, G. Koch and R. Strom, eds. (1978), *Systems Theory in Immunology*, Lect. Notes in Biomath. (New York: Springer-Verlag), vol. 32.
5. Cantrell, D. A., and K. A. Smith (1983), "Transient Expression of Interleukin 2 Receptors: Consequences for T Cell Growth," *J. Exp. Med.* **158**, 1895–1911.
6. Cantrell, D. A., and K. A. Smith (1984), "The Interleukin-2 T-Cell System: a New Cell Growth Model," *Science* **224**, 1312–1316.
7. DeLisi, C. (1983), "Mathematical Modeling in Immunology," *Ann. Rev. Biophys. Bioeng.* **12**, 117–138.
8. Doedel, E. J. (1981), "AUTO: a Program for the Automatic Bifurcation Analysis of Autonomous Systems," *Cong. Num.* **30**, 265–284.
9. Farmer, J. D., N. H. Packard, and A. S. Perelson (1986), "The Immune System, Adaptation, and Machine Learning," *Physica* **22D**, 187–204.
10. Fujii, M., K. Sugamura, K. Sano, M. Nakai, K. Sugita and Y. Hinuma (1986), "High-Affinity Receptor-Mediated Internalization and Degradation of Interleukin 2 in Human T Cells," *J. Exp. Med.* **163**, 550–562.
11. Fujita, T., C. Takaoka, H. Matsui and T. Taniguichi (1983), "Structure of the Human Interleukin-2 Gene," *Proc. Natl. Acad. Sci. USA* **80**, 7437–7441.
12. Golubitsky, M., and D. Schaeffer (1985), *Singularities and Groups in Bifurcation Theory* (New York: Springer-Verlag).
13. Greene, W. C., and W. J. Leonard (1986), "The Human Interleukin-2 Receptor," *Ann. Rev. Immunol.* **4**, 69–95.
14. Guckenheimer, J., and P. Holmes (1983), *Nonlinear Oscillations, Dynamical Systems and Bifurcation of Vector Fields* (New York: Springer-Verlag).
15. Herzenberg, L. A., S. J. Black and L. A. Herzenberg (1980), "Regulatory Circuits and Antibody Responses," *Eur. J. Immunol.* **10**, 1-11.
16. Hoffmann, G. W. (1975), "A Theory of Regulation and Self-Nonself Discrimination in an Immune Network," *Eur. J. Immunol.* **5**, 638-647.
17. Hoffmann, G. W., and T. Hraba, eds. (1985), *Immunology and Epidemiology*, Lect. Notes in Biomath. (Springer-Verlag, New York), vol. 65.
18. Holbrook, N., K. A. Smith, A. J. Fornace, C. Comean, R. L. Wiskocil, and G. R. Crabtree (1984), "T-Cell Growth Factor: Complete Nucleotide Sequence and Organization of the Gene in Normal and Malignant Cells," *Proc. Natl. Acad. Sci. USA* **81**, 1634–1638.

19. Hooten, J. W. L., C. Gibbs and V. Paetkau (1985), "Interaction of Interleukin 2 with Cells: Quantitative Analysis of Effects," *J. Immunol.* **135**, 2464-2473.
20. Jenkins, M. K., D. M. Pardoll, J. Mizuguchi, T. M. Chused and R. H. Schwartz (1987), "Molecular Events in the Induction of a Nonresponsive State in Interleukin 2-Producing Helper T-Lymphocyte Clones," *Proc. Natl. Acad. Sci. USA* **84**, 5409–5413.
21. Kath, W. L., and J. D. Murray (1985), "Analysis of a Model Biochemical Switch," *SIAM J. Appl. Math.* **45**, 943–955.
22. Kubicek, M., and M. Marek (1983), *Computational Methods in Bifurcation Theory and Dissipative Structure* (New York: Springer-Verlag).
23. Marchuk, G. I. (1983), *Mathematical Models in Immunology* (New York: Optimization Software).
24. Meinhardt, H. (1982), *Models of Biological Pattern Formation* (New York: Academic Press).
25. Meuer, S. C., R. E. Hussey, D. A. Cantrell, J. C. Hodgdon, S. F. Schlossman, K. A. Smith and E. L. Reinherz (1984), "Triggering of the T3-Ti Antigen-Receptor Complex Results in Clonal T-Cell Proliferation through an Interleukin 2-Dependent Autocrine Pathway, *Proc. Natl. Acad. Sci. USA* **81**, 1509–1513.
26. Mizel, S. B. (1982), "Interleukin 1 and T Cell Activation," *Immunol. Rev.* **63**, 51–71.
27. Morgan, D. A., F. W. Ruscetti, and R. C. Gallo (1976), "Selective *in Vivo* Growth of T Lymphocytes from Normal Human Bone Marrow," *Science* **193**, 1007–1008.
28. Perelson, A. S., M. Mirmirani, and G. F. Oster (1976), "Optimal Strategies in Immunology. I. B-Cell Differentiation and Proliferation," *J. Math. Biol.* **3**, 325–367.
29. Perelson, A. S., M. Mirmirani, and G. F. Oster (1978), "Optimal Strategies in Immunology. II. B Memory Cell Production," *J. Math. Biol.* **5**, 213–256.
30. Perelson, A. S., and C. A. Macken (1984), "Kinetics of Cell-Mediated Cytotoxicity: Stochastic and Deterministic Multistage Models," *Math. Biosci.* **70**, 161–194.
31. Rabinowitz, P. H. (1977), *Applications of Bifurcation Theory* (New York: Academic Press).
32. Smith, K. A. (1980), "T-Cell Growth Factor," *Immunol. Rev.* **51**, 337–357.
33. Smith, K. A. (1984a), "Interleukin 2," *Ann. Rev. Immunol.* **2**, 319–333.
34. Smith, K. A. (1984b), "Lymphokine Regulation of T Cell and B Cell Function," *Fundamental Immunology*, Ed. W. E. Paul (New York: Raven Press), 559–576.
35. Smith, K. A. (1987), "The Two-Chain Structure of High-Affinity IL-2 Receptors," *Immunol. Today* **8**, 11–13.
36. Smith, K. A., and D. A. Cantrell (1985), "Interleukin 2 Regulates Its Own Receptors," *Proc. Natl. Acad. Sci. USA* **82**, 864–868.

37. Stern, J. B., and K. A. Smith (1986), "Interleukin-2 Induction of T-Cell G_1 Progression and c-*myb* Expression," *Science* **233**, 230–206.
38. Taniguichi, T., H. Matsui, T. Fugita, C. Takoka, N. Kashima, R. Yoshimoto, and J. Hamuro (1983), "Structure and Expression of a Cloned cDNA for Human Interleukin-2," *Nature* **302**, 305–310.
39. Teshigawara, K., H.-M. Wang, K. Kato, and K. A. Smith (1987), "Interleukin 2 High-Affinity Receptor Expression Requires Two Distinct Binding Proteins," *J. Exp. Med.* **165**, 223–238.

M. KAUFMAN
Faculté des Sciences, Université Libre de Bruxelles, Service de Chimie-Physique II, Campus Plaine C.P. 231, Bd. du Triomphe, B-1050 Brussel, Belgium

Role of Multistability in an Immune Response Model: A Combined Discrete and Continuous Approach

1. INTRODUCTION

In this article, we present an attempt at developing a theoretical framework for the study of complex regulatory systems, and at integrating the various processes known to regulate antibody production into a coherent ensemble that accounts for some main features of the humoral immune response.

A novel aspect of our approach is the combined use of two methods of description: a logical or discrete formalisation, and a more classical analysis in terms of differential equations.

The discrete approach typically is a qualitative treatment. It is characterized by the fact that the variables, which represent the level or concentration of each component, and the corresponding functions, which describe the rate of production of each component, take only a limited number of values 0, 1, 2, ...[1,2] This all-or-none character gives a somewhat caricatural image of the system, but greatly simplifies its description and analysis. This logical method has been essential to determine the minimum complexity required for a feasible model as it provides the main qualitative features to be expected for a given interaction diagram, without having to specify the details of the cellular interactions or to introduce a great number of parameters. The essence of the method consists in encoding into logical relations the main regulatory interactions characterizing the system, and deriving from these relations the stable regime states, if any, and the possible dynamic pathways. The simplicity of such an analysis allows one to extract the key elements responsible for a given behaviour and may even suggest minimum conditions that must be met for a better alternative.[3]

The continuous description leads to a more refined and quantitative analysis. It requires the specification of the form of the mathematical functions describing the cellular interactions, but permits the relaxation of the strong non-linearity inherent in the discrete description. Indeed, it has been shown that the qualitative picture obtained with the logical method remains valid in the corresponding continuous treatment for a wide range of nonlinearity. Moreover, the time evolution of the antigen concentration is more naturally included in the continuous description, thus allowing for a more precise study of the system response as a function of the amounts of antigen injected or of the way in which antigen is administered (single or repeated injections).[4]

Our purpose has been to investigate to what extent one may account for some important aspects of immune regulation by simple models. Starting with a small number of essential components (antigen, B, T helper, T suppressor cells of the relevant specificity, and the corresponding antibodies) and a few experimentally documented interactions between them, we have developed a class of models which predict a high multiplicity of regime states, both in the absence of antigen and for constant antigen levels. This high multistationarity determines the dynamic behaviour of the system upon antigenic stimulation. Antigen injections provoke switches among the regime states which might be related to the different modes of the immune response, and depend on how much and in what way antigen is administered, and on the previous antigenic history of the system.[3,4]

2. DESCRIPTION OF THE MODELS

The principal features of the regulatory schemes that we have analyzed are:

- the well-established negative feedback loops between T helper and suppressor lymphocytes;[5,6]
- the occurrence of T-T interactions both at the level of helpers and suppressors populations (for these interactions we have taken into account the experimental facts in favour of the idea that each of these compartments of T lymphocytes exerts a positive effect on its own development);[7,8] and
- the fact that immature B cells are highly susceptible to negative signalling.[9,10]

Our starting point is shown on Figure 1. A detailed description can be found in previous publications.[3,4)] The central part of this model is formed by a triple loop

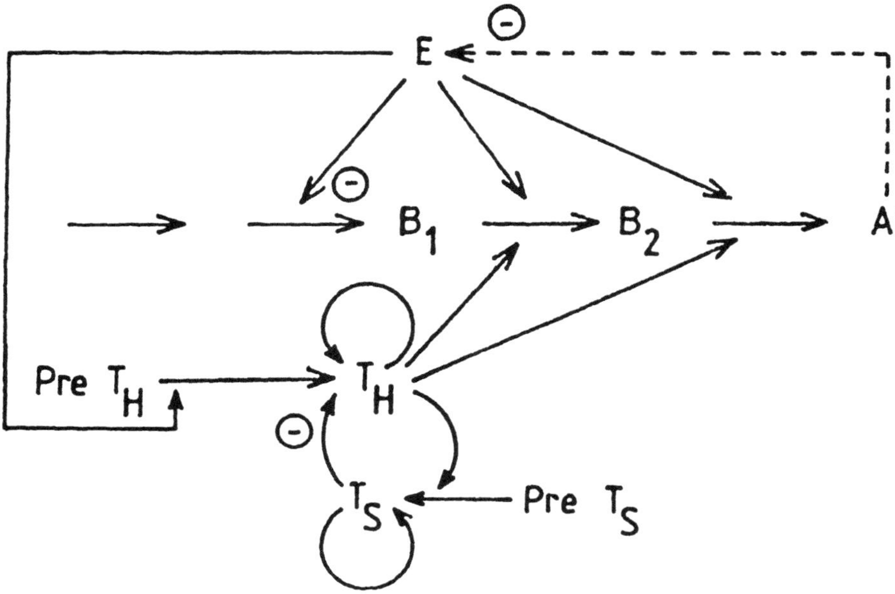

FIGURE 1 Schematic diagram of the interactions. E: antigenic determinant or epitope; Ab: antibody; B_1: virgin B cells; B_2: more mature B cells; T_H: T helper cells; T_S: T suppressor cells. Unless otherwise specified, the interactions are positive.

involving the regulatory T cells. The positive regulations at the level of helper and suppressor lymphocytes are *schematically* represented by autocatalytic loops. The complete model includes differentiation and maturation of the B lymphocytes into antibody secreting cells under the concerted action of antigen and T_H cells of the required specificity. The lymphocytes are assumed to come into close contact either through idiotype–anti-idiotype interactions[11] or through antigen bridges.[12] In particular, for the T-T interactions we have reasoned in terms of antigen-independent interactions. The stimulatory or suppressive character of the transmitted signals, however, are defined by the nature of the interacting partners. Suppression is considered to be mainly exerted on the T helper cells and also, we temporarily reason as if antigen is not involved in the development of the suppressor population. We will come back to this point in Sections 3.2 and 4.3.

Figure 2 shows an example of a somewhat more detailed version of the model. Its core comprises cross-interactions between idiotypic and anti-idiotypic related T_H and T_S cells.[11,13–15] The main point being that the cross-interactions between similar subsets of lymphocytes again form positive loops. Antigen, and possibly the B cells, are here explicitly involved in the development of both compartments of T lymphocytes.

These interaction diagrams show which components influence each other, but do no specify how these interactions are connected. In the absence of definite experimental data, several possibilities have been considered.[3] As will be discussed later, some choices appear to be crucial in order to observe an appropriate behaviour, others not.

We will focus here on the first and simpler network, and only mention some results that have been obtained with the more elaborate version of the model, a detailed analysis of which will be presented elsewhere.

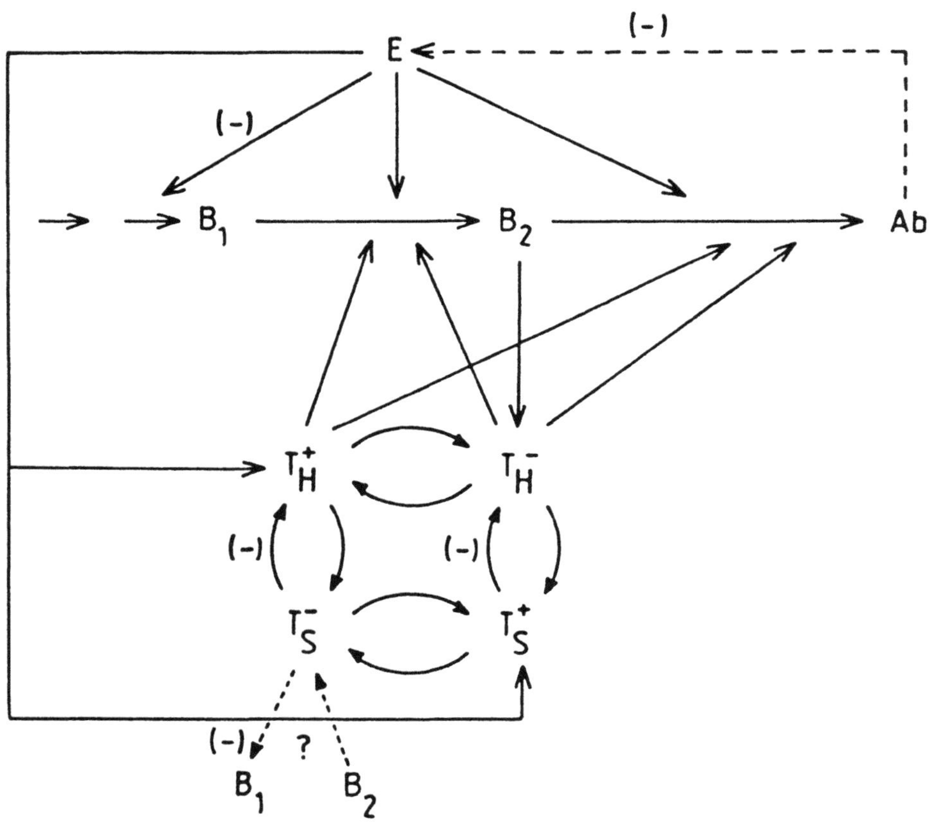

FIGURE 2 Extended interaction diagram involving pairs of mutually specific helper T cells and pairs of mutually specific suppressor T cells. T^+: idiotypic; T^-: autoanti-idiotypic. The meaning of the symbols is as in Figure 1. Unless otherwise specified, the interactions are positive.

3. DISCRETE ANALYSIS

3.1. NAIVE BOOLEAN ANALYSIS

In applying the discrete approach one may consider two levels of description. Let us first treat the system in a very naive way and reason as if the level of each component is either negligible or functional, and in the same way, that the functions describing their rate of production are either negligible or significant. Both variables and functions, thus, take the values 0 or 1. Similarly, there is no interaction or strong interaction between the components. Starting from the ensemble of control loops that are assumed to play a role in the process to be analyzed, complemented with a choice of connections between them, one derives logical equations which relate, at any time, the logical values of the functions to the logical values of the variables.[1] Considering the triple loop involving the T_H and T_S cells in Figure 1, the logical equations may be written as :

$$H = e \cdot \overline{s} + h \tag{1}$$

or

$$H = (e + h) \cdot \overline{s} \tag{1'}$$

and

$$S = h + s \tag{2}$$

where h and s denote the levels of T_H and T_S cells respectively, H and S their rate of production, and e is an input variable describing the antigen concentration. Precursor cells of each required type are assumed to be constantly present, their level, thus, being equal to one. The symbols $(\cdot)$ and $(+)$ correspond to the logical connectives AND and inclusive OR, respectively. The overbar denotes the NOT operation.

Eq. (1) means that precursor T_H cells are activated when antigen is present *and* the level of activated T_S cells is negligible, *or* when activated T_H cells are already present. In the alternative Eq. (1′) inhibition by the T_S cells is considered for both processes. Eq. (2) means that the development of T_S cells from a precursor pool requires either the presence of T_H cells *or* of already activated T_S cells. From these equations, one derives the state table, Table 1, giving the stable and transient states. The stable regime states are those where the value of each function agrees with the corresponding variable: there is no order to change.

As shown in Table 1, in the absence of antigen ($e = 0$) the first set of equations leads to three stable regime states, whereas for the second set of equations, there is no stable state with helper cells present in significant amounts. At this point, one would be tempted to reject the second choice of logical connections since it cannot account for a positive memory of previous antigenic challenges. In the presence of antigen ($e = 1$), the trivial state (00) is no longer a stable state.

TABLE 1 State Tables Corresponding to Eqs. (1)–(2) and (1′)–(2) in the Absence ($e = 0$) or Presence ($e = 1$) of Antigen[1]

$e = 0$		$e = 1$		$e = 0$		$e = 1$	
h s	H S	h s	H S	h s	H S	h s	H S
(0 0)	0 0	$\bar{0}$ 0	1 0	(0 0)	0 0	$\bar{0}$ 0	1 0
(0 1)	0 1	(0 1)	0 1	(0 1)	0 1	(0 1)	0 1
(1 1)	1 1	(1 1)	1 1	$\bar{1}$ 1	0 1	$\bar{1}$ 1	0 1
1 $\bar{0}$	1 1	1 $\bar{0}$	1 1	1 $\bar{0}$	1 1	1 $\bar{0}$	1 1

[1] Here the dashes indicate which state variables have an order to change. The stable states are circled.

The logical analysis of the complete network shown in Figure 1 with Eqs. (1)–(2) for the development of the T_H and T_S populations, has been presented in detail in a previous publication.[3] The main conclusions are that in the absence of antigen, the system may persist depending on its past history, in any one of three stable regime states: a "virgin" state in which only virgin B_1 cells are present, a "memory" state in which B_1 cells and both T_H and T_S cells are present and a "suppressed" state where besides B_1 cells, T_S cells are also present. For a constant non-zero antigen level, one observes a paralyzed and a suppressed state with negligible levels of each type of B cells. Besides the regime states, one may also deduce the possible temporal sequences of states upon antigen injection. A detailed study of these pathways predicts a reasonable dynamics for primary and secondary responses upon single antigen injections and the establishment of paralysis for persisting antigen. The suppressed state, however, can never be reached starting from the virgin state and the establishment of suppression is beyond the scope of this simple boolean analysis.

3.2 GENERALIZED DISCRETE ANALYSIS

The logical analysis which has briefly been described above gives a very crude idea of the behaviour of the system, in the presence and absence of antigen. We have extended our logical treatment so as to include two important aspects that were disregarded until now, namely:

- the strength of the various interactions, and

- the fact that when a component exerts a control on the development of more than one element, its significant level is not necessarily the same for the different interactions.

This leads to a major generalization of the discrete approach through the introduction of "logical parameters" and, if necessary, of a multileveled logic. A detailed presentation of this generalized method which has been developed by E. Snoussi can be found elsewhere.[16] Here, let us consider again the triple loop on Figure 1, taking into account that the functional thresholds for the T_S–T_H and T_S–T_S interaction are not the same, and similarly for the T_H–T_S and T_H–T_H interactions. Several situations are now possible depending on which threshold is taken to be the lowest for each component. Given a choice of the relative position of the thresholds, one associates to the real variables X_1 and X_2 describing the concentration of T_H and T_S cells, the discrete variables q_1, q_2 defined by:

$$q_1 = d_1(X_1), \quad q_2 = d_2(X_2)$$

where the subscripts 1 and 2 for the discretisation operator d indicate that the discretisation scales of X_1 and X_2 are based on two different sets of thresholds. q_1, q_2 thus take the value 0, 1 or 2 depending on the relative location of the concentrations X_1 and X_2 with respect to their thresholds values. It has been shown[16] that the dynamic behaviour of the system as a function of q_1, q_2 is given by the discretised value of the pseudo-boolean expression which may be deduced, for each component, directly from the interation diagram. Thus, the two alternative sets of equations become, for example:

TABLE 2 State Table Corresponding to Eq. (3)

q_1	q_2	f_1	f_2
0	0	K_1	0
0	1	K_1	K_3
0	2	0	K_3
1	0	K_1	K_2
1	1	K_1	K_6
1	2	0	K_6
2	0	K_5	K_2
2	1	K_5	K_6
2	2	K_4	K_6

TABLE 3 State Table Corresponding to Eq. (4)

q_1	q_2	f_1	f_2
0	0	K_1	0
0	1	K_1	K_3
0	2	0	K_3
1	0	K_5	0
1	1	K_5	K_3
1	2	0	K_3
2	0	K_5	K_2
2	1	K_5	K_6
2	2	0	K_6

$$f_1(q_1, q_2) = d_1\left(a_1 \cdot \overline{q}_2{}^{(2)} + a_4 \cdot q_1{}^{(2)}\right) \tag{3.1}$$

$$f_2(q_1, q_2) = d_2\left(a_2 \cdot q_1{}^{(1)} + a_3 . q_2{}^{(1)}\right) \tag{3.2}$$

and

$$f_1(q_1, q_2) = d_1\left(a_1 \cdot \overline{q}_2{}^{(2)} + a_4 \cdot q_1{}^{(1)} \cdot \overline{q}_2{}^{(2)}\right) \tag{4.1}$$

$$f_2(q_1, q_2) = d_2\left(a_2 \cdot q_1{}^{(2)} + a_3 \cdot q_2{}^{(1)}\right) \tag{4.2}$$

where the parameters a_1 are real numbers describing the strength of each interaction and the plus sign refers to the usual arithmetic sum. (α) denotes the order of the threshold which is involved in a given interaction, with the notation:

$$q_i{}^{(\alpha)} \in \{0, 1\}, \quad q_i{}^{(\alpha)} = 0 \iff q_i < \alpha \quad \text{and} \quad \overline{q}_i{}^{(\alpha)} = 1 - q_i{}^{(\alpha)}$$

The logical parameters are defined by:

$$K_1 \equiv d_1(a_1), \quad K_2 \equiv d_2(a_2), \quad K_3 \equiv d_2(a_3), \quad K_4 \equiv d_1(a_4),$$
$$K_5 \equiv d_1(a_1 + a_4), \quad K_6 \equiv d_2(a_2 + a_3)$$

and the antigen concentration is now included in parameters K_1 and K_5. Even if a component is significantly present, its contribution to the development of a given population will depend on the relative strength of interaction. As shown in the corresponding state Tables 2 and 3, the stable and transient states this time are determined by the discrete values of the logical parameters K.

Figure 3 illustrates the behaviour of Eq. (3) for different antigen levels, upon introducing a choice of logical parameter values in the corresponding state table. In the absence of antigen ($K_1 = 0$), the conclusions remain the same as before. There are three stable regime states: a virgin state (00), a suppressed state (02) and a memory state (22). When antigen is present the virgin state is no longer a stable state. In this description however, for low antigen levels ($K_1 = 1$), only the suppressed state can be reached from the zero state as indicated by the arrows on Figure 3(A2). Whereas for higher antigen concentrations ($K_1 = 2$), the memory state also becomes accessible (Figure 3(A3)) and which state will be arrived at depends on the relative values of the establishment time delays of both variables.

In the case of Eq. (4), a main difference with respect to the simple boolean analysis already appears in the absence of antigen: depending on the choices of threshold and parameter values, the system now exhibits, besides the virgin (00) and suppressed (01) states, a third stable state (21) corresponding to a high level of helpers cells and a medium level of suppressors cells (Figure 4(A1)),and this alternative should thus not be excluded. In addition, as shown for a different set of the logical parameters on Figure 4(A2), this memory state may correspond to a *cycle* and be responsible for oscillations at the level of the T cells.

It can be shown that the logical parameters which appear in this discrete description are linked to the kinetic constants of homologous continuous systems in which the regulatory interactions are described by combinations of sums and products of Hill functions:

$$F_j{}^+(X_1/\theta_{ij};n) = \frac{X_i{}^n}{\theta_{ij}^n + X_i{}^n} \qquad \text{for activation} \tag{5.1}$$

and

$$F^- = 1 - F^+ \qquad \text{for inhibition} \tag{5.2}$$

Where n is the Hill number and θ_{ij} a threshold parameter for the regulation of the j-th variable by component i. In the case of Eqs. (3) and (4), the corresponding continuous equations are given respectively by:

$$\frac{dX_1}{dt} = k_1 \cdot F_1{}^-\left(X_2/\theta_{21};n_1\right) + k_4 \cdot F_1{}^+\left(X_1/\theta_{11};n_4\right) - d_1X_1 \tag{6.1}$$

$$\frac{dX_2}{dt} = k_2F_2{}^+\left(X_1/\theta_{12};n_2\right) + k_3 \cdot F_2{}^+\left(X_2/\theta_{22};n_3\right) - d_2X_2 \tag{6.2}$$

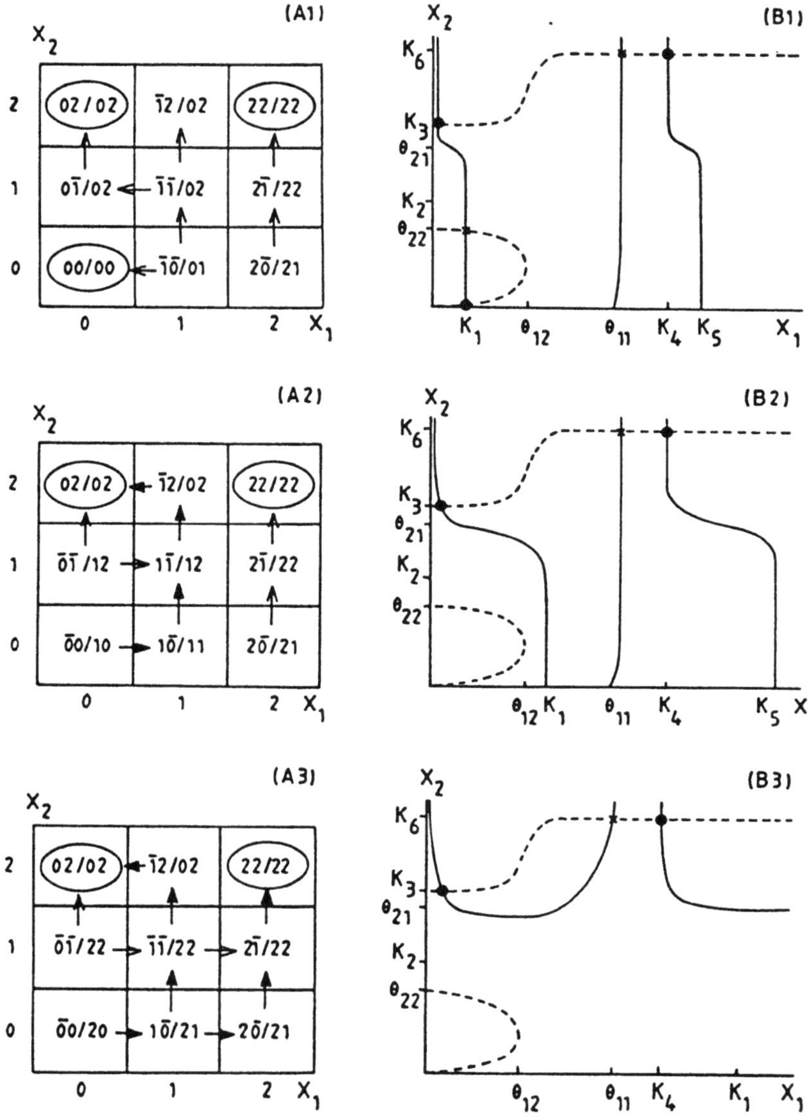

FIGURE 3 A. State tables corresponding to Eq. (3) for the logical parameter values $K_2 = 1$, $K_3 = K_4 = K_5 = K_6 = 2$, and different antigen levels. (A1): $K_1 = 0$; (A2): $K_1 = 1$ and (A3): $K_1 = 2$. The stable states are circled. The dashes indicate which variables have an "order" to change. The arrows give the possible temporal sequences of states. B. Schematic representation in the phase plane (X_1, X_2) of the isoclines corresponding to the continuous equations (6.1) and (6.2), for increasing antigen levels. The values of the kinetic parameters $K = k/d$ are comparable to the logical parameter values. (B1): $K_1 = 0$; (B2): $\theta_{12} < K_1 < \theta_{11}$; (B3): $K_1 > \theta_{11}$.

with $\theta_{11} > \theta_{12}$, $\theta_{22} < \theta_{21}$; and:

$$\frac{dX_1}{dt} = \{k_1 + k_4 F_1{}^+(X_1/\theta_{11}; n_4)\} \cdot F_1{}^-(X_2/\theta_{21}; n_1) - d_1 X_1 \tag{6.3}$$

$$\frac{dX_2}{dt} = k_2 F_2{}^+(X_1/\theta_{12}; n_2) + k_3 F_2{}^+(X_2/\theta_{22}; n_3) - d_2 X_2 \tag{6.4}$$

with $\theta_{11} < \theta_{12}$, $\theta_{22} < \theta_{21}$.

The antigen concentration $[E]$ is included in k_1:

$$k_1 = k'_1 \cdot \frac{[E]^n}{\theta_{E1}^n + [E]^n}$$

It has been mathematically proven that the discrete stable states are approximations of stable nodes in the homologous continuous system, whereas a cycle indicates the existence of a focus.[16] This correspondence which is strictly valid for infinite Hill numbers, however, remains applicable for a wide range of values of n. To illustrate the correspondence between discrete and continuous approaches, the isoclines of the homologous continuous equations are presented schematically, for each situation, on Figures 3(B) and 4(B). The continuous description gives, in addition, the location of the unstable steady states.

The same approach has been applied to the core of the more extended model of Figure 2. One finds easily conditions for a multiplicity of stable states similar to the ones of the simpler scheme. However, related to the mediation of antigen and B cells in the development of the suppressor population, a new possibility appears here: when the antigen concentration is further increased, the immune state with high level of helper cells disappears and a new stable state with high suppressors and low helpers level is established. Therefore, at high antigen concentration, the system may again be suppressed and this could play a role for low responsiveness at high antigen doses.

The discrete formalism, thus, provides detailed information on the multiplicity of stable states that can be obtained for a given interaction diagram. It gives a good idea of the relative parameter ranges that have to be considered to observe appropriate stable states and a suitable dynamics upon antigen administration. But the antigen is treated in a primitive way: for the sake of simplicity antigen is considered as an external variable in spite of the fact that other components retroact on its concentration.

4. CONTINOUS ANALYSIS

4.1. KINETIC EQUATIONS

The continuous approach naturally incorporates an additional equation which describes in a self-consistent way the disappearance of antigen by complex formation with antibody and by slower spontaneous decay. This leads to a more precise insight in the predictions of the models. The following set of differential equations have been used to describe model 1[4]:

$$\frac{dX_1}{dt} = k_1 F_1{}^-(X_6) - k_2 X_1 X_3 F_2{}^+(X_6) - d_1 X_1 \tag{7.1}$$

$$\frac{dX_2}{dt} = k_2 X_1 X_3 F_2{}^+(X_6) + m_2 X_2 F^+(X_6) X_3 - k_5 X_2 F^+(X_6) X_3 - d_2 X_2 \tag{7.2}$$

$$\frac{dX_3}{dt} = k_3 F_3{}^+(X_6) F_3{}^-(X_4) + m_3 F_3{}^+(X_3) + m_3' F_3{}^+(X_6) X_3 - d_3' X_3 X_4 - d_3 X_3 \tag{7.3}$$

$$\frac{dX_4}{dt} = k_4 X_3 + m_4 F_4{}^+(X_4) + k_4' F_4{}^+(X_3) X_4 - d_4 X_4 \tag{7.4}$$

$$\frac{dX_5}{dt} = k_5 X_2 F^+(X_6) X_3 - p k_6 X_5{}^p X_6{}^q - d_5 X_5 \tag{7.5}$$

$$\frac{dX_6}{dt} = -q k_6 X_5{}^p X_6{}^q - d_6 X_6 \tag{7.6}$$

where the X_i's, $i = 1, 6$, are respectively the concentrations of the components B_1, B_2, T_H, T_S, Ab, and E, and the $F_i(X_j)$'s are Hill functions defined by Eqs. (5.1) and (2). Antigen bridging together with stimulation is represented by the saturation function $F^+(X_6)$. The existence of a constant level of precursor cells of each required type is included in the rate constants. p and q are the stochiometric coefficients of the antigen-antibody reaction. A term of spontaneous decay, $-d_i X_i$, is considered for each cellular or molecular species.

Compared to the discrete treatment (see also [4]) the nonlinearity is importantly reduced by considering a low degree of cooperativity (i.e., values of the Hill numbers ranging between 2 and 5) and also by retaining sigmoids only for a small number of interactions. Moreover, in order to fit more closely the experimental data some extra terms are inserted into the continuous equations, such as the loss of virgin B_1 cells by differentiation into more mature B_2 cells and the explicit proliferation and maturation of the B_2 cells; an antigen-dependent proliferation of the T_H cells and T_H dependent proliferation of the T_S cells; and possibly, a killing effect of the T helper by the T suppressor cells. It turns out that these additional features merely act as modulators and do not introduce qualitative changes.

4.2. STEADY STATE ANALYSIS

A detailed steady state analysis is presented in [4]. The agreement with the predictions of the discrete analysis is here illustrated on Figure 5 which shows the isoclines corresponding to Eqs. (7.3) and (7.4) for various antigen concentrations.

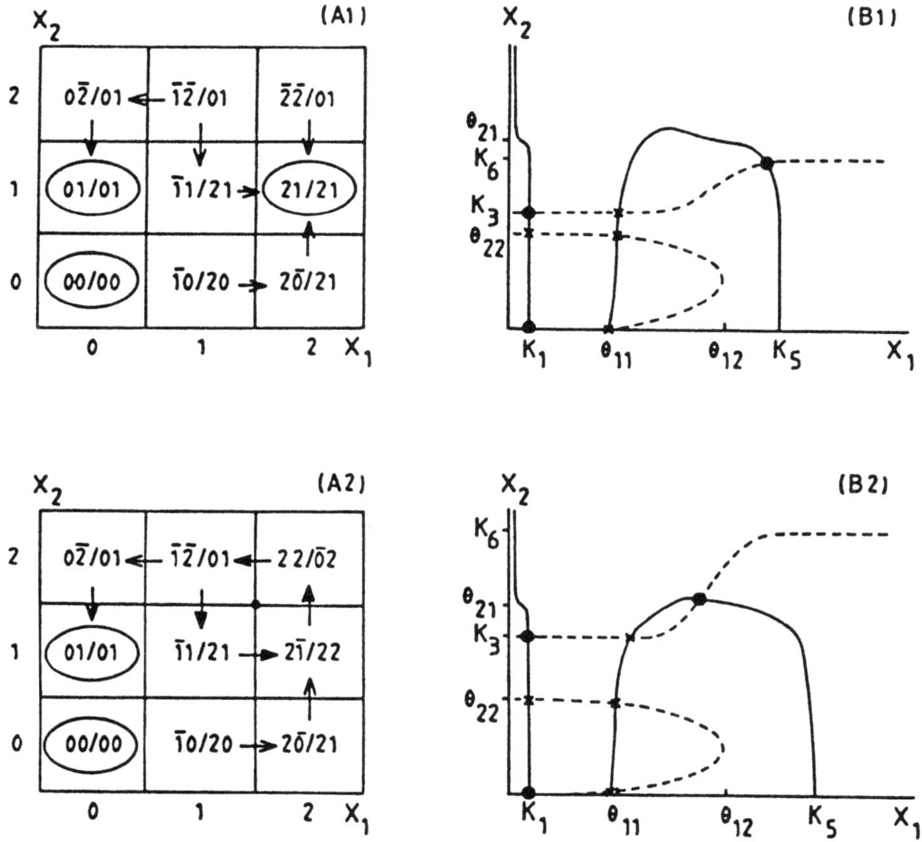

FIGURE 4 A. State tables corresponding to Eq. (4) in the absence of antigen. (A1): $K_1 = 0$, $K_2 = K_3 = K_6 = 1$, $K_5 = 2$; (A2): $K_1 = 0$, $K_2 = K_3 = 1$, $K_5 = K_6 = 2$. The stable states are circled. The dashes indicate which variables have an "order" to change. The arrows give the possible temporal sequences of states.
B. Schematic representation in the phase plane (X_1, X_2) of the isoclines corresponding to the continuous equations (6.3) and (6.4). The values of the kinetic parameters $K = k/d$ are comparable to the logical parameter values. (B1): $K_6 < \theta_{21}$; (B2): $K_6 > \theta_{21}$.

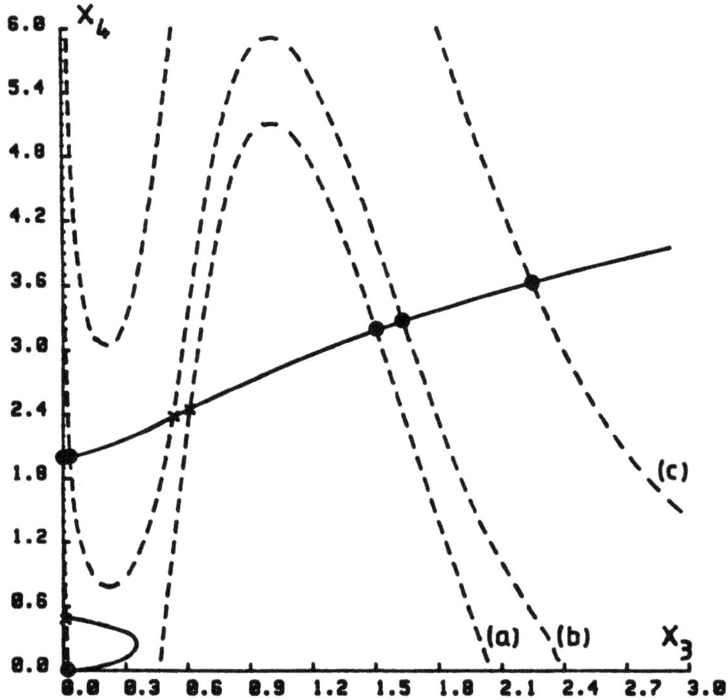

FIGURE 5 Isoclines in the phase plane (X_3, X_4). The parameter values are those listed in Section 4.3 except for $k_4 = k_4{}' = .05$, $m_3{}' = .1$ and $d_3{}' = .01$. Broken lines correspond to $\dot{X}_3 = 0$, solid lines to $\dot{X}_4 = 0$. Curves (a), (b) and (c) correspond respectively to the antigen levels $X_6 = .01,.03$ and $.09$. The dots and crosses denote respectively the stable and unstable steady states. For increasing antigen concentration, the virgin and suppressed states disappear successively.

Figures 6 and 7 give the trajectories around the "suppressed" and "memory" states for two antigen levels. These figures should be compared with Figures (3(A2)) and (3(A3)). One observes that the size of the domains of attraction of the stable steady states strongly depends on the antigen level.

4.3. DYNAMIC BEHAVIOUR

In the following the role of the steady states for the response of the system as a function of the doses of antigen that are injected or of the way in which antigen is administered will be illustrated by considering the response of the system as described by Eq. (7) for two typical situations: after pretreatment with different antigen doses and upon single antigen injections of different sizes. The dynamic

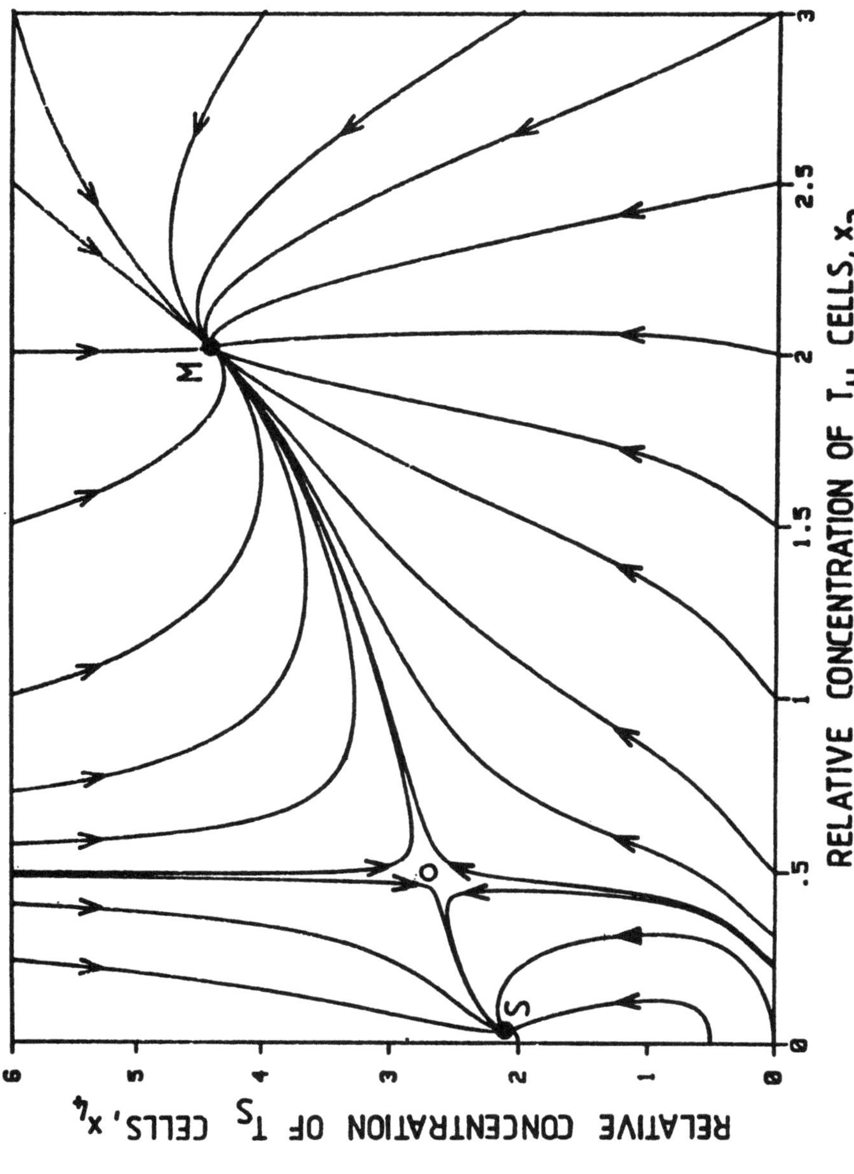

FIGURE 6 Phase plane diagram in the presence of a constant level of antigen $X_6 = .03$. The parameter values are listed in Section 4.3. Only the suppressed state (S) can be reached from the trivial state.

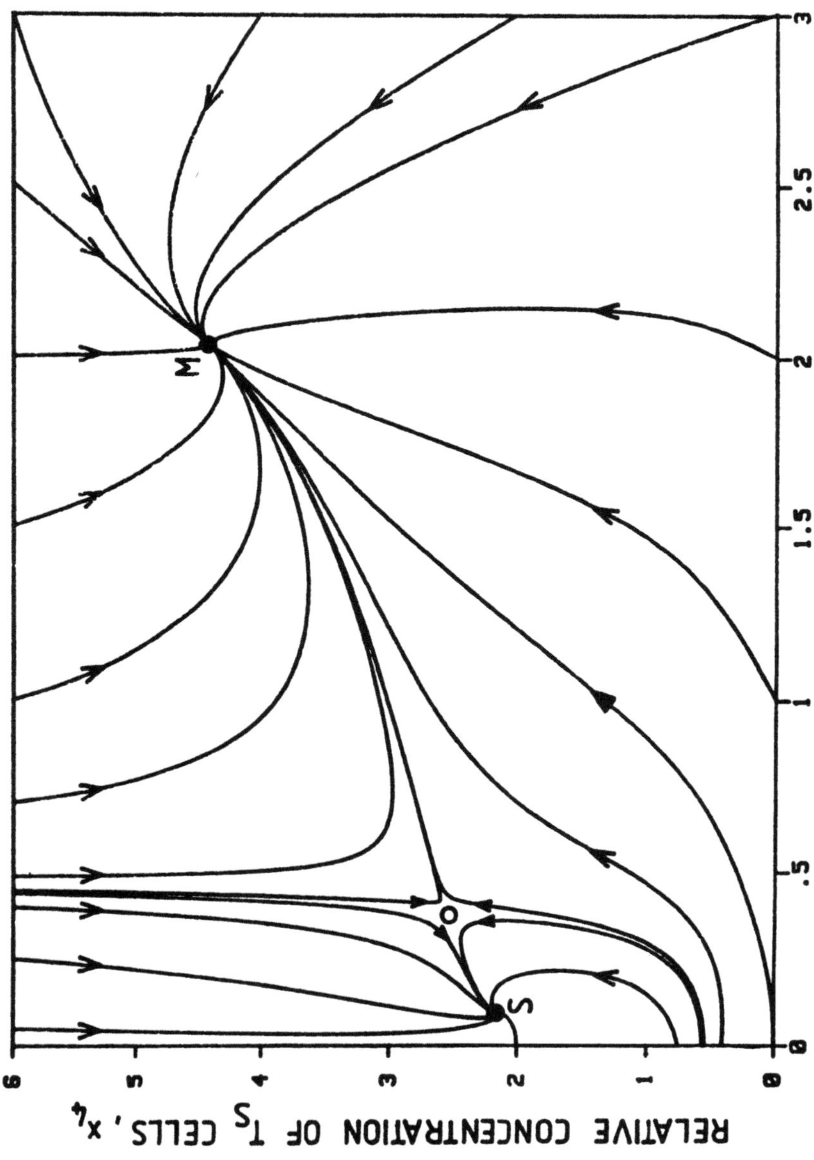

FIGURE 7 Phase plane diagram in the presence of a constant level of antigen $X_6 = .05$. The parameter values are listed in Section 4.3. Only the memory state (M) can be reached from the trivial state.

behaviour is obtained by numerical integration of the equations with the following parameter values:

$$k_1 = .6 \quad k_2 = .1 \quad k_3 = k_5 = m_2 = .4 \quad k_4 = k_6 = .2 \quad m_3 = m_4 = .5$$
$$d_1 = d_2 = .1 \quad d_3 = d_4 = d_5 = .2 \quad d_6 = .05 \quad k_4' = m_3' = d_3' = 0$$
$$\theta_{44}/\theta_{43} = \theta_{33}/\theta_{34} = 1 \quad \theta_{61}/\theta_{62} = \theta_{63}/\theta_{62} = \theta/\theta_{62} = .1$$

$n = 2$ for all the sigmoid functions. X_6, X_3 and X_4 are reduced to their threshold value θ_{62}, θ_{33} and θ_{44} respectively.

The results which are presented below are qualitative results since the parameter values were not adjusted in order to take quantitative data into account.

DYNAMIC BEHAVIOUR AFTER PRETREATMENT. Pretreatment here consists of maintaining a given antigen level in the system during a time interval of 50 days. This can equally be simulated by repeated antigen injections at sufficiently short time intervals, or simply by maintaining a constant antigen concentration in the system.The initial state is the virgin state. The response to an antigen challenge which normally evokes a response in a non-pretreated system is then recorded. The behaviour after pretreatment is summarized on Figure 8 showing the size of the response after a normal antigen dose. The curve which is obtained is in very good qualitative agreement with experimental data as reported, for instance, in [17].

Depending on the doses that are used for pretreatment one distinguishes four regions.[4]

- A "*virgin*" region, I, where pretreatment leaves the system almost unaffected: it remains in the vicinity of the virgin state and the response to a next antigenic challenge is of the primary type (Figure 9, curve (a)).
- A "*low-dose-paralysis*" region, II, where pretreatment drives the system into the suppressed state (see Figure 6). A subsequent normal stimulation results in low amounts of antibody (Figure 9, curve (b)).
- An "*immunity*" region, III, where pretreatment induces the memory state (see Figure 7) without producing antibodies. The next antigenic challenge provokes a secondary type of response (Figure 9, curve (c)).
- A "*high-dose-paralysis*" region, IV, where pretreatment now leads the system towards a paralyzed state characterized by high levels of T_H and T_S cells, but strong depletion of the B cells. A next antigen injection fails to evoke a response. This paralysis is gradually lost when antigen disappears by spontaneous decay.

High and low dose paralysis are, thus, associated in our model with different steady state branches which implies that the underlying mechanisms are different. Low dose paralysis is related to an inefficient activation of the B cells: the suppressor T cells inhibit the development of the T helper population which is required at several stages of differentiation and maturation of the antibody producing cells.

High dose paralysis mainly results from clonal depletion of the B_1 cells in the continuous presence of antigen. The two regions of paralysis are separated by an immunity zone where evolution towards the "T memory" state prevails.

DYNAMIC BEHAVIOUR UPON ISOLATED ANTIGENIC STIMULATIONS. Figure 10 summarizes schematically the situation after single injections of antigen of different sizes. With respect to the curve obtained after pretreatment, one observes here a shift towards higher antigen concentrations.

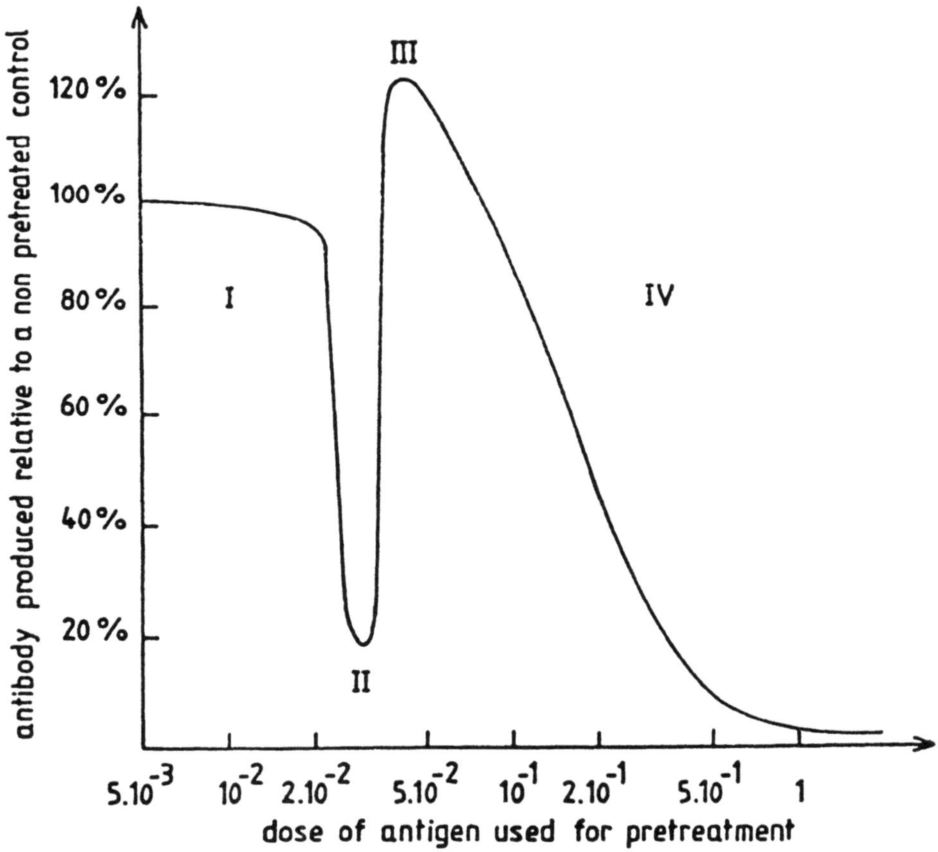

FIGURE 8 The effect of pretreatment with different antigen doses is reflected in the size of the secondary response relative to a non-pretreated system.

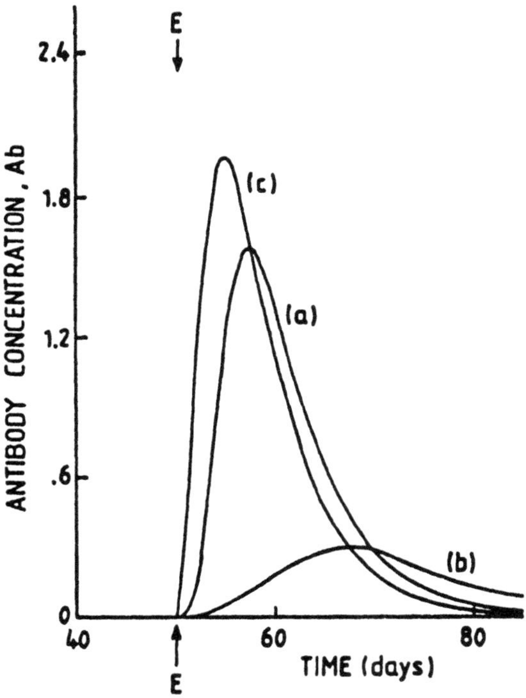

FIGURE 9 Antibody response after pretreatment.

Again there are different regions:

1. The system is not sensible to a very low dose (region I) and a second stronger injection will provoke a primary response.
2. An antigen dose corresponding to region II brings the system, without antibody production, into the suppressed state from where it responds poorly to a second optimal dose. This low responsiveness may be reversed by adequate subsequent antigen challenges which lead the system to a fully responsive state.
3. A somewhat higher antigen dose (region III) induces the phenomenon of priming without antibody production: without a good primary response a high level of both T_H and T_S cells is built up, and the system responds more rapidly and stronger to a next antigen injection than starting from the virgin state. If one includes the production of B memory cells in our description, no B memory cells are produced during priming.
4. An antigen injection corresponding to region IV leads to a significant primary response during which B memory cells are produced, if included in the model. A next stimulation is followed by a secondary response.

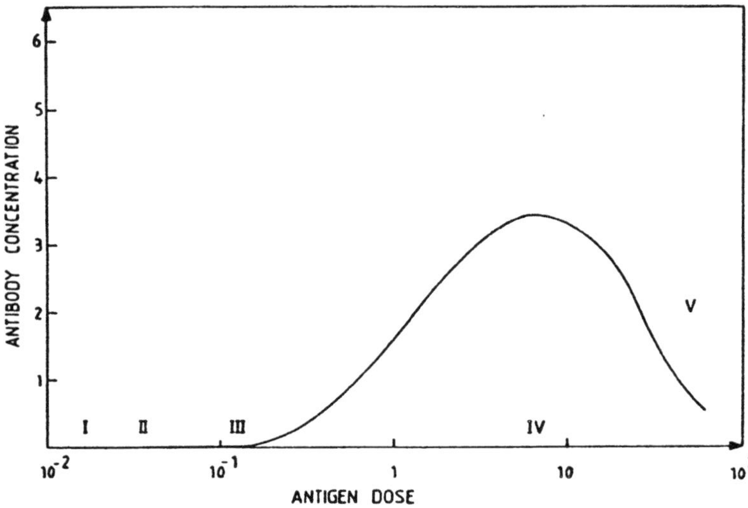

FIGURE 10 Antibody response obtained upon a single injection of antigen as a function of the antigen dose.

5. When the antigen dose is further increased (region V) the amplitude of the antibody response declines and becomes negligible. Here again, rapid but temporary depletion of the B cells is observed. For the simple interaction scheme that we have analyzed, both T suppressor and T helper cells are activated in this region. However, in the case of the extended model of Figure 2, which includes the stimulation of idiotypic T suppressor cells by antigen, the existence of a suppressed state at high antigen concentrations might also play a role in suppressing the response for increasing antigen doses.

5. CONCLUSIONS

A formal analysis of the regulation of the humoral immune response has been developed which is based on a small number of well-established facts about lymphocyte interactions and some reasonable assumptions.

The discrete approach proved to be extremely useful in developing and analyzing the models, by itself and in conjunction with a classical continuous analysis. The generalisation of the logical method improved strongly the predictability of the continuous system from its logical caricature. A similar type of approach combining boolean and continuous methods has recently been developed by Eisenfeld et al.[18,19]

The core of the models that emerge from our study is formed by interactions between T helper and T suppressor cells. The assumptions mainly consist in the connections that are proposed for the various interactions and the role ascribed to antigen at several levels of the network. In particular, the T_H–T_H and T_S–T_S interactions are considered to be antigen-independent, and this choice is essential in order to account for three biologically relevant regime states in the absence of antigen. If, on the other hand, the stimulation of the suppressor T cells by the helper T cells is assumed to be antigen-dependent, an additional regime state appears corresponding to a significant level of T_H, but low level of T_S cells. It appears clearly from the discrete analysis, and this as well for the interaction schemes that are presented here as for other variants that have been studied, that the presence of positive loops is fundamental to observe a multiplicity of stable states and that this multiplicity may increase with the number of positive loops that are involved. The important point is the positive character of the loops and not so much the detail of their structure: they may include several steps, with possibly an even number of negative interactions. The negative loops, on the other hand, are responsible for the fact that the steady states may correspond to a cycle or focus, which implies the possibility of damped and even sustained oscillations at the level of the T cells, and presently we study the conditions in which these T cells oscillations may induce oscillatory antibody production.

Depending on the range of antigen doses or on the way in which antigen is injected, the system that we have presented stabilizes in different regime states. This, in turn, determines the dynamic behaviour upon a next antigenic stimulation. Some of the transitions that are observed are irreversible: once the system has encountered antigen, it never goes back to the virgin state; others are temporary, such as the induction of the suppressed state.

The permanent T memory state which is created after a sufficiently strong challenge with antigen, is responsible for rapid secondary responses. The differences in magnitude between primary and secondary responses, however, do not compare with those that can be observed experimentally. Even if this point can be improved by including in the model the fact that B memory cells are produced after a successful immunization, this aspect still needs further investigation.

Our description certainly is over-simplified, but might help to clarify some aspects of the interplay between antigenic and idiotypic regulation.

ACKNOWLEDGMENTS

This work has been performed in close collaboration with J. Urbain and R. Thomas at the Department of Molecular Biology of the Université Libre de Bruxelles. I thank E. Snoussi for fruitful discussions. Financial support has been received from the "Instituts Internationaux de Physique et de Chimie (Fondés par E. Solvay)" and from the Commission of the European Communities (Actions de Stimulation n° STI-074-J-CD).

REFERENCES

1. Thomas, R. (1979), ed., *Kinetic Logic A Boolean Approach to the Analysis of Complex Regulatory Systems* (Berlin-New York: Springer Verlag), 29.
2. Thomas, R. (1984), "Logical Description, Analysis and Synthesis of Biological and Other Networks Comprising Feedback Loops," *Adv. Chem. Phys.* **55**, 247.
3. Kaufman, M., J. Urbain, and R. Thomas (1985), "Towards a Logical Analysis of the Immune Response," *J. Theor. Biol.* **114**, 527.
4. Kaufman, M., and R. Thomas (1987), "Model Analysis of the Bases of Multistationarity in the Humoral Immune Response," *J. Theor. Biol.*, in press.
5. Eardley, D.D. (1980), "Feedback Suppression: An Immunoregulatory Circuit," *Fed. Proc.* **39**, 3114.
6. Green, D.R., P. M. Flood, and R. K. Gershon (1983), "Immunoregulatory T-Cell Pathways," *Ann. Rev. Immunol.* **1**, 439.
7. Tada, T., and K. Okumura (1980), "The Role of Antigen-Specific T-Cell Factors in the Immune Response," *Adv. Immunol.* **28**, 1.
8. Hoffmann, G.W. (1980), "On Network Theory and H-2 Restriction," *Contemp. Topics in Immunobiology*, Ed. N. L. Warner (New York: Plenum Press), vol. 11, 185.
9. Lederberg, J. (1959), "Genes and Antibodies," *Science* **129**, 1649.
10. Bruyns, C., G. Urbain-Vansanten, C. De Vos-Cloetens, and J. Urbain (1976), "Ontogeny of Mouse B Lymphocytes and Inactivation by Antigen of Early B Lymphocytes," *Proc. Natl. Acad. Sci. USA* **73**, 2462.
11. Urbain, J., C. Wuilmart, and P. A. Cazenave (1981), "Idiotypic Regulation in Immune Networks," *Contemp. Topics in Molecular Immunol.*, Eds. F. P. Inman and W. J. Mandy (New York: Plenum Press), **8**, 113.
12. Mitchison, N.A. (1971), "The Carrier Effect in Secondary Response to Hapten-Protein Conjugates. II. Cellular Cooperation," *Eur. J. Immunol.* **1**, 18.
13. Urbain, J., C. Collignon, J. D. Franssen, B. Mariamé, O. Léo, G. Urbain-Vansanten, P. Van de Walle, M. Wikler, and C. Wuilmart (1979), "Idiotypic Networks and Self-Recognition in the Immune System," *Ann. Immunol. (Inst. Pasteur)* **130C**, 281.
14. Eichmann, K., I. K. Falk, and K. Rajewsky (1975), "Antigen-Independent Cooperation between T and B Lymphocytes that Possess Similar and Complementary Idiotypes," *Europ. J. Immunol.* **5**, 661.
15. Hetzelberger, D., and K. Eichmann (1978), "Idiotypic Selectivity in the Cooperation between T and B Lymphocytes," *Europ. J. Immunol.* **8**, 846.
16. Snoussi, E., submitted to *Dynamics and Stability of Systems.*
17. Shellam, G.R., and G. J. V. Nossal (1968), "Mechanisms of Induction of Immunological Tolerance," *Immunology* **14**, 273.
18. Eisenfeld, J. (1987), "Compatibility Tests of Immunological Control Loops," *Ann. N.Y. Acad. Sci.* **504**, 132.

19. Eisenfeld, J., and P. Prueitt, "Systematic Approach to Modeling Immune Response," this volume.

J. EISENFELD and P. PRUEITT
Department of Mathematics, The University of Texas at Arlington, Arlington, Texas 76019

Systematic Approach to Modeling Immune Response

SUMMARY

Consideration of the interactions among cells and cell products involved in regulating antibody responses suggests that such interactions are organized into integrated circuits. To illustrate how a circuit-based regulatory system could be expected to operate, various laboratories have introduced a multitude of models; however, when these models are tested against observations regarding memory and low- and high-dose paralysis, none are completely satisfactory.

This paper applies the boolean approach to obtain a step-by-step mathematical procedure for developing an immune response model. The procedure begins by listing a complete set of constraints that the model is required to satisfy and culminates in the development of a system of differential equations that is guaranteed to satisfy those constraints.

This procedure is used to obtain a minimal two-lymphocyte model that simulates a set of nine immune response patterns. Then, it is illustrated how this minimal model serves as a core regulatory circuit which can be developed into more detailed models by adding or expanding circuits.

Theoretical Immunology, Part One, SFI Studies in the Sciences of Complexity
Ed. A. S. Perelson, Addison-Wesley Publishing Company, 1988

1. INTRODUCTION

Numerous laboratories are actively seeking to elucidate the nature and extent of the interactions that regulate the quantity of antibody-secreting cells produced in response to antigen. Various mathematical models have been introducted with an eye toward capturing salient aspects of immune response phenomena, suggesting new experiments, and perhaps most importantly, developing a systematic method for building models so that, as more detailed information becomes available, promising models can be easily updated.

Classical experiments already testify to the enormous complexity involved in the immune response. The amount of antibody produced at a given time after immunization is a biphasic function of antigen dose, the response being weak at low and high doses, and rising through a maximum at some intermediate level. Moreover, if the initial exposure is followed with a second dose of the same antigen, the secondary response will be greatly enhanced when the primary exposure was optimal, and markedly reduced if the primary exposure was either low or high (Mitchison, 1964; Golub, 1981; Bellanti, 1985). The immune system's memory of previous encounters with antigen is, thus, both qualitative and quantitative, and stores information on the amount and on the molecular details of the stimulating molecule. The enhanced response is referred to simply as memory, whereas the suppressed response, which also involves information storage, is referred to as high or low zone tolerance, depending on the level of the primary dose.

The above observations suggest (Hoffmann,1980; Herzenberg et al., 1980; Kaufman et al., 1985; Eisenfeld and DeLisi, 1985; Eisenfeld, 1987) that a model of immune response mechanism should incorporate at least three stable states referred to as memory, tolerance, and the virgin state. An immune system is said to be in it's virgin state if it has no memory of a previous encounter with the antigen in question, whereas, if it remembers a single encounter with a more or less optimal antigen dose, it is said to be in its memory state. Similarily, if it stores information on an encounter with a sub-optimal or supra-optimal antigen dose, it is said to be in its tolerant state. The term "tolerance" as it is used here, connotes a state of the system which allows diminished or suppressed responses rather than no response. The system may be driven away from the memory state (or the tolerant state) by further encounters with the antigen, but it returns to this state. Information storage in a tolerance response state is transient, of the order of days; whereas in a memory state information storage is long lived and might last the life time of the individual (see Bellanti, 1985 and references therein for further discussions). The above remarks are summarized in Figure 1.

The above response characteristics indicate that the state of the immune system, i.e., the particular mixture of the various types of cells and molecules, is dependent on the history of antigenic exposure. Antigen drives transitions between

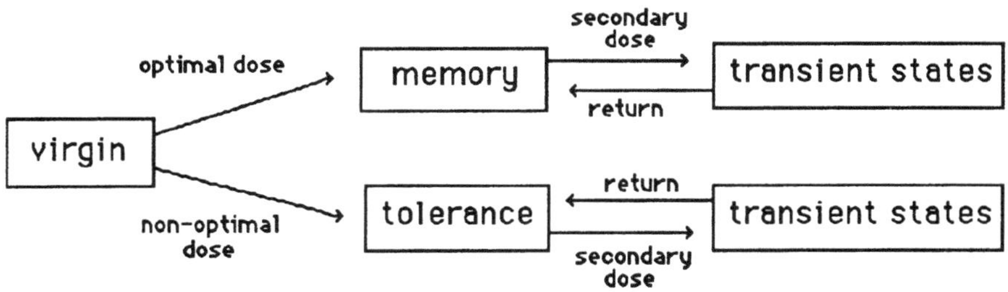

FIGURE 1 Immune responses. The system is driven from the virgin state to memory by an optimal dose. A secondary or a further encounter with the specific antigen drives the system through a number of transient states from which it returns to the memory state. A sub-optimal or supra-optimal dose drives the system from the virgin state to tolerance from which it is driven, by a secondary encounter, through transient states and then it returns to the tolerant state.

states by stimulating some yet-to-be-elucidated sequence of biochemical events that ultimately lead to shifts in the population balance by turning some cell types on, and turning others off. The complex homeostatic mechanisms regulating the quantity of serum antibody finds its origin in the vast web of interactions amoung the cells and molecules of the immune system. Herzenberg et al. (1980) suggested that the core regulatory circuit of the immune network is composed of the interactions between the T helper cells, which amplify the response to the antigenic challange, and the T suppressor cells, which limit the proliferation of T helper cells. Immunologic memory is characterized by high concentration of T helper cells as compared with the immunologic state of tolerance, which is characterized by high concentration of T suppressor cells (Herzenberg et al., 1980; Kaufman et al., 1985; Eisenfeld and DeLisi, 1985; Eisenfeld, 1987). The virgin state is neutral, dominated neither by T-helpers or T suppressors.

Herzenberg et al. also introduced the notion that both types of T lymphocytes operate on two levels, a high level in which the lymphocyte is sufficiently established so as to invoke a response and a low level in which the lymphocyte lies dormant. This prompted Kaufman et al. (1985) to introduce a formal procedure for modeling immune response phenomena which is based on boolean network logic.

Both Herzenberg et al. and Kaufman et al. introduced several minimal models of the immune response phenomena. These models are illuminating and stimulating, yet they are incomplete. The Herzenberg models do not incorporate a virgin state; these investigators were primarily concerned with the memory and tolerance states and the possible mechanisms that might lead to the establishment of one of those states over the other. Moreover, the kinetic interplay between antigen and the various cell types is described only diagramatically, not in terms of differential equations. The Kaufman models are much more detailed, replete with differential equations, elegantly derived, and with computer simulations. The major shortcoming of their (best) model, by the authors' own admission, is that there is no "escape mechanism" for the T helper cells and the T suppressor cells in the sense that, once established, these populations persist indefinitely. As a consequence, any level of antigen dose, even non-optimal, leads eventually to the memory state.

Prompted by the work of Herzenberg et al. and Kaufman et al., we seek to develop a minimal model that could replicate all of the responses described in Figure 1. The model to be presented consists of only two populations aside from the antigenic challange (compare with the Kaufman six-population model or the even more complex Herzenberg models). All cell types involved in degradation of antigen, including the T helper cells, are comglomerated into a single population referred to simply as helpers. The other population, referred to as suppressors, are characterized by their role as suppressors of helpers; this class is represented by the T suppressor cells. These two populations, known to play an essential role in antibody production, are minimal components that must be included in any model of the immune system.

Having obtained a minimal model, it is not difficult to expand it to a more detailed model while preserving the core regulatory circuit. As an illustration, we present (Section 7) an expanded model containing B cells and antibody populations.

The simplicity of the model building procedure to be presented here is, in and of itself, a significant contribution, perhaps more important then the models it generated. The systematic method, an extension of the formalism set forth by Herzenberg et al. and Kaufman et al., will facilitate the construction of more detailed models, involving larger numbers of components, which can be developed by this method as more information about the underlying cell dynamics becomes available.

The model building procedure adopted here belongs to the class of methods often referred to as the "boolean" (or "logical network") approach (see Section 8C). Our adoption of the boolean approach differs from that used by Kaufman et al. (see Section 8D).

2. DESCRIPTION OF IMMUNOLOGICAL STATES IN TERMS OF BOOLEAN CODE

We consider a two-lymphocyte model consisting of helpers and suppressors as described above. We wish to formulate a system of differential equations involving these two populations and antigen, which replicates in a continuous fashion the qualitive responses described in Figure 1, by a step-by-step procedure. The first step is to describe the various immunologic states of the system in boolean code.

A boolean variable is a variable that takes on one of two values, zero or one. With respect to the bi-level concept discussed above, the levels of the helper populations are described in terms of the boolean variable h, where $h = 0$ denotes low helper concentration and $h = 1$ denotes high helper concentration. Similarly, $s = 0$ denotes low suppressor concentration and $s = 1$ denotes high supressor concentration. Antigen can be introduced at three levels, $e = 1$ denotes low or sub-optimal, $e = 2$ denotes medium or optimal, and $e = 3$ denotes high or supra-optimal. Moreover, $e = 0$ denotes that the antigen is absent or not present to a significant degree. Since e is not a boolean variable, having four values rather than two, antigen is represented by two boolean variables $e = (e_2, e_1)$. In terms of the binary number system, we have $(0,0) = 0$, $(0,1) = 1$, $(1,0) = 2$ and $(1,1) = 3$. Of the sixteen states represented by the boolean vector (h, s, e_2, e_1), three of these states, memory, tolerance and the virgin state, are construed as steady states while the rest are transient states.

In the immunologic state of memory, the immune system is primed to respond to the antigenic challenge. This state, characterized by a high concentration of helpers and a low concentration of suppressors, is represented by (1,0,0,0). The tolerance state, the antithesis of memory characterized by low helper and high suppressor, is represented by (0,1,0,0). Considering that helpers and suppressors are established in response to the presence of antigen, the virgin state should be represented by (0,0,0,0). The immune responses (Figure 1) are initiated by introducing antigen at one of three possible levels when the system is in one of the three steady states, memory, tolerance or the virgin state. This gives nine initial states. The fact that a state is an initial state for one type of response sequence does not prevent it from being an intermediate state along another response sequence. The 16 boolean states are described in Table 1. For the sake of brevity of notation, commas separating the zeros and ones are omitted.

3. BOOLEAN DESCRIPTION OF THE SEQUENCE OF EVENTS LEADING EITHER TO MEMORY OR TOLERANCE

Before seeing antigen, one thinks of the immune system as being in the virgin state; the antigen then drives the system to memory or to tolerance, depending on the dose. One thinks of memory, tolerance and the virgin state as steady states.

There are nine immune responses depicted in Figure 1, arising from the three possible ways the three steady states may be perturbed by introducing antigen. The initial state and the terminal state for each sequence is described in Table 1. The problem now is to determine the intermediate states. For this purpose we take into account a number of constraints listed below.

TABLE 1 Description of Boolean States

Number	State	Immunologic Description
1	0000	virgin (steady state)
2	1000	memory (steady state)
3	0100	tolerance (steady state)
4	1100	transient state
5	0010	medium-dose introduced at virgin, leads to memory
6	1010	medium-dose introduced at memory, leads to memory
7	0110	medium-dose at introduced at tolerance, leads to tolerance
8	1110	transient state
9	0001	low-dose introduced at virgin, leads to tolerance
10	1001	low-dose introduced at memory, leads to memory
11	0101	low-dose introduced at tolerance, leads to tolerance
12	1101	transient state
13	0011	high-dose introduced at virgin, leads to tolerance
14	1011	high-dose introduced at memory, leads to memory
15	0111	high-dose introduced at tolerance, leads to tolerance
16	1111	transient state

CONSTRAINT 1: Any state on a sequence leading to memory (resp. tolerance) can not occur on a sequence leading to tolerance (resp. memory) and certainly the virgin state can not occur on such sequences.

CONSTRAINT 2: Memory, tolerance and the virgin states, being steady states, lead only to themselves.

CONSTRAINT 3: Both suppressors and helper are established only in the presence of antigen ($e = 1, 2$, or 3).

CONSTRAINT 4: Antigen is degraded in the presence of, and only in the presence of, helpers ($h = 1$).

CONSTRAINT 5: Helpers are degraded in the presence of, and only in the presence of, suppressors.

CONSTRAINT 6: The role of the suppressors, i.e., to degrade the helpers, suggests that the suppressor population would not be established in the absence of helpers; however, the two populations may be synthesized simultaneously.

CONSTRAINT 7: Perturbation of the virgin state by a medium antigenic dose must lead to memory. Perturbation of the virgin state by a low or a high antigenic dose must lead to tolerance. Perturbation of the memory (resp. tolerance) state by any antigenic dose (low, medium or high) must lead back to memory (resp. tolerance).

We consider first the establishment of memory. Suppose that antigen is introduced into the virgin state at the medium level. The initial state is (0010). The event should result in memory, (1000). The presence of antigen should stimulate the proliferation of helpers and possibly also suppressors, but, if suppressors were to be established first, they would suppress the helpers. Moreover, there is no mechanism to deplete suppressors once they are established. The process by which memory is achieved must occur in two steps, first the proliferation of helpers (0010) $\longrightarrow$ (1010) followed by the degradation of antigen, (1010) $\longrightarrow$ (1000). The reduction of antigen actually occurs in two steps (1010) $\longrightarrow$ (1001) $\longrightarrow$ (1000). However, later on in the development of the model, the two antigen components will be catenated into a single continuous variable, in which case the obvious constraint, that the antigenic level must pass through "low" on its way from "medium" to "zero," will be automatically satisfied. In the present boolean description, it suffices to express antigen degradation by a single step.

Now, if the memory state is perturbed by introducing antigen at one of the three levels, then the response should be to catabolize the antigen. As in the above sequence, suppressors can not be established. The three sequences are listed in Table 2.

TABLE 2 Description of the Response Sequences in Boolean Code

Event	Sequence
Establishment of memory	(0010) $\longrightarrow$ (1010) $\longrightarrow$ (1000)
Low-dose perturbation of memory	(1001) $\longrightarrow$ (1000)
Med-dose perturbation of memory	(1010) $\longrightarrow$ (1000)
High-dose perturbation of memory	(1011) $\longrightarrow$ (1000)
Low zone tolerance	(0001) $\longrightarrow$ (1101) $\longrightarrow$ (1100) $\longrightarrow$ (0100)
High zone tolerance	(0011) $\longrightarrow$ (1111) $\longrightarrow$ (1100) $\longrightarrow$ (0100)
Low-dose perturbation of tolerance	(0101) $\longrightarrow$ (1101) $\longrightarrow$ (1100) $\longrightarrow$ (0100)
Med-dose perturbation of tolerance	(0110) $\longrightarrow$ (1110) $\longrightarrow$ (1100) $\longrightarrow$ (0100)
High-dose perturbation of tolerance	(0111) $\longrightarrow$ (1111) $\longrightarrow$ (1100) $\longrightarrow$ (0100)

Next we consider the establishment of tolerance. Suppose that antigen is introduced at a low level, (0001), stimulating a sequence culminating in tolerance, (0100). (This event is called low zone tolerance.) The presence of antigen should stimulate the growth of both helpers (to degrade the antigen) and suppressors (to degrade the helpers). The question is which population is established first or are they established simultaneously? The possibility of the helpers being established first, (0001) $\longrightarrow$ (1001), is ruled out since (1001) leads to memory (constraint 1). The possibility of the suppressors being established first is ruled out by constraint 6. There remains the simultaneous establishment of both populations, (0001) $\longrightarrow$ (1101). Once established, the suppressors will degrade the helpers, but not before the helpers degrade the antigen, (1101) $\longrightarrow$ (1100) $\longrightarrow$ (0100).

High zone tolerance is described in the same fashion only now the process begins with high antigen dose, (0011). As before, helpers and suppressors are established simultaneously, then helpers degrade antigen, followed by degradation of helpers by suppressors, (0011) $\longrightarrow$ (1111) $\longrightarrow$ (1100) $\longrightarrow$ (0100).

Finally, if the tolerance state is perturbed by introducing antigen at one of the three levels, then the response should be to degrade the antigen. In view of the constraint 4, helpers must be established (in spite of the presence of suppressors already established), and then the helps must be degraded by the suppressors. These three sequences, along with the other six, are listed in Table 2.

4. THE NEXT-STEP FUNCTIONS

We wish to determine the analytical formulas that generate the sequences listed in Table 2. That is, we wish to find the functions $f_i, i = 1,2,3,4$, such that, if $(h(t_n), s(t_n), e_2(t_n), e_1(t_n))$ is a state occurring along any one of these sequences, at time t_n, then the next state along that particular sequence will be given by

$$h(t_{n+1}) = f_1\big(h(t_n), s(t_n), e_2(t_n), e_1(t_n)\big) \tag{1}$$

$$s(t_{n+1}) = f_2\big(h(t_n), s(t_n), e_2(t_n), e_1(t_n)\big), \tag{2}$$

$$e_2(t_{n+1}) = f_3\big(h(t_n), s(t_n), e_2(t_n), e_1(t_n)\big), \tag{3}$$

$$e_1(t_{n+1}) = f_4\big(h(t_n), s(t_n), e_2(t_n), e_1(t_n)\big), \tag{4}$$

For this purpose we follow a procedure elegantly presented in the book by Gelfand and Walker (1984). This method is illustrated by the computation of f_1 .

The function f_1 (or any boolean function of (h, s, e_2, e_1)) is given by

$$\begin{aligned} f_1(h, s, e_2, e_1) = {} & c_1 + c_2 h + c_3 s + c_4 hs + c_5 e_2 + c_6 he_2 + c_7 se_2 + c_8 hse_2 + c_9 e_1 + c_{10} h \\ & + c_{11} se_1 + c_{12} hse_1 + c_{13} e_2 e_1 + c_{14} he_2 e_1 + c_{15} se_2 e_1 + c_{16} hse_2 e_1, \end{aligned}$$

where the $c_k, k = 1$ to 16, are coefficients. Let $C = (c_1, c_2, \ldots, c_{16})$ denote the coefficient vector. This vector is obtained by inverting a linear equation, $CG_4 = V$; the vector V and the matrix G_4 are given below.

To obtain V, one denotes the boolean states as $s_1, s_2, \ldots, s_{16}$ in lexicographic order (the order given in Table 1). For example, $s_1 = (0000)$, $s_2 = (1000), \ldots, s_{16} = (1111)$. The kth component of V is denoted by $v_k = f_1(s_k)$. Since s_k are steady states for $k = 1, 2, 3$, $f_1(s_k)$ is the first component of s_k; i.e., $f_1(0000) = 0 = v_1$, $f_1(1000) = 1 = v_2$ and $f_1(0100) = 0 = v_3$. To obtain $f_1(s_k)$ for $k = 4, \ldots, 16$, we locate s_k in Table 2. For instance, $s_4 = (1100)$ occurs along the last five sequences; its next state is (0100). Thus, $v_4 = f_1(s_k) = 0$. In this manner, we obtain the result, $V = (0100111111111111)$.

In general, the matrix G_n depends only on the number n of boolean variables ($n = 1, 2, \ldots$). Moreover, G_{n+1} is obtained from G_n by the recursion formula,

$$G_{n+1} = \begin{bmatrix} G_n & G_n \\ 0 & G_n \end{bmatrix}, \quad n = 1, 2, \ldots .$$

Starting with

$$G_1 = \begin{bmatrix} 1 & 1 \\ 0 & 1 \end{bmatrix},$$

the recursion formula generates G_n for arbitrary n. For the case $n = 4$, we have

$$G_4 = \begin{bmatrix}
1 & 1 & 1 & 1 & 1 & 1 & 1 & 1 & 1 & 1 & 1 & 1 & 1 & 1 & 1 & 1 \\
1 & 0 & 1 & 0 & 1 & 0 & 1 & 0 & 1 & 0 & 1 & 0 & 1 & 0 & 1 & 0 \\
0 & 0 & 1 & 1 & 0 & 0 & 1 & 1 & 0 & 0 & 1 & 1 & 0 & 0 & 1 & 1 \\
0 & 0 & 0 & 1 & 0 & 0 & 0 & 1 & 0 & 0 & 0 & 1 & 0 & 0 & 0 & 1 \\
0 & 0 & 0 & 0 & 1 & 1 & 1 & 1 & 0 & 0 & 0 & 0 & 1 & 1 & 1 & 1 \\
0 & 0 & 0 & 0 & 0 & 1 & 0 & 1 & 0 & 0 & 0 & 0 & 0 & 1 & 0 & 1 \\
0 & 0 & 0 & 0 & 0 & 0 & 1 & 1 & 0 & 0 & 0 & 0 & 0 & 0 & 1 & 1 \\
0 & 0 & 0 & 0 & 0 & 0 & 0 & 1 & 0 & 0 & 0 & 0 & 0 & 0 & 0 & 1 \\
0 & 0 & 0 & 0 & 0 & 0 & 0 & 0 & 1 & 1 & 1 & 1 & 1 & 1 & 1 & 1 \\
0 & 0 & 0 & 0 & 0 & 0 & 0 & 0 & 0 & 1 & 0 & 1 & 0 & 1 & 0 & 1 \\
0 & 0 & 0 & 0 & 0 & 0 & 0 & 0 & 0 & 0 & 1 & 1 & 0 & 0 & 1 & 1 \\
0 & 0 & 0 & 0 & 0 & 0 & 0 & 0 & 0 & 0 & 0 & 1 & 0 & 0 & 0 & 1 \\
0 & 0 & 0 & 0 & 0 & 0 & 0 & 0 & 0 & 0 & 0 & 0 & 1 & 1 & 1 & 1 \\
0 & 0 & 0 & 0 & 0 & 0 & 0 & 0 & 0 & 0 & 0 & 0 & 0 & 1 & 0 & 1 \\
0 & 0 & 0 & 0 & 0 & 0 & 0 & 0 & 0 & 0 & 0 & 0 & 0 & 0 & 1 & 1 \\
0 & 0 & 0 & 0 & 0 & 0 & 0 & 0 & 0 & 0 & 0 & 0 & 0 & 0 & 0 & 1
\end{bmatrix}$$

The inverse of G_n may be obtained by the recursion formula

$$G_{n+1}^{-1} = \begin{bmatrix} G_n^{-1} & -G_n^{-1} \\ 0 & G_n^{-1} \end{bmatrix}, \quad n = 1, 2, \ldots, \qquad G_1^{-1} = \begin{bmatrix} 1 & -1 \\ 0 & 1 \end{bmatrix}.$$

For the case of interest,

$$G_4^{-1} = \begin{bmatrix}
1 & -1 & -1 & 1 & -1 & 1 & 1 & -1 & -1 & 1 & 1 & -1 & 1 & -1 & -1 & 1 \\
0 & 1 & 0 & -1 & 0 & -1 & 0 & 1 & 0 & -1 & 0 & 1 & 0 & 1 & 0 & -1 \\
0 & 0 & 1 & -1 & 0 & 0 & -1 & 1 & 0 & 0 & -1 & 1 & 0 & 0 & 1 & -1 \\
0 & 0 & 0 & 1 & 0 & 0 & 0 & -1 & 0 & 0 & 0 & -1 & 0 & 0 & 0 & 1 \\
0 & 0 & 0 & 0 & 1 & -1 & -1 & 1 & 0 & 0 & 0 & 0 & -1 & 1 & 1 & -1 \\
0 & 0 & 0 & 0 & 0 & 1 & 0 & -1 & 0 & 0 & 0 & 0 & 0 & -1 & 0 & 1 \\
0 & 0 & 0 & 0 & 0 & 0 & 1 & -1 & 0 & 0 & 0 & 0 & 0 & 0 & -1 & 1 \\
0 & 0 & 0 & 0 & 0 & 0 & 0 & 1 & 0 & 0 & 0 & 0 & 0 & 0 & 0 & -1 \\
0 & 0 & 0 & 0 & 0 & 0 & 0 & 0 & 1 & -1 & -1 & 1 & -1 & 1 & 1 & -1 \\
0 & 0 & 0 & 0 & 0 & 0 & 0 & 0 & 0 & 1 & 0 & -1 & 0 & -1 & 0 & 1 \\
0 & 0 & 0 & 0 & 0 & 0 & 0 & 0 & 0 & 0 & 1 & -1 & 0 & 0 & -1 & 1 \\
0 & 0 & 0 & 0 & 0 & 0 & 0 & 0 & 0 & 0 & 0 & 1 & 0 & 0 & 0 & -1 \\
0 & 0 & 0 & 0 & 0 & 0 & 0 & 0 & 0 & 0 & 0 & 0 & 1 & -1 & -1 & 1 \\
0 & 0 & 0 & 0 & 0 & 0 & 0 & 0 & 0 & 0 & 0 & 0 & 0 & 1 & 0 & -1 \\
0 & 0 & 0 & 0 & 0 & 0 & 0 & 0 & 0 & 0 & 0 & 0 & 0 & 0 & 1 & -1 \\
0 & 0 & 0 & 0 & 0 & 0 & 0 & 0 & 0 & 0 & 0 & 0 & 0 & 0 & 0 & 1
\end{bmatrix}$$

Thus, $C = VG_4{}^{-1} = (0\ 1\ 0\ \text{-}1\ 1\ \text{-}1\ 0\ 1\ 1\ \text{-}1\ 0\ 1\ \text{-}1\ 1\ 0\ \text{-}1)$, yielding

$$f_1 = h - hs + e_2 - he_2 + hse_2 + e_1 - he_1 + hse_1 - e_2e_1 + he_2e_1 - hse_2e_1,$$

After simplifying, we obtain

$$f_1 = e_1 + e_2 - e_1e_2 + \overline{e}_1\overline{e}_2\overline{s}\, h, \tag{5}$$

where we adopt the notation $\overline{b} = 1 - b$, for any boolean variable b. In similar fashion, we derive the other three boolean functions:

$$f_2 = s + e_1\overline{s}h, \tag{6}$$
$$f_3 = e_2\overline{h}, \tag{7}$$
$$f_4 = e_1\overline{h}. \tag{8}$$

It should be noted that the above application of the next-step-function algorithm is slightly modified from that given in Gelfand and Walker (1984). (See Section 8C.)

5. THE CONTINUOUS MODEL

The model described by Eqs. (1–8) is discrete both with respect to time and population levels. The first step is to extend the discrete model to a homologous continuous model. We wish to carry out this procedure in a systematic manner so that the method may apply in full generality. The following procedure was suggested by Glass (1975) and by Glass and Pasternak (1978) who address the problem of writing systems of ordinary differential equations which approximate boolean networks.

In the discrete model, h and s are boolean variables taking values in the discrete set 0,1. In the continuous analog, the continuous variables h and s take values in subintervals of the interval [0,1] . The high helper level, $h = 1$, corresponds in the continuous model to the subinterval, $\eta_2 \leq h \leq 1$, where η_2 is the threshold value at which the system senses that the helper population is no longer at the low level and recognizes a high concentrations of helpers. Similarly, as the helper concentration decreases from high to low, the level of helpers is not recognized as being low until h decreases below a low threshold, η_1. Of course $\eta_1 < \eta_2$. Similarly, $s = 0$ is associated with a subinterval, $0 \leq s \leq \eta_3$ and $s = 1$ is associated with a subinterval, $\eta_4 \leq s \leq 1$, where $\eta_3 < \eta_4$ are the low and high thresholds, respectively, for s.

TABLE 3 Association between Boolean States and Regions

Number	State, b_i boolean	decimal	Region, R_i
1	0 0 0 0	(000)	$[0, \eta_1] \times [0, \eta_3] \times [0, \eta_5]$
2	0 0 0 1	(001)	$[0, \eta_1] \times [0, \eta_3] \times [\eta_6, \eta_7]$
3	0 0 1 0	(002)	$[0, \eta_1] \times [0, \eta_3] \times [\eta_8, \eta_9]$
4	0 0 1 1	(003)	$[0, \eta_1] \times [0, \eta_3] \times [\eta_{10}, \infty]$
5	0 1 0 0	(010)	$[0, \eta_1] \times [\eta_4, 1] \times [0, \eta_5]$
6	0 1 0 1	(011)	$[0, \eta_1] \times [\eta_4, 1] \times [\eta_6, \eta_7]$
7	0 1 1 0	(012)	$[0, \eta_1] \times [\eta_4, 1] \times [\eta_7, \eta_8]$
8	0 1 1 1	(013)	$[0, \eta_1] \times [\eta_4, 1] \times [\eta_9, \infty]$
9	1 0 0 0	(100)	$[\eta_2, 1] \times [0, \eta_3] \times [0, \eta_5]$
10	1 0 0 1	(101)	$[\eta_2, 1] \times [0, \eta_3] \times [\eta_6, \eta_7]$
11	1 0 1 0	(102)	$[\eta_2, 1] \times [0, \eta_3] \times [\eta_8, \eta_9]$
12	1 0 1 1	(103)	$[\eta_2, 1] \times [0, \eta_3] \times [\eta_{10}, \infty]$
13	1 1 0 0	(110)	$[\eta_2, 1] \times [\eta_4, 1] \times [0, \eta_5]$
14	1 1 0 1	(111)	$[\eta_2, 1] \times [\eta_4, 1] \times [\eta_6, \eta_7]$
15	1 1 1 0	(112)	$[\eta_2, 1] \times [\eta_4, 1] \times [\eta_8, \eta_9]$
16	1 1 1 1	(113)	$[\eta_2, 1] \times [\eta_4, 1] \times [\eta_{10}, \infty]$

In the discrete model, the four levels of antigen are represented by two boolean variables: zero (or very small) by $(e_2, e_1) = (0,0)$ or $e = 0$, low by $(e_2, e_1) = (0,1)$ or $e = 1$, medium by $(e_2, e_1) = (1,0)$ or $e = 2$, high by $(e_2, e_1) = (1,1)$ or $e = 3$. In the continuous model, these four antigenic levels are represented by disjoint intervals $[0, \eta_5]$, $[\eta_6, \eta_7]$, $[\eta_8, \eta_9]$, $[\eta_{10}, \infty]$, where $\eta_6 < 1 < \eta_7$, $\eta_8 < 2 < \eta_9$ and $\eta_{10} < 3$. As in the case of helpers and suppressors, the boundaries of these intervals represent threshold values for sensing the antigenic levels. The association between boolean states and regions is shown in Table 3. The boolean states, along with their associated regions, are numbered for notational convenience associated to the decimal representation and are not in lexiograghic order.

In the discrete model, the evolution of h is governed by Eq. (1), where f_1 is given by Eq. (5). We can express this equation in the equivalent form $\Delta h(n) = f_1 - h(n)$, where $\Delta h(n) = h(n+1) - h(n)$ is the finite difference operator. (For brevity sake, we omitted the arguments of f_1.) The corresponding equation in the continuous model has the form $dh/dt = F_1 - h$, where F_1 is a function homologous to f_1, defined in the region

$$\mathcal{R} = \bigcup_{i=1}^{16} R_1, \tag{9}$$

by the rule,

$$F_1(h, s, e) = f_1(b_i), \tag{10}$$

whenever (h, s, e) belongs to the region R_i that is associated with the boolean state b_i, $i = 1, \ldots, 16$. If the η_i are chosen so that

$$\begin{aligned} &\eta_1 < .5 < \eta_2,\ \eta_3 < .5 < \eta_4, \\ &\eta_5 < .5 < \eta_6 < 1 < \eta_7 < 1.5 < \eta_8 < 2 < \eta_9 < 2.5 < \eta_{10} < 3, \end{aligned} \tag{11}$$

then the formula giving F_1 can be obtained from that giving f_1, i.e., Eq. (5), by a simple device. We will show that, in terms of the step function,

$$\mathcal{S}(x) = \begin{cases} 0 & \text{if } x < .5 \\ 1 & \text{if } x > .5 \end{cases}, \tag{12}$$

$$F_1(h, s, e) = \mathcal{S}(e) + \big(1 - \mathcal{S}(e)\big)\mathcal{S}(\overline{s})\mathcal{S}(h) \tag{13}$$

for $(h, s, e) \varepsilon \mathcal{R}$ (recall the notation, $\overline{s} = 1 - s$). To see why this is so, observe that $f_1 = e_1 + e_2 - e_1 e_2 + \overline{e}_1 \overline{e}_2 \overline{s} h$ has the value one if either e_1 or e_2(or both) $= 1$, or if $e_1 = e_2 = 0$, while $h = 1$ and $s = 0$; otherwise $f_1 = 0$. Thus, $F_1 = 1$ (on $\mathcal{R}$) if either $e > .5$ or if $e < .5$ while $h > .5$ and $s < .5$; otherwise $F_1 = 0$. This is precisely the statement of Eq. (13) if the threshold values are chosen so as to satisfy the inequalities (11).

Similarly, the differential equation governing the evolution of s is $ds/dt = F_2 - s$ where F_2 is defined on $\mathcal{R}$ by the rule

$$F_2(h, s, e) = f_2(b_i), \tag{14}$$

whenever $(h, s, e) \varepsilon R_i$, $i = 1, \ldots, 16$. Now $f_2 = s + she_1$ is one if either $s = 1$ or if $s = 0$ while $h = 0$ and $e_1 = 1$; otherwise $f_2 = 0$. Now $e_1 = 1$ occurs at two antigenic levels, low and high, corresponding to, in the continuous model, $\eta_6 < e < \eta_7$ and $e > \eta_{10}$. Thus, keeping in mind inequalities (11) and the fact that F_2 is defined only on $\mathcal{R}$, $F_2 = 1$ if either $s > .5$ or if $s < .5$ while $h < .5$ and either $.5 < e < 1.5$ or $e > 2.5$. In terms of the step function $\mathcal{S}$ given by Eq. (12),

$$F_2(h, s, e) = \mathcal{S}(s) + \mathcal{S}(\overline{s})\mathcal{S}(\overline{h})\left[\mathcal{S}(e)\mathcal{S}(2 - e) + \mathcal{S}(e - 2)\right] \tag{15}$$

The antigenic evolution is given in the discrete model by $\Delta e_i = e_i h - e_i$, $i = 1, 2$. If $h = 0$, then $\Delta e_i = 0$ and the antigenic level is maintained. If $h = 1$, the $\Delta e_i = -e_i$ and the antigen decreases exponentially. The homologous dynamics in the continuous model are: $de/dt = 0$ if $h < \eta_1$, and $de/dt = -e$ if $h > \eta_2$ for $0e < \eta_5$, $\eta_6 < e < \eta_7$, $\eta_8 < e < \eta_9$ or $e > \eta_{10}$. With the threshold values satisfying inequality (11), the dynamical equation governing e may be expressed as

$$de/dt = -e\mathcal{S}(h). \tag{16}$$

Eq. (16), coupled with the other two dynamical equations derived above,

$$dh/dt = F_1(h, s, e) - h, \tag{17}$$

$$ds/dt = F_2(h, s, e) - s, \tag{18}$$

govern the evolution of the state (h, s, e), when and only when $(h, s, e)\varepsilon\mathcal{R}$, i.e., when and only when the level of helpers, suppressors and antigen are clearly recognized. Eqs. (17) and (18) are of the generalized Lotka-Volterra form $dx_i/dt = f_i(x_1, x_2, \ldots, x_n) - g_i(x_1, x_2, \ldots, x_n)$ (May, 1973) which are used extensively to model the dynamical behavior of interacting populations. $f_i \geq 0$ is the function that represents the help given to x_i by other populations and the intrinsic birth rate. $g_i \geq 0$ represents the suppression of x_i by other populations and the intrinsic death rate. Eq. (16) has the same form, but there is no help term present; there is only a degradation term. This reflects the case, such as with chemical antigens, where antigen is not produced in the system. A central focus of the modeling procedure is to suggest the kinetics by which the transitory presence of antigen drives the immune system to either memory or tolerance.

Eq. (16) may be written as

$$de/dt = F_3 - e, \tag{19}$$

where

$$F_3(h,s,e) = e(1 - \mathcal{S}(h)) \tag{20}$$

The functions F_i $i = 1,2,3$, are defined only on $\mathcal{R}$. If the state (h,s,e) does not belong to $\mathcal{R}$, then this state is said to be ambiguous. When traversing through the region of ambiguous states, the response functions F_i, $i = 1,2,3$, remain unchanged from their values upon entering the ambiguous region. In other words, if the trajectory $(h(t), s(t), e(t))$ enters the ambiguous region at some instant t_0, then it is governed by the equations $dh/dt = G_1 - h$, $ds/dt = G_2 - s$, $de/dt = G_3 - e$, where

$$G_i = lim_{t \to t_o -} F_i(h(t), s(t), e(t)),\ i = 1,2,3, \tag{21}$$

until the trajectory leaves the ambiguous region. At this instant, a new unambiguous level state is recognized and the system is again governed by Eqs. (16)–(18). We may say that the system "remembers" its last unambiguous level state and reacts accordingly.

It may be shown that starting in any region R_i, the trajectory will remain in R_i if b_i is a steady state of the discrete model, or it will leave R_i and enter R_k ,where b_k, follows b_i in the discrete model. In other words, the continuous model completely replicates the discrete analog in establishing the patterns given in Table 2. The analytic results supporting this conclusion will be presented in a follow-up paper devoted entirely to the formal aspects involved in deriving a continuous analog to a discrete boolean model. For the present, we will demonstrate the validity of the method through computer simulations (Section 6).

The model may be simplified while maintaining the same qualitive patterns by replacing the step function $\mathcal{S}$ by the sigmoidal function of the Hill type

$$\mathcal{S}_n(x) = \begin{cases} (2x)^n/(1+(2x)^n) & \text{when } x \geq 0 \\ 0 & \text{when } x \leq 0 \end{cases} \tag{22}$$

The choice of the threshold values η_i and the positive integer n will be discussed in Section 8B.

6. NUMERICAL SIMULATIONS

We now examine the evolution of the immune responses as predicted by the model,

$$dh/dt = \mathcal{S}_n(e) + (1 - \mathcal{S}_n(e))\mathcal{S}_n(\overline{s})\mathcal{S}_n(h) - h, \tag{23}$$

$$ds/dt = \mathcal{S}_n(s) + \mathcal{S}_n(\overline{s})\mathcal{S}_n(\overline{h})[\mathcal{S}_n(e)\mathcal{S}_n(2-e) + \mathcal{S}_n(e-2)] - s, \tag{24}$$

$$de/dt = -\,e\mathcal{S}_n(h), \tag{25}$$

where $\mathcal{S}_n$ is the sigmoidal function given in Eq. (22).

For simplicity and convenience, we set the threshold values in terms of a single parameter δ. The η_i are given by,

$$\eta_1 = \eta_3 = \eta_5 = .5 - \delta, \eta_2 = \eta_4 = \eta_6 = .5 + \delta, \eta_7 = 1.5 - \delta,$$
$$\eta_8 = 1.5 + \delta, \eta_9 = 2.5 - \delta, \eta_{10} = 2.5 + \delta \qquad (26)$$

Observe that, for $0 < \delta < .5$, the threshold values given in Eqs. (26) are consistent with inequalities (11).

The following computer simulations, Figures 2–10, were run on a Macintosh computer using a second order Runge-Kutta method, with $n = 30$ and $\delta = 0.035$. In each of the figures, 2-10, h is designated by a "o" placed on its trajectory every 40 steps, s by "▪" every 40 steps and e by "•" every 80 steps. The time intervals spent in the ambiguous region are displayed below the graphs. The boolean level states that occur between these intervals are also shown. The order of the boolean states and the length of the stay in the ambiguous region is determined by the computer program and is printed as data output below the graphs.

Figure 2 depicts the establishment of memory. Introducing antigen at the medium level in the virgin state, $(h, s, e) = (0, 0, 2)$, helpers are established. As helpers become more plentiful, antigen is degraded. Helpers and antigen both evolve to steady state values, helpers to a high level (actually the steady-state helper value is x_{15}, see Section 8B) and antigen to zero. Throughout the process, $s = 0$. The kinetics emulate the discrete process: $(002) \longrightarrow (102) \longrightarrow (101) \longrightarrow (100)$.

Figure 3 depicts low zone tolerance. Introducing antigen at the low level in the virgin state $(h, s, e) = (0, 0, 1)$, both helpers and suppressors are established. As helpers become established, antigen is degraded, as in the establishment of memory, but now the helpers are degraded in the presence of antigen. Helpers decay exponentially to zero. At the point at which the helper population is recognized as low, antigen is no longer degraded and survives at a very low level. Meanwhile, suppressors continue to evolve exponentially to a high steady-state level. The kinetics emulates the discrete process: $(001) \longrightarrow (111) \longrightarrow (110) \longrightarrow (010)$.

Figure 4 depicts high zone tolerance. The kinetic behavior is similar to that observed in low zone tolerance except that the initial antigen level is at a high level, $e = 3$. Observe the resemblance to the discrete pattern : $(003) \longrightarrow (113) \longrightarrow (112) \longrightarrow (111) \longrightarrow (110) \longrightarrow (010)$.

Figures 5–7 depict the immune responses that occur when antigen is introduced in the memory state. In each case the antigen level is reduced to zero while the helper and suppressor levels remain unchanged.

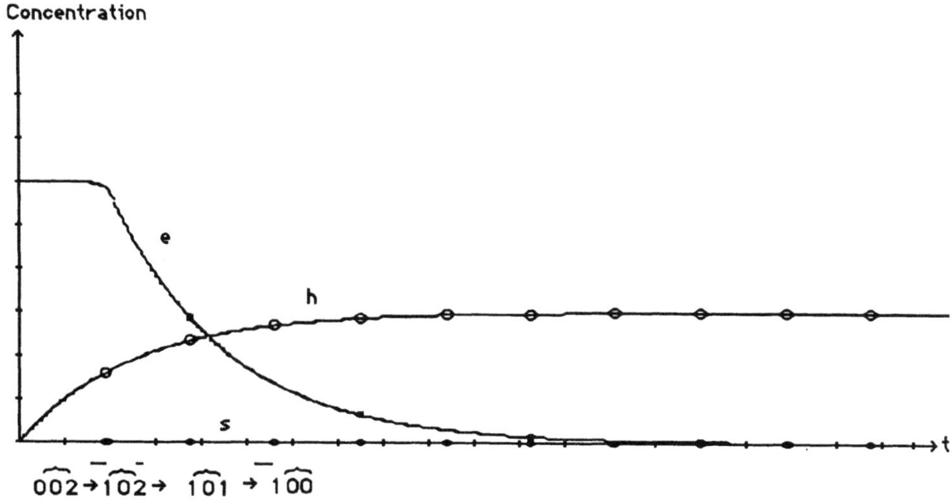

FIGURE 2 Medium-dose establishment of memory.

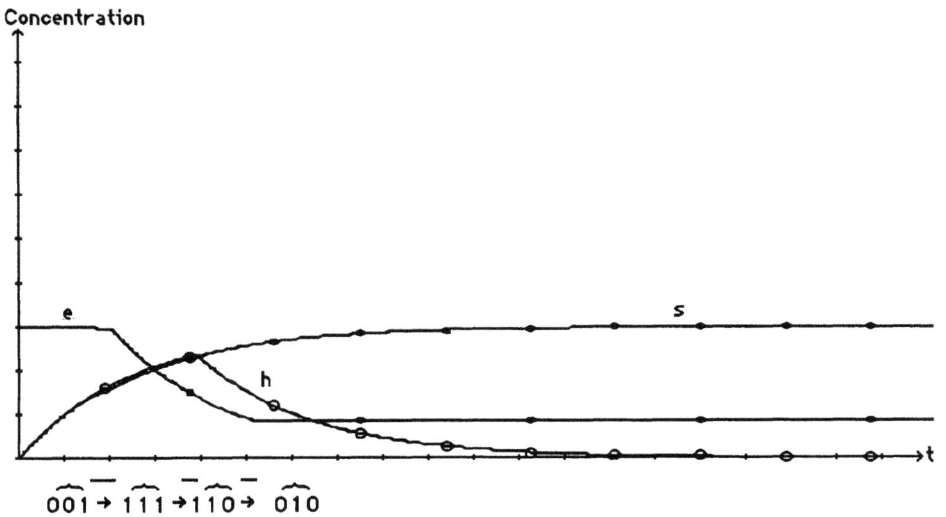

FIGURE 3 Low-dose establishment of low zone tolerance.

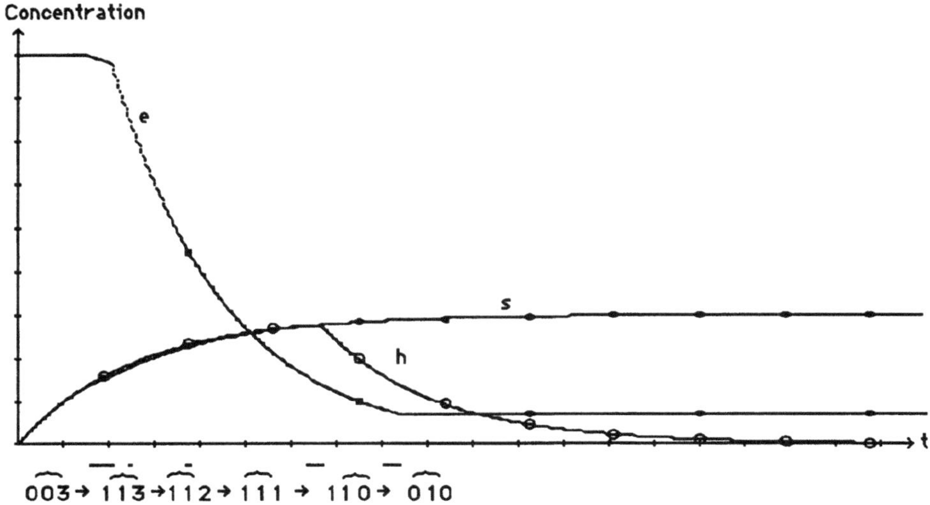

FIGURE 4 High-dose establishment of high zone tolerance.

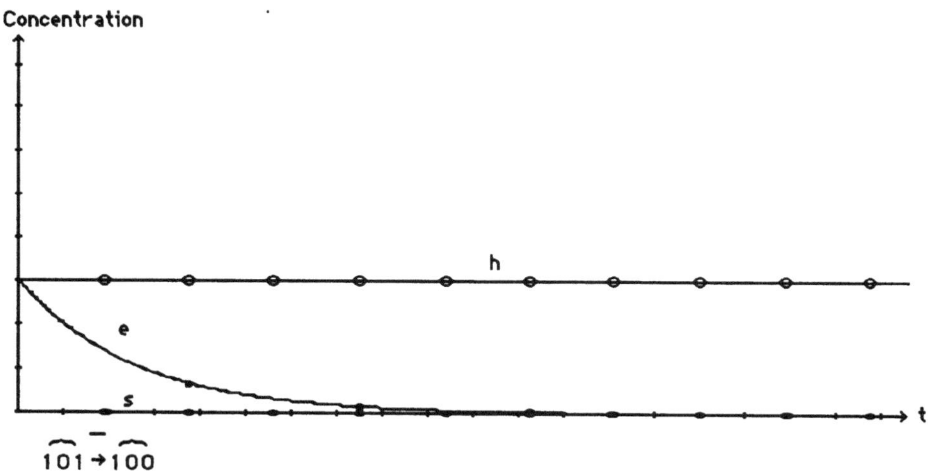

FIGURE 5 Low-dose perturbation of memory.

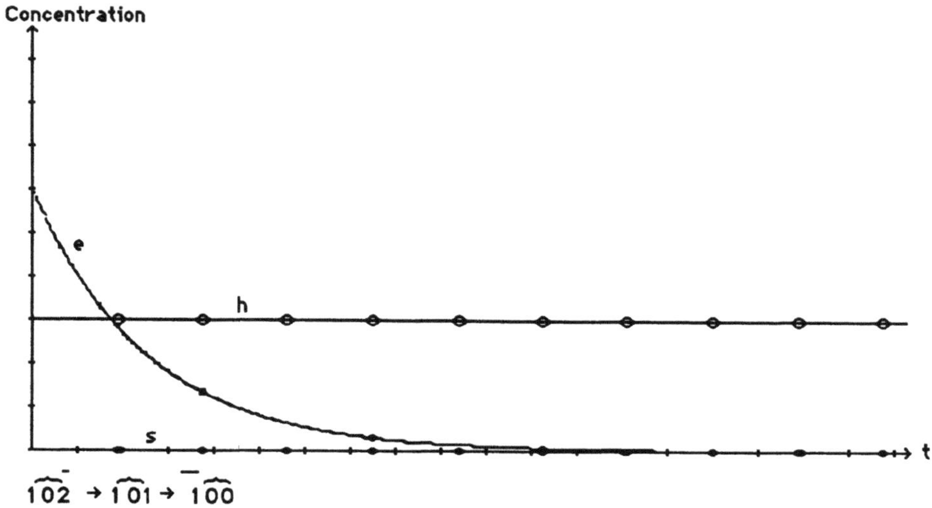

FIGURE 6 Medium-dose perturbation of memory.

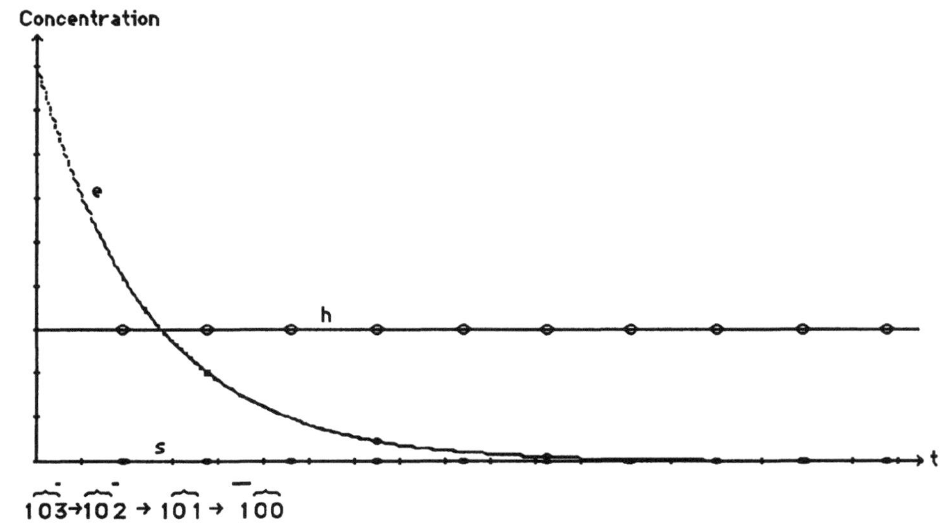

FIGURE 7 High-dose perturbation of memory.

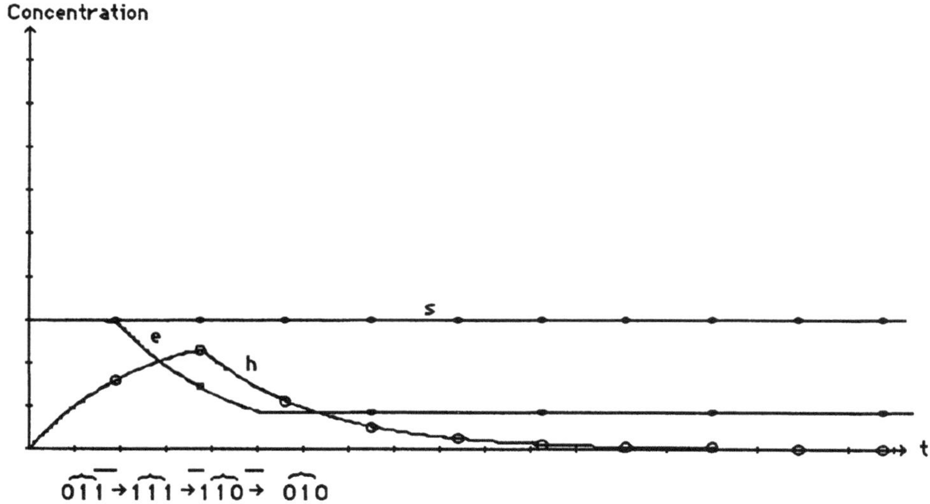

FIGURE 8 Low-dose perturbation of tolerance.

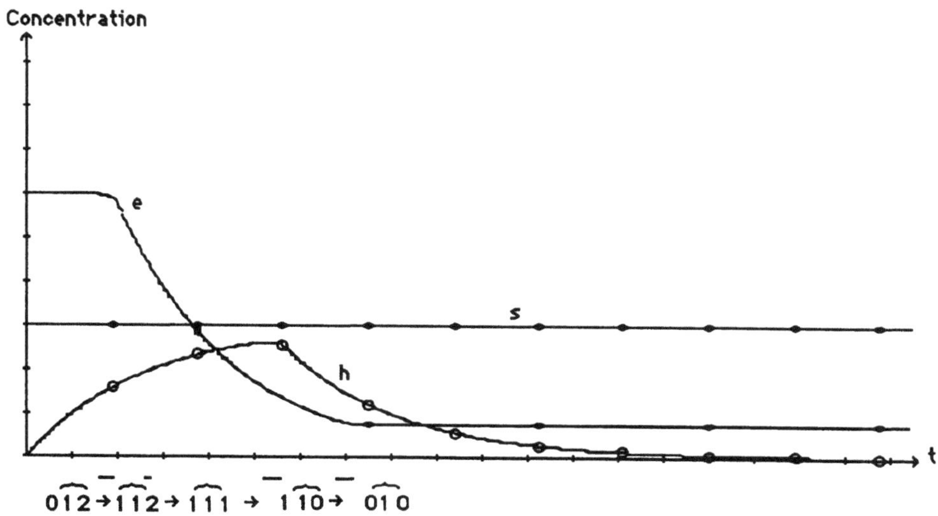

FIGURE 9 Medium-dose perturbation of tolerance.

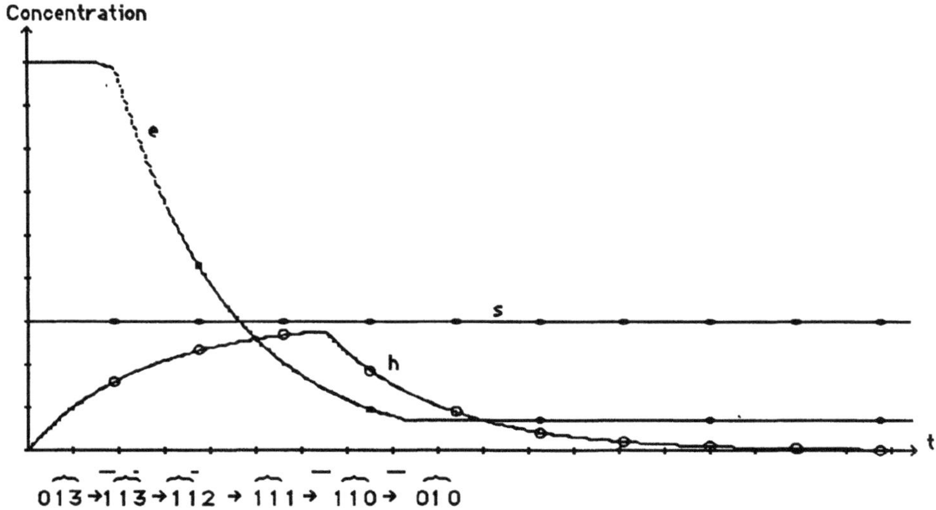

FIGURE 10 High-dose perturbation of tolerance.

Figures 8-10 depict the immune responses that occur when antigen is introduced in the tolerance state. The presence of antigen stimulates the development of helpers, which, in turn, degrades the antigen. Once the antigenic level is reduced below the low level, the helpers begin to disappear. Throughout the process the suppressor level remains unchanged.

7. Model expansion

The major advantage of the minimal model is that it provides a possible mechanism between helpers and suppressors so as to effect the nine immune responses that are described in Figure 1 and are specified in Table 2. The relationships between helpers, suppressors and antigen that are found in the minimal model may be expressed diagrammatically in Figure 11(A). We refer to these relationships as the core regulatory circuit (CRC). (After Herzenberg et al.,1980, although their core regulatory circuit is different.)

To refine the minimal model, it may not be necessary to repeat the step-by-step process as long as the CRC is not significantly altered. Suppose, for instance, we wish to distinguish B cells (b) and antibody (a) as separate populations. Then, instead of the helpers inhibiting antigen directly the process occurs in three steps as shown in Figure 11(B).

To generate the expanded model, the equations (both boolean and differential) pertaining to the helper and suppressor populations remain unchanged from the previous version. To obtain the equations (boolean and differential) pertaining to antigen, we merely replace "h" by "a" whenever the symbol h appears (since a is now the antigen inhibitor rather than h). Since helpers stimulate B cells while B cells produce antibody, both processes occurring in the presence of antigen, the next-step functions pertaining to those variables are:

$$f_4 = (e_1 + e_2 - e_1 e_2)h, \qquad f_5 = (e_1 + e_2 - e_1 e_2)b$$

respectively. Thus, the system of differential equations governing the expanded model is:

$$de/dt = -\, e\mathcal{S}_n(a), \tag{27}$$
$$db/dt = \mathcal{S}_n(e)\mathcal{S}_n(h) - b, \tag{28}$$
$$da/dt = \mathcal{S}_n(e)\mathcal{S}_n(b) - a, \tag{29}$$

in conjunction with Eqs. (23)–(24).

As is true in the case of the other populations, b and a are recognized in distinct levels: low in the interval $[0, 0.5 - \delta]$and high in the interval $[0.5 + \delta, 1]$.

The following computer simulations, Figures 12–20, were run on a Macintosh computer using a second-order Runge-Kutta method, with $n = 30$ and $\delta = 0.035$. In each figure, b is designated by "o" placed every 80 steps (the circles are spaced further apart than that designating h), a is designated by having no markings and h, s and e are drawn as before (Figures 2–10). The boolean level states of the vector (h, s, e, b, a) that occur between these intervals are also shown.

It is interesting to compare each figure with its minimal-model analog. Consider Figure 12; its boolean sequence is 00200 ⟶ 10200 ⟶ 10210 ⟶ 10111 ⟶ 10011 ⟶ 10001 ⟶ 10000. When the last two numbers (representing b and a) are deleted, we obtain: 002 ⟶ 102 ⟶ 102 ⟶ 101 ⟶ 100 ⟶ 100 ⟶ 100. Omitting consecutive repetitions, this truncated sequence becomes 002 ⟶ 102 ⟶ 101 ⟶ 100, which is precisely the boolean sequence that is associated with the analog (Figure 2). This observation reflects the fact that medium-dose

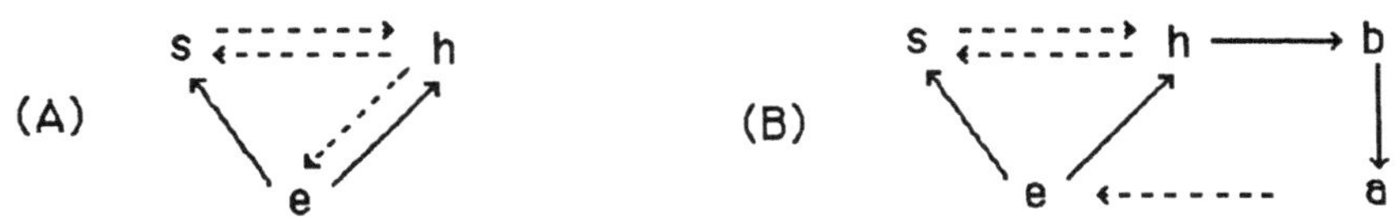

FIGURE 11 (A) Minimal model: helpers inhibit antigen directly. (B) Expanded model: helpers stimulate B cells, B cells produce antibody, and antibody inhibit antigen. (Dashed arrows indicate a negative effect; solid arrows indicate a positive effect.)

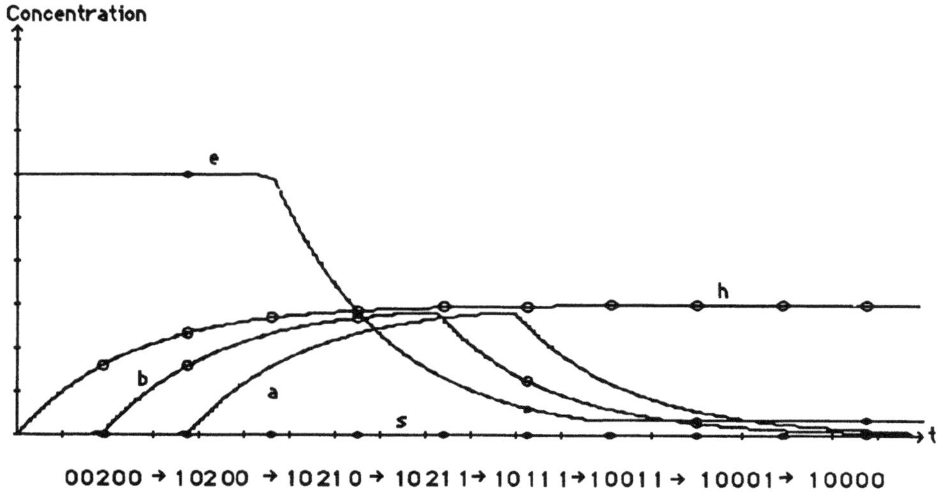

FIGURE 12 Medium-dose establishment of memory.

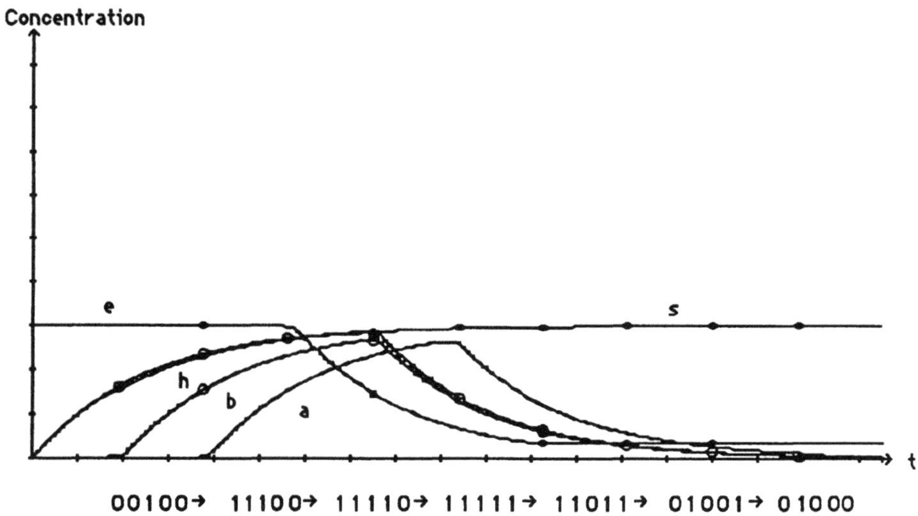

FIGURE 13 Low-dose establishment of low zone tolerance.

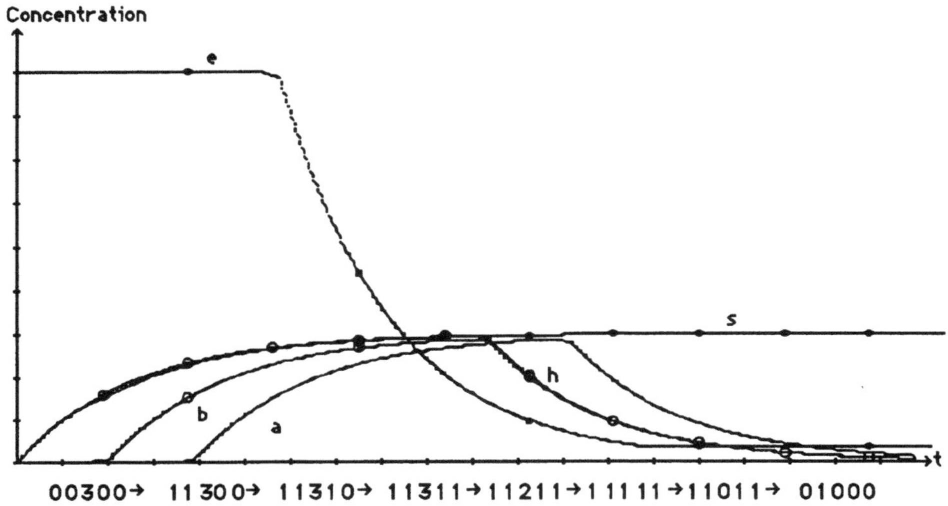

FIGURE 14 High-dose establishment of high zone tolerance.

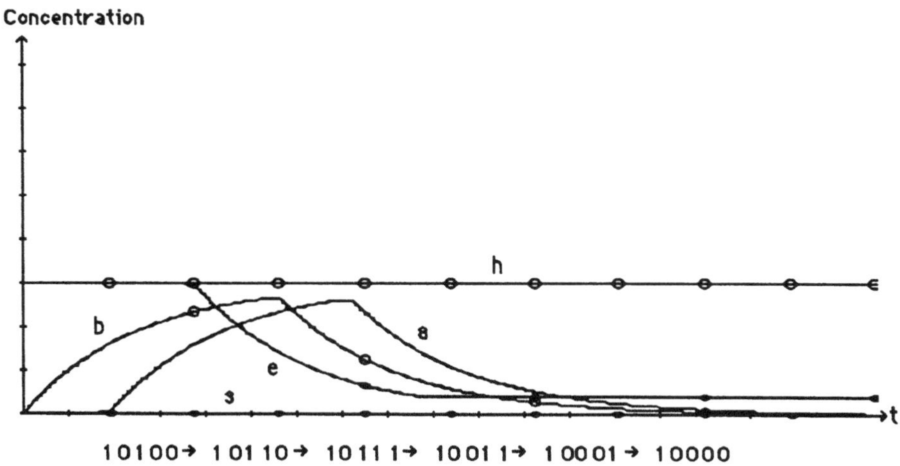

FIGURE 15 Low-dose perturbation of memory.

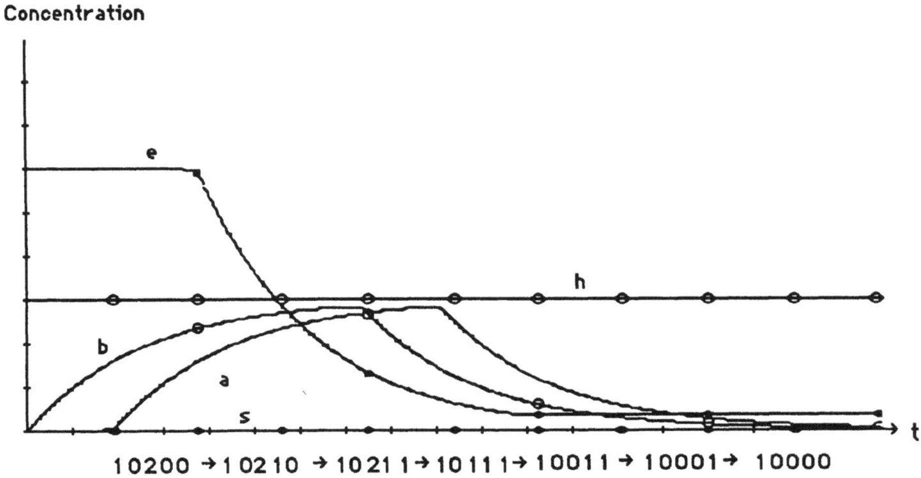

FIGURE 16 Medium-dose perturbation of memory.

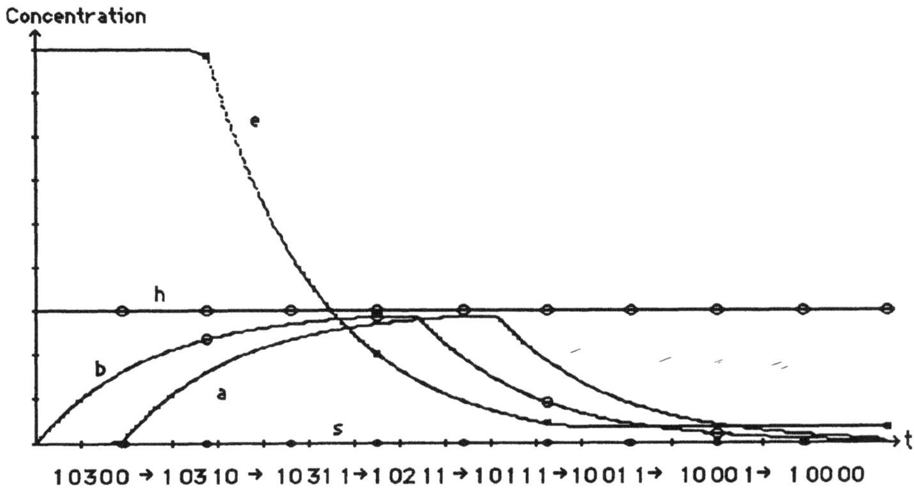

FIGURE 17 High-dose perturbation of memory.

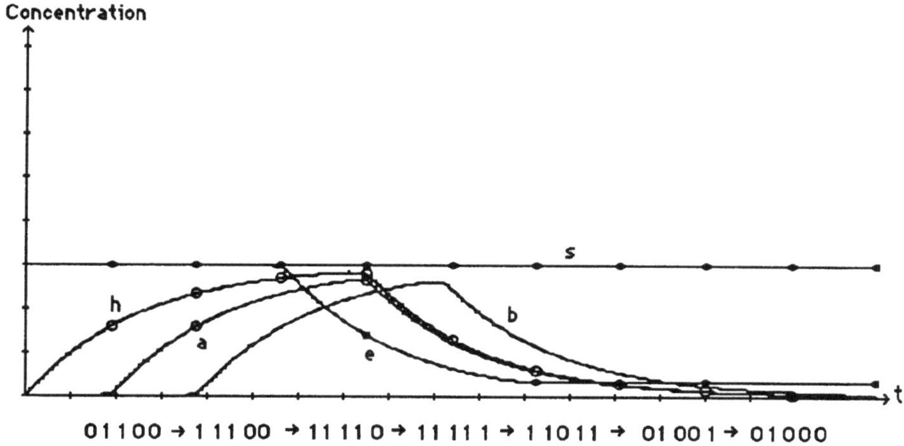

FIGURE 18 Low-dose perturbation of tolerance.

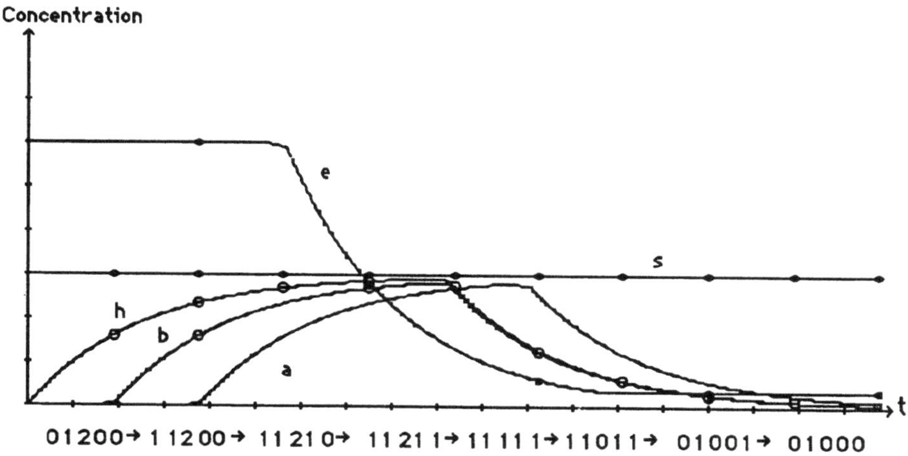

FIGURE 19 Medium-dose perturbation of tolerance.

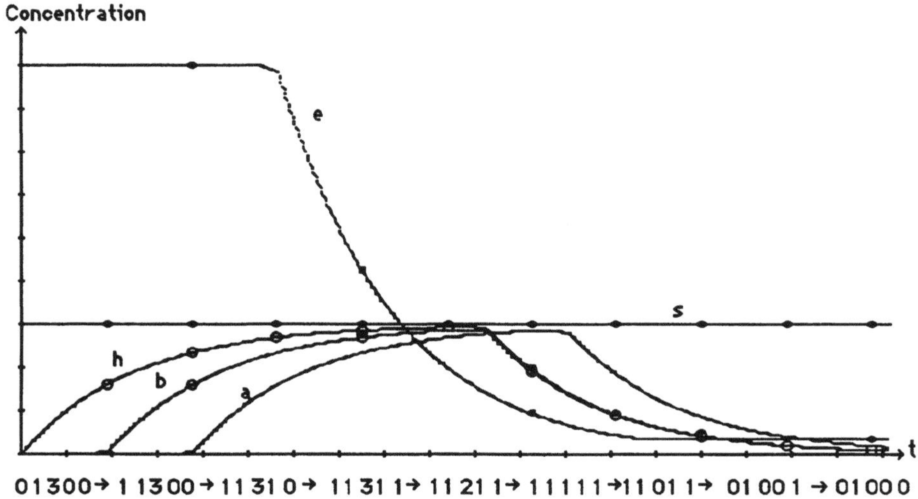

FIGURE 20 High-dose perturbation of tolerance.

establishment of memory is basically the same in the expanded model as in the minimal model except that there is a time delay in antigen destruction owing to the addition of two more steps in the process (Figure 11). This observation about the boolean sequences applies to each of the nine immune responses. Considering that the expanded model is an extension of the minimal model and that it gives essentially the same results, it is not necessary to discuss each of the figures, 12–20, individually.

8. DISCUSSION

A. INTERPRETATION OF THE SYSTEM OF DIFFERENTIAL EQUATIONS

Since the system of differential Eqs. (23)–(25), was derived from immune response patterns depicted in Figure 1, each term appearing in this system must relate to these response patterns.

We consider the helper Eq. (23). The term $\mathcal{S}_n(e)$ represents the establishment of helpers in response to the presence of antigen. In fact, if n is sufficiently large and $e > .5$, then $\mathcal{S}_n(e) \sim 1$ while $(1 - \mathcal{S}_n(e)) \sim 0$. Thus, $dh/dt \sim 1 - h$. Hence, $h(t)$ tends exponentially to a high level near unity. The term $(1-\mathcal{S}_n(e))\mathcal{S}_n(\overline{s})\mathcal{S}_n(h)$ represents perseverance of helper in the absence of both antigen and suppressor; this is necessary for the establishment of memory. Observe that if $e \sim 0$, $s \sim 0$ and $h \sim 1$, then $dh/dt \sim 0$ so the high level of helper is maintained. The remaining term, $-h$, effects exponential decay or exponential growth. Since $dh/dt = F_1 - e$, h

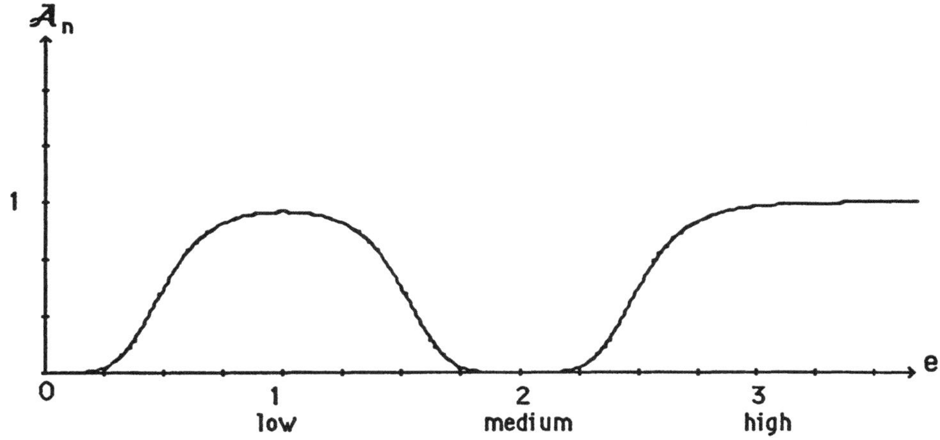

FIGURE 21 The sensitivity of suppressor to antigen (at $n = 5$).

decays exponentially to zero in a region where $F_1 \sim 0$ (this occurs when s is high and e is at its lowest level) and h grows exponentially when $F_1 \sim 1$ (as already noted above).

Next we consider the suppressor equation, (24). The term $\mathcal{S}_n(s)$ represents the perseverance of suppressor once this population is established so as to maintain tolerance. The term $\mathcal{A}_n(e)\mathcal{S}_n(\overline{s})\mathcal{S}_n(\overline{h})$, where

$$\mathcal{A}_n(e) = \mathcal{S}_n(e)\mathcal{S}_n(2-e) + \mathcal{S}_n(e-2), \tag{30}$$

represents the establishment of suppressor in the presence of antigen. This occurs only when both helper and suppressor are at low levels. The function $\mathcal{A}_n$, which represents the sensitivity of suppressor to antigen, expresses the biphasic nature of the immune response phenomenon since $\mathcal{A}_n(e)$ peaks at low and high levels of e while falling to a minimum at the medium antigenic level (see Figure 21).

The graph of $\mathcal{A}_n$ might be compared with the classical description of immune response, Figure 22 (e.g., see p. 195 in Bellanti, 1985). Observe the inverse relationship between Figures 21 and 22.

The antigen equation (27) is easy to interpretate. At high antibody levels, $de/dt \sim -e$, so e decays exponentially, while at low antibody levels $de/dt \sim 0$ so that the antigenic level is maintained. Observe that in Figures 12–20, the antigenic level is maintained once "a" decreases to a low level; consequently, a residual amount of antigen persists. This theoretical result is consistant with observation made by Tew & Phipps (1982). The remaining equations (28)–(29) are self-evident.

B. SYSTEMIC PROCEDURE FOR MODEL DEVELOPMENT

Our model was developed in a systematic fashion and reflects a descriptive paradigm of general immune response phenomena. We believe that this systematic procedure, in and of itself, is a significant step forward in the continuing search for a theoretical basis underlying immunological patterns. Therefore, it is important to analyze this procedure in a general setting which we will do in a forthcoming paper. However, some aspects of the procedure of a technical nature will be discussed below.

We consider first the choice of the parameter δ which controls the size of the ambiguous region. One might ask why there should be an ambiguous region in the first place. The ambiguous region is a consequence of the bi-level concept (see Introduction). Given the hypothesis that the immune system recognizes certain cell types and molecules at distinct levels of concentration, then these levels must be adequately separated; for, otherwise, the system would be unstable. That is, whenever a population in question nears the thresholds separating distinct levels, then, if these thresholds are too close, a small perturbation might cause a sharp switch to a different recognition of levels; thereby resulting in an entirely different response. This instability, due to a too small level separation, is detectable in the model as well. For instance, if the process representing low zone tolerance is initiated

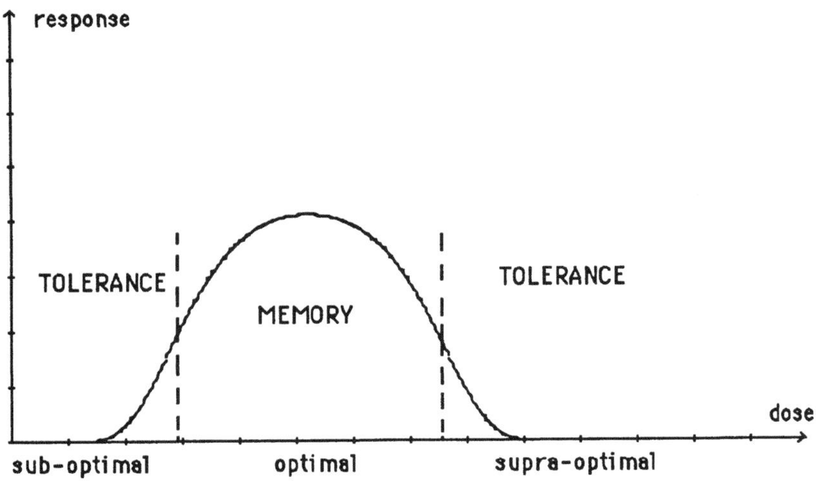

FIGURE 22 Qualitative description of immune response phenomena. Immune response is strong at optimal antigenic levels and weak at sub-optimal and supra-optimal levels.

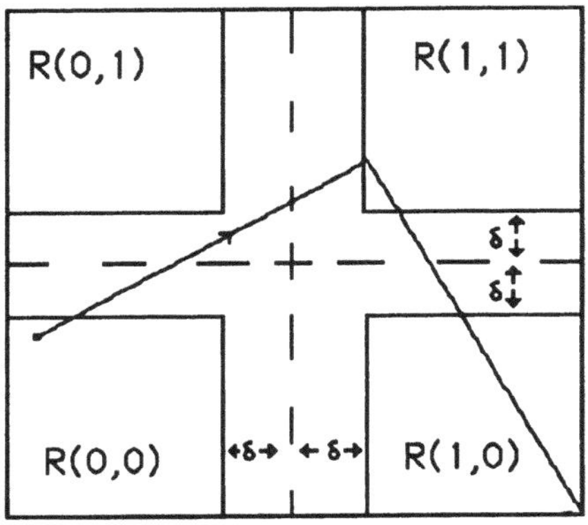

FIGURE 23 Ambiguous region. The discrete sequence $(00) \to (11) \to (10)$ corresponds to a continuous trajectory traversing $R(0,0) \to R(1,1) \to R(1,0)$. Upon leaving $R(0,0)$, the trajectory enters the ambiguous region. It is programmed to retain its course until it enters another unambiguous region which is, if δ is sufficiently large, the region $R(1,1)$. However, if δ is too small, then the trajectory enters $R(0,1)$ and its course may be altered.

with h slightly positive instead of h equal to zero; then, if δ is chosen too small, the system will evolve to the memory state rather than to tolerance. By giving h a "head start," h will be established while s is still low. The system then recognizes a level state which leads to memory. As illustration of the ambiguous region is shown in Figure 23.

Another reason why the ambiguous region is important is that the continuous system will, in general, have unstable steady states. These unstable steady states have no counterpart in the discrete system and they can alter a trajectory from the path that was programmed into the discrete system. However, these unstable steady states always lie in the ambiguous region. The trajectory, being "blindfolded" in the ambiguous region, does not "see" the unstable steady states; it passes through them or near them unhampered.

It can be shown that the choice $\delta > (\sqrt{5} - 2)/2 \simeq 0.12$ insures the proper dynamical behavior patterns no matter how h, s and e are initiated, as long as the initial values are in the appropriate ranges for the level state in question, provided n is sufficiently large. Besides guaranteeing stability, the larger δ values have the advantage of making the extrema in the curves $h(t)$ and $s(t)$ more pronounced. For instance, if $\delta = 0.1$, then a low h can be as high as 0.4 and a high h can be as low as 0.6, while if (to take as extreme case) $\delta = 0.4$, then h is not recognized as low until its value is below 0.1 and it is not recognized as high until its value exceeds 0.9.

The replacement of the step function $\mathcal{S}$ by a smooth function $\mathcal{S}_n$ was introduced for several reasons. Smooth functions are preferable in general and the Hill functions $\mathcal{S}_n$ have been used extensively in modeling biochemical phenomena. Technically,

$\mathcal{S}_n(x)$ converges uniformly to $\mathcal{S}(x)$ in the range $\mid x - 0.5 \mid > \delta$ so that, if n is chosen sufficiently large (with this choice depending upon δ), then $\mathcal{S}_n(x)$ is a very close approximation to $\mathcal{S}(x)$. For instance, if $\delta > 0.12$ and $n > 22$, then the graphs of $\mathcal{S}$ and $\mathcal{S}_n$ are within .01 of each other; too close to be distinguished if the graphs are drawn on standard size paper. From the biochemical point of view, a step function represents a phenomena in which all members of a given population react spontaneously in essentially the same fashion resulting in a sharp change while a smooth function represents an accumulative average of events occurring at different times. Therefore, it is interesting to experiment with n, as a model parameter, to see how the value of n alters the dynamics of the model. However, technical difficulties may arise by choosing n too low. For instance, the high steady-state value of either h or s is the positive root, x_n, of the polynomial $P_n(x) = (2x)^{n-1} - (2x)^{n-2} - \ldots - 2x - 1$. For large n, $x_n \sim 1 - 2^{-n}$. As h or s converges to x_n, its value will be below x_n in the high-level range. Since $x_2 = 0.5$, $n = 2$ is ruled out since the high range begins above 0.5. Since $x_3 = (1+\sqrt{5})/4$, δ must be chosen smaller than $(1+\sqrt{5})/4 - 0.5 \sim 0.309$ to allow $n = 3$ to be feasible. In general, x_n must be in the high range, i.e., $x_n > 0.5 + \delta$. Since x_n increases monotonically to unity as $n \rightarrow \infty$, the larger n the greater the flexibility for choosing δ.

There is a curious interplay between δ and n. The effects of varying these model parameters will be studied in a forthcoming paper.

C. THE BOOLEAN APPROACH

The qualitative treatment presented here is based on the boolean approach made in Gelfand & Walker (1984) and Thomas (1979), which has been applied extensively in biology beginning with the early works of Rashevsky (1960), Sugita (1963) and Kauffman (1969). Our model-building procedure employs or modifies the works of several authors. The notion of representing the four-level antigen variable as a binary boolean vector was suggested by Van Ham (1979), as was the use of the parameter δ to distinquish ambiguous regions. A qualitative agreement between the boolean and the continuous models has been observed with the use of piecewise linear functions (Glass and Pasternack,1978) and with the use of steep sigmoids (Glass and Kauffman,1973). The algorithm for generating the next-step functions given in Gelfand & Walker (1984) was modified so as to yield coefficients suitable for extension to a continuous homologous model. The algorithm presented by Gelfand & Walker obtains the next-step functions in binary arithmetic $(1+1=0)$. The binary formalism is not conducive when the variables are allowed to vary continuously. The only departure from the Gelfand-Walker algorithm is the computation of the matrix $G_n{}^{-1}$. In binary, $G_n{}^{-1} = G_n$, while in real arithmetic $G_n{}^{-1}$ is determined by the recursion formula given in Section 4.

D. COMPARISON WITH KAUFMAN ET AL.

The approach used by Kaufman et al. (1985) may seem similar to ours, but there are notable differences. Both use the boolean approach; however, Kaufman et al. began by putting together feedback loops to postulate a network model. They then test their model by considering whether or not the model-generated boolean sequences are consistent with certain biological constraints. Our approach reverses the order of these steps. We begin with a set of biological constraints to obtain a set of boolean sequences (e.g., Table 2). Then we employ the modified Gelfand-Walker algorithm to obtain the next-step functions.

The first step, conversion from boolean to continuous, is also present in Kaufman et al. (1985), but their procedure is not completely formalistic. Some terms in their differential equations are unexplainably non-sigmoidal. Their motivation, different from ours, seems to be to test the range of nonlinearity, rather than to present a systematic procedure which is guaranteed to work for all models.

The models that we obtained differ from the Kaufman models. As noted earlier, the Kaufman models always lead to the memory state.

ACKNOWLEDGMENT

The authors wish to thank Dr. Charles DeLisi for his helpful discussions and for his calling their attention to instrumental reference material and to Dr. Leon Glass for proofreading the paper and making suggestions.

REFERENCES

1. Bellanti, J. A. (1985), *Immunology: Basic Processes*, Philadelphia: Saunders.
2. Eisenfeld, J. (1987), "Compatibility Tests on Immunological Control Loops," *Ann. N.Y. Acad. Sci.* **504**, 132–150.
3. Eisenfeld, J., and C. DeLisi (1985), "On Conditions for Qualitative Instability of Regulatory Circuits and Application to Immunological Control Loops," *Mathematics and Computers in Biomedical Applications*, Eds. J. Eisenfeld and C. DeLisi (New York: Elsevier), 39–53.
4. Gelfand, A. E., and C. C. Walker (1984), *Ensemble Modeling* (New York: Marcel Dekker).
5. Glass, L. (1975), "Combinatorial and Topological Methods in Nonlinear Chemical Kinetics," *J. Chem. Phys.* **63**, 1325–1335.
6. Glass, L., and S. A. Kauffman (1973), "The Logical Analysis of Continuous Nonlinear Biochemical Control Networks," *J. Theor. Biol.* **39**, 103–129.
7. Glass, L., and J. S. Pasternack (1978), "Prediction of Limit Cycles in Mathematical Models of Biological Oscillations," *Bull. Math. Biol.* **40**, 27.
8. Golub, E. S. (1981), *The Cellular Basis of the Immune Response* (Sunderand, Mass.: Sinauer Associates).
9. Herzenberg, L. A., S. J. Black, and L. A. Herzenberg, (1980), "Regulatory Circuits and Antibody Responses," *Eur. J. Immunol.* **10**, 1–11
10. Hoffmann, G. W. (1980), "On Network Theory and H-2 Restriction," *Contemporary Topics in Immunology*, Ed. N.L. Warner, vol.II, 185–226.
11. Kauffman, S. A. (1969), "Metabolic Stability and Epigenesis in Randomly Constructed Genetic Nets," *J. Theor. Biol.* **114**, 527–561.
12. Kaufman, M., J. Urbain, and R. Thomas (1985), *J. Theor. Biol.* **114**, 527–561.
13. May, R. M. (1973), *Stability and Complexity in Ecosystems* (Princeton, NJ: Princeton University Press).
14. Mitchison, N. A. (1964), "Introduction of Immunological Paralysis in Two Zones of Dosage," *Proc. Roy. Soc. Ser. B.* **161**, 275–292.
15. Rashevsky, N. (1960), *Mathematical Biophysics*, Vol. 2, New York: Dover.
16. Sugita, M. (1963), "Functional Analysis of Chemical Systems in Vivo Using a Logical Circuit Equivalent, II. The Idea of Molecular Automation," *J. Theor. Biol.* **4**, 179–192.
17. Tew, J. G., and P. Phipps (1982), "Cyclic Antibody Production: Role of Antigen Retaining Follicular Denritic Cells and Antibody Feedback Regulation," *Regulation of Immune Response Dynamics*, Eds. C. DeLisi and J. R. J. Hiernaux (Boca Raton, Florida: CRC Press), Vol. I, 28–41.
18. Thomas, R. (1979), *Kinetic Logic a Boolean Approach to the Analysis of Complex Regulatory Systems [Lecture Notes Biomath, 29]* (New York: Springer-Verlag).

19. Van Ham, P. (1979), "How to Deal with More Than Two Levels," *Kinetic Logic a Boolean Approach to the Analysis of Complex Regulatory Systems* [Lecture Notes Biomath., 29], Ed. R. Thomas (New York: Springer-Verlag), 326–343.

R. B. KING
Department of Chemistry, University of Georgia, Athens, Georgia 30602, U.S.A.

The Flow Topology of the Herzenberg Immunological Control Networks

ABSTRACT

Immunological control networks can be modeled by bipartite, signed, directed graphs called influence diagrams. Possible flow topologies around unstable steady states in such systems can be determined by kinetic logic based on the directed Boolean cubes of switching circuit theory. An example of an immunological control network is the Herzenberg core regulatory circuit (CRC) which consists of a single positive circuit of length 4 with alternating positive and negative edges. The flow topology of this CRC indicates locking into one of two stable configurations corresponding to "help" and "suppression." Addition of a single vertex to the Herzenberg CRC so as to preserve the bipartite nature of the signed, directed graph leads to a network with five independent internal variables and two connected feedback circuits, either positive and negative circuits of length 4 or positive circuits of lengths 4 and 2. The flow topology of either of these systems indicates that addition of the fifth vertex switches the basis four-vertex CRC into one of its two stable configurations.

1. INTRODUCTION

The general objective of this paper is the examination of immunological control networks using topological and graph theoretical methods similar to those previously used for chemical reaction networks.[1,2,3] Extension of these methods to immunological control networks is not trivial since their general patterns are significantly different from those of chemical reaction networks. The scope of this paper is limited to the important topological features of the reported immunological control networks with particular emphasis on the models of Herzenberg, Black, and Herzenberg.[4] Discussion of experimental data leading to the postulation of specific immunological control networks is thus considered beyond the scope of this paper.

Two research groups have recently published work closely related to the work discussed in this paper (see Kaufman, this volume, and Eisenfeld and Prueitt, this volume). Kaufman, Urbain, and Thomas[5] have used a similar, but not identical, flow topological analysis based on the ideas of kinetic logic[6] to determine the steady states of a model of an immunological control network shown in the present paper not to contain all of the important features of the Herzenberg models. Eisenfeld and DeLisi[7,8] have analyzed the qualitative stability of immunological control networks using network diagrams to depict essential topological relationships. This paper extends the approach of Eisenfeld and DeLisi by considering the flow topologies rather than the stabilities of immunological control networks and treats the Herzenberg models rather than the models of Kaufman, Urbain, and Thomas.[5] Such topological models of immunological control networks provide guidance for much more complicated continuous analyses using specific systems of differential equations.

2. BACKGROUND

An immunological control network can be represented by a bipartite, signed, directed graph. The vertices of this graph represent distinguishable subsets of lymphocytes. The lymphocytes encountered in the models in this paper are helper and suppressor thymus-derived (T) lymphocyte (Th and Ts, respectively) and the so-called B lymphocytes. Each lymphocyte has an associated idiotype which may be positive or negative corresponding to a coloring of the vertices of the graph in one of two colors. In the graphs used in this paper, vertices associated with positive and negative idiotypes are represented by squares or rectangles and circles or ellipses, respectively.

The directed edges of the immunological control network represent interactions between lymphocytes in different subsets based on complementary idiotype/anti-idiotype interactions.[9] Thus, edges can be directed only from a positive to a negative vertex or from a negative to a positive vertex. Edges between pairs of vertices of the same sign (i.e., positive to positive or negative to negative) are not

FIGURE 1 A strong vertex.

FIGURE 2 A weak vertex can either be (a) a sink or (b) a source.

possible. An immunological control network, thus, can be represented by a bipartite graph. The directed edges are given positive or negative signs depending upon whether they represent help or suppression, respectively.

The vertices of an immunological control network can be classified as either strong or weak vertices. A *strong vertex* has at least one edge directed towards it and at least one edge directed away from it, e.g., Figure 1. A *weak vertex* either has only edges directed towards it (a *sink*, Figure 2a) or only edges directed away from it (a *source*, Figure 2b). Strong vertices represent internal variables and weak vertices represent external variables. A strong vertex with two or more edges directed *towards* it is called a *turbulent vertex.*

The relationships between the internal variables are of interest in the treatment of immunological control networks in this paper as well as in the previous work on chemical reaction networks.[1,2,3,] Such relationships are represented by signed, directed graphs known as *influence diagrams.* An influence diagram is obtained from the corresponding immunological control network by deleting all of the weak vertices and any edges connected to weak vertices until only strong vertices remain. Such deleted weak vertices represent external variables such as antigens (sources, e.g., Figure 2b) or antibodies (sinks, e.g., Figure 2a). Conversion of an immunological control network into the corresponding influence diagram can require several stages if removal of a weak vertex and its associated edges converts a strong vertex into a weak vertex.

In the simplest approach, loops representing self-interactions are deleted when converting the immunological control network to the corresponding influence diagram. More accurate, but much more complicated models can be obtained by replacing such self-interaction loops by a positive circuit between two complementary idiotypes as suggested by Kaufman, Urbain, and Thomas.[5] However, this approach adds one variable and at least one circuit to the influence diagram for each loop

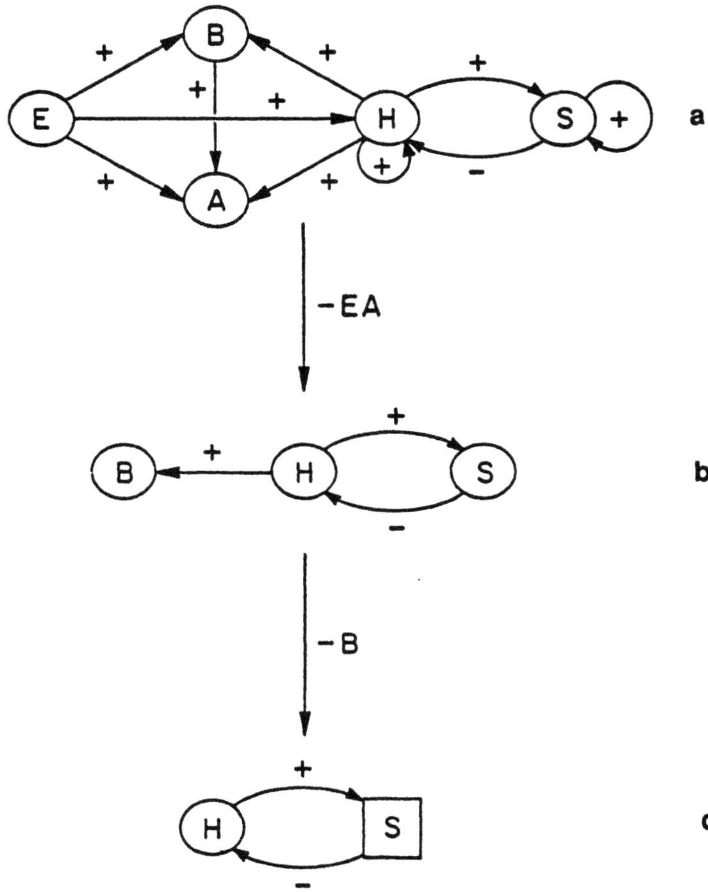

FIGURE 3 The immunological control network of Kaufman, Urbain, and Thomas. (1) the original five-vertex, signed, directed graph; (b) removal of the weak vertices E and A and the associated edges $E \to A$, $E \to H$, $E \to B$, $B \to A$, $H \to A$ as well as the loops at H and S to give a three-vertex, signed, directed graph; (c) removal of the weak vertex B and the edge $H \to B$ from the three-vertex, signed, directed graph to give a (bipartite) two-vertex negative circuit; the vertices H and S correspond to helper and suppressor T cells, respectively.

deleted, thereby leading to a considerably more complicated analysis of the corresponding flow topology. For this reason, this latter approach will not be considered further in this paper.

This procedure is illustrated by the conversion of the immunological control network studied by Kaufman, Urbain, and Thomas[5] to the corresponding influence diagram (Figure 3). In the original network (Figure 3a), the vertices A and E are weak vertices and B, H, and S are strong vertices. Removal of the vertices A and E and the associated five edges $E \rightarrow B$, $E \rightarrow H$, $E \rightarrow A$, $B \rightarrow A$, and $H \rightarrow A$ as well as the loops at vertices H and S leads to a network with the three vertices B, H, and S (Figure 3b). In this stage, B becomes a weak vertex because of the removal of the only edge directed to B, namely $E \rightarrow B$. Thus, a second stage involving the removal of the vertex B and the associated edge $H \rightarrow B$ is necessary in order to obtain the two-vertex influence diagram (Figure 3c) corresponding to the original five-vertex immunological control network of Kaufman, Urbain, and Thomas (Figure 3a). In this case, the influence diagram has far less structure than the original immunological control network indicating the relatively minor role played by the internal variables in this model and the significant effect of the loops at vertices H and S.

The circuits in the influence diagrams are important in the treatment of this paper. A *circuit* in an influence diagram consists of a path which starts with a given vertex and follows various edges in the directions of the arrows until the original vertex is reached again. The *length* of a circuit is the number of edges in the circuit that must be traversed from a given vertex until that same vertex is reached again. A circuit is *negative* if it has an odd number of negative edges and *positive* if it has an even number of negative edges or no negative edges. Following previous practice,[1,2,3] positive and negative circuits of length n will be called B_n and C_n circuits, respectively.

Now, consider the qualitative dynamics of a system depicted by a given influence diagram as represented by the flow topology around an unstable equilibrium point. Possible flow topologies include one or more stable steady states as well as oscillations or chaos. The earlier work on chemical reaction networks focused on networks exhibiting periodic oscillations[1] or aperiodic chaos.[3] The current work on immunological control networks, on the other hand, is concerned with networks proceeding from an initial virgin state by exposure to antigen to give one or more steady states corresponding, for example, to suppression or help configurations.[4] The flow topologies exhibited by some fundamental types of influence diagrams are discussed in a previous paper.[2]

The procedure for determining the flow topology corresponding to a given influence diagram uses methods based on switching circuit theory[10,11] as adapted by Glass and collaborators[12–17] and which have been termed *kinetic logic*.[6] In the logical description used in this approach, time may be involved either synchronously[12–14] or asynchronously[5,18] Since the special features of the more complicated asynchronous version (e.g., the effects of time delays) are not necessary for this work, the simpler synchronous version will be used throughout this paper.

Consider the immunological control network represented by an influence diagram as a synchronous switching network in which time is quantized so that the signs of the first time derivatives of the internal variables at time $t + 1$ are determined by their signs at time t.[13] The switching state at any time of such a network containing n internal variables can be represented by an n-dimensional vector of 1's and 0's corresponding to positive and negative time derivatives, respectively, of the n internal variables. Such an n-vector is called a *state vector*. The number of different possible state vectors in an n-dimensional system is 2^n and these state vectors can be represented by the 2^n vertices of an n-dimensional cube. The possible transitions from states at synchronous time t to those at time $t + 1$ may then be represented by arrows directed along the edges of the n-dimensional cube. In such transitions, the value of exactly one component of the state vector changes. Furthermore, in this synchronous treatment, the discrete time scale t is chosen so that it advances one unit each time a single component of the state vector changes. The resulting n-dimensional cube with directed edges is called a *state transition diagram*. The center of the n-dimensional cube representing a state transition diagram corresponds to a steady state in which all of the first time derivatives of the internal variables are zero and, therefore, do not correspond to either Boolean variable 0 or 1. If this steady state is unstable, the transitions represented by the directed edges of the state transition diagram define the fundamental topology of the flow in the neighborhood of the unstable steady state. If the steady state is stable, the state transition diagram is not relevant to the qualitative dynamics of the system since it is not possible to leave the stable steady state to exhibit the flow topology indicated by the state transition diagram.

The calculation of a state transition diagram corresponding to a given turbulent influence diagram can be performed by the following five-step procedure:

1. A logical relation (OR or AND) is assigned between each pair of edges directed towards a given turbulent vertex. An OR relationship between arrows from vertices X and Y both going to a turbulent vertex Z implies that the internal variables X and Y affect the level of Z in separate processes corresponding to an equation of the general type:

$$dZ/dt = aX^r + bY^s. \tag{1}$$

   An AND relationship between arrows from vertices X and Y both going to a turbulent vertex Z implies that the internal variables X and Y affect the level of Z in the same process corresponding to an equation of the type:

$$dZ/dt = aX^r Y^s. \tag{2}$$

   The OR relationship at turbulent vertices appears to be more realistic for influence diagrams based on immunological control networks and is used in this paper. In the case of influence diagrams modelling chemical reaction networks, either OR or AND relationships at turbulent vertices may be realistic depending upon the nature of the chemical system.
2. A *local truth table* is generated for each circuit in the influence diagram to indicate possible transitions between state vectors. In order to see how such a local truth table is generated, consider possible effects of one internal variable X on a second internal variable Y. If X helps Y as indicated by a positive arrow from X to Y in the circuit under consideration, then in the local truth

table the values for Y in each possible state vector at time $t+1$ will correspond to the values for X at time t. If X suppresses Y as indicated by a negative arrow from X to Y in the circuit under consideration, then in the local truth table the values for Y in each possible state vector at time $t+1$ will be the opposite of the values for X at time t (i.e., a 0 for X at time t leads to a 1 for Y at time $t+1$ and vice versa).

3. The local truth tables for each of the circuits in the influence diagram are combined to give a *global truth table* using the logical relationships at the turbulent vertices as follows:

   a. If any of the local truth tables show a value of 0 for a variable represented by an AND turbulent vertex, then the global truth table for that variable also shows a value of 0.

   b. If any of the local truth tables show a value of 1 for a variable represented by an OR turbulent vertex, then the global truth table for that variable also shows a value of 1.

4. Each state vector in the *global* truth table at time $t+1$ is compared with the corresponding state vector at time t. If any such pair of state vectors are the same, then that state vector corresponds to a stable configuration of the immunological control network modeled by the influence diagram.
5. If no stable configurations are found, then possibilities for oscillatory and/or chaotic behavior are explored by using the global truth table to determine the corresponding state transition diagram as discussed in a previous paper.[3] Since this situation does not arise in the immunological control networks discussed in this paper, the relevant algorithm will not be repeated here.

3. THE HERZENBERG MODELS

The Herzenberg models for immunological control networks[4] use five independent variables: B cells, two types of helper T cells ($Th1$ and $Th2$), and two types of suppressor T cells ($Ts1$ and $Ts2$). Helper Th cells carry anti-idiotypic V_H structures and, thus, correspond to negative vertices in the bipartite immunological control networks (ellipses in Figure 4). Suppressor Ts cells and B cells carry idiotypic V_H structures which are complementary to the anti-idiotypic V_H structures and, thus, correspond to positive vertices in the bipartite immunological control networks (rectangles and squares in Figure 4). The soluble products of the Herzenberg models[4] are viewed as dependent rather than independent variables with their levels being so tightly coupled to B cells that they do not need to be considered as

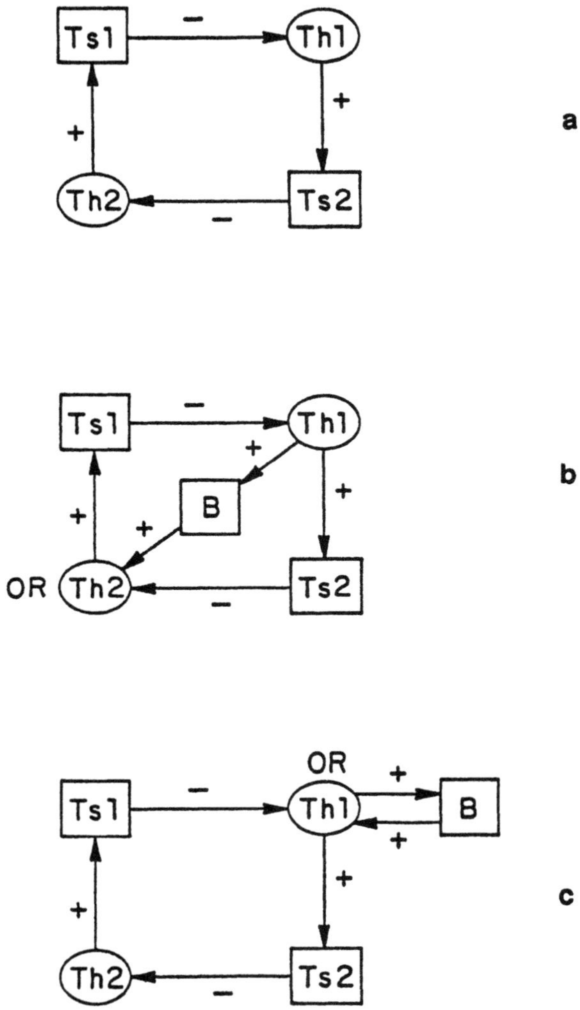

FIGURE 4 Influence diagrams corresponding to the Herzenberg models. (a) the four-vertex CRC; (b) addition of a fifth vertex (B) to give the $B_4 + C_4$ system stabilizing the "suppression" configuration; (c) addition of a fifth vertex (B) to give the $B_4 + B_2$ system stabilizing the "help" configuration.

separate variables. In this way, the number of variables in the Herzenberg models considered in this paper can be limited to five.

The Herzenberg models consist of a bistable core regulatory circuit (CRC) which can be maintained stably in either the help or suppression mode. One or more auxiliary regulatory circuits (ARC's) are linked to this CRC and can switch the CRC between the help and suppression modes.

The Herzenberg CRC is depicted in Figure 4a. It represents the simplest way of linking helper and suppressor *T* cells to satisfy the following conditions:

1. Anti-idiotypic cells can only interact with idiotypic cells or vice versa so that only bipartite graphs are allowed, thereby excluding graphs containing circuits of odd lengths.
2. The *Th* and *Ts* cells are differentiated so that the populations of target *Th* and helper *Th* for a given *Ts* are drawn from different *Th* populations.

The Herzenberg CRC (Figure 4a) is, thus, seen to consist of a single positive circuit of length 4, called a fundamental quadrilateral.[2] By a theorem in a previous paper (Theorem 8, in ref. 2), a positive fundamental quadrilateral must have two attracting regions or stable configurations exactly as suggested in the Herzenberg analysis.

TABLE 1 Truth Table for the Herzenberg CRC

Time t				Time $t+1$				
Ts1	Th1	Ts2	Th2	Ts1	Th1	Ts2	Th2	
0	0	0	0	0	1	0	1	
0	0	0	1	1	1	0	1	
0	0	1	0	0	1	0	0	
0	0	1	1	1	1	0	0	
0	1	0	0	0	1	1	1	
0	1	0	1	1	1	1	1	
0	1	1	0	0	1	1	0	stable configuration ("help")
0	1	1	1	1	1	1	0	
1	0	0	0	0	0	0	1	
1	0	0	1	1	0	0	1	stable configuration ("suppression")
1	0	1	0	0	0	0	0	
1	0	1	1	1	0	0	0	
1	1	0	0	0	0	1	1	
1	1	0	1	1	0	1	1	
1	1	1	0	0	0	1	0	
1	1	1	1	1	0	1	0	

TABLE 2 Truth Table for a Herzenberg Model with An Auxiliary Circuit Establishing the Suppression Configuration

Time t					Local (B_4) Time $t+1$					Local (C_4) Time $t+1$					Global Time $t+1$				
Ts1	Th1	Ts2	Th2	B	Ts1	Th1	Ts2	Th2	B	Ts1	Th1	Ts2	Th2	B	Ts1	Th1	Ts2	Th2	B
0	0	0	0	0	0	1	0	1	-	0	1	-	0	0	0	1	0	1	0
0	0	0	0	1	0	1	0	1	-	0	1	-	1	0	0	1	0	1	0
0	0	0	1	0	1	1	0	1	-	1	1	-	0	0	1	1	0	1	0
0	0	0	1	1	1	1	0	1	-	1	1	-	1	0	1	1	0	1	0
0	0	1	0	0	0	1	0	0	-	0	1	-	0	0	0	1	0	0	0
0	0	1	0	1	0	1	0	0	-	0	1	-	1	0	0	1	0	1	0
0	0	1	1	0	1	1	0	0	-	1	1	-	0	0	1	1	0	0	0
0	0	1	1	1	1	1	0	0	-	1	1	-	1	0	1	1	0	1	0
0	1	0	0	0	0	1	1	1	-	0	1	-	0	1	0	1	1	1	1
0	1	0	0	1	0	1	1	1	-	0	1	-	1	1	0	1	1	1	1
0	1	0	1	0	1	1	1	1	-	1	1	-	0	1	1	1	1	1	1
0	1	0	1	1	1	1	1	1	-	1	1	-	1	1	1	1	1	1	1
0	1	1	0	0	0	1	1	0	-	0	1	-	0	1	0	1	1	0	1
0	1	1	0	1	0	1	1	0	-	0	1	-	1	1	0	1	1	1	1
0	1	1	1	0	1	1	1	0	-	1	1	-	0	1	1	1	1	0	1
0	1	1	1	1	1	1	1	0	-	1	1	-	1	1	1	1	1	1	1
1	0	0	0	0	0	0	0	1	-	0	0	-	0	0	0	0	0	1	0
1	0	0	0	1	0	0	0	1	-	0	0	-	1	0	0	0	0	1	0
1	0	0	1	0	1	0	0	1	-	1	0	-	0	0	1	0	0	1	0[1]
1	0	0	1	1	1	0	0	1	-	1	0	-	1	0	1	0	0	1	0
1	0	1	0	0	0	0	0	0	-	0	0	-	0	0	0	0	0	0	0
1	0	1	0	1	0	0	0	0	-	0	0	-	1	0	0	0	0	1	0
1	0	1	1	0	1	0	0	0	-	1	0	-	0	0	1	0	0	0	0
1	0	1	1	1	1	0	0	0	-	1	0	-	1	0	1	0	0	1	0

1	1	0	0	0	0	0	1	1	-	0	0	-	0	1	0	0	1	1	1
1	1	0	0	1	0	0	1	1	-	0	0	-	1	1	0	0	1	1	1
1	1	0	1	0	1	0	1	1	-	1	0	-	0	1	1	0	1	1	1
1	1	0	1	1	1	0	1	1	-	1	0	-	1	1	1	0	1	1	1
1	1	1	0	0	0	0	1	0	-	0	0	-	0	1	0	0	1	0	1
1	1	1	0	1	0	0	1	0	-	0	0	-	1	1	0	0	1	1	1
1	1	1	1	0	1	0	1	0	-	1	0	-	0	1	1	0	1	0	1
1	1	1	1	1	1	0	1	0	-	1	0	-	1	1	1	0	1	1	1

[1] stable configuration "suppression"

TABLE 3 Truth Table for a Herzenberg Model with An Auxiliary Circuit Establishing the Help Configuration

Time t					Local (B_4) Time $t+1$					Local (B_2) Time $t+1$					Global Time $t+1$				
Ts1	Th1	Ts2	Th2	B	Ts1	Th1	Ts2	Th2	B	Ts1	Th1	Ts2	Th2	B	Ts1	Th1	Ts2	Th2	B
0	0	0	0	0	0	1	0	1	-	-	0	-	-	0	0	1	0	1	0
0	0	0	0	1	0	1	0	1	-	-	1	-	-	0	0	1	0	1	0
0	0	0	1	0	1	1	0	1	-	-	0	-	-	0	1	1	0	1	0
0	0	0	1	1	1	1	0	1	-	-	1	-	-	0	1	1	0	1	0
0	0	1	0	0	0	1	0	0	-	-	0	-	-	0	0	1	0	0	0
0	0	1	0	1	0	1	0	0	-	-	1	-	-	0	0	1	0	0	0
0	0	1	1	0	1	1	0	0	-	-	0	-	-	0	1	1	0	0	0
0	0	1	1	1	1	1	0	0	-	-	1	-	-	0	1	1	0	0	0
0	1	0	0	0	0	1	1	1	-	-	0	-	-	1	0	1	1	1	1
0	1	0	0	1	0	1	1	1	-	-	1	-	-	1	0	1	1	1	1
0	1	0	1	0	1	1	1	1	-	-	0	-	-	1	1	1	1	1	1
0	1	0	1	1	1	1	1	1	-	-	1	-	-	1	1	1	1	1	1
0	1	1.	0	0	0	1	1	0	-	-	0	-	-	1	0	1	1	0	1
0	1	1	0	1	0	1	1	0	-	-	1	-	-	1	0	1	1	0	1[1]
0	1	1	1	0	1	1	1	0	-	-	0	-	-	1	1	1	1	0	1
0	1	1	1	1	1	1	1	0	-	-	1	-	-	1	1	1	1	0	1
1	0	0	0	0	0	0	0	1	-	-	0	-	-	0	0	0	0	1	0
1	0	0	0	1	0	0	0	1	-	-	1	-	-	0	0	1	0	1	0
1	0	0	1	0	1	0	0	1	-	-	0	-	-	0	1	0	0	1	0
1	0	0	1	1	1	0	0	1	-	-	1	-	-	0	1	1	0	1	0
1	0	1	0	0	0	0	0	0	-	-	0	-	-	0	0	0	0	0	0
1	0	1	0	1	0	0	0	0	-	-	1	-	-	0	0	1	0	0	0
1	0	1	1	0	1	0	0	0	-	-	0	-	-	0	1	0	0	0	0
1	0	1	1	1	1	0	0	0	-	-	1	-	-	0	1	1	0	0	0

1	1	0	0	0	0	0	1	1	-	-	0	-	-	1	0	0	1	1	1
1	1	0	0	1	0	0	1	1	-	-	1	-	-	1	0	1	1	1	1
1	1	0	1	0	1	0	1	1	-	-	0	-	-	1	1	0	1	1	1
1	1	0	1	1	1	0	1	1	-	-	1	-	-	1	1	1	1	1	1
1	1	1	0	0	0	0	1	0	-	-	0	-	-	1	0	0	1	0	1
1	1	1	0	1	0	0	1	0	-	-	1	-	-	1	0	1	1	0	1
1	1	1	1	0	1	0	1	0	-	-	0	-	-	1	1	0	1	0	1
1	1	1	1	1	1	0	1	0	-	-	1	-	-	1	1	1	1	0	1

[1] stable configuration "help"

In order to determine the nature of these two stable configurations, the truth table for the Herzenberg CRC is determined (Table 1). The stable "help" configuration corresponds to a ($Ts1$, $Th1$, $Ts2$, $Th2$) state vector of (0, 1, 1, 0) or $dTs1/dt < 0$, $dTh1/dt > 0$, $dTs2/dt > 0$, and $dTh2/dt < 0$. In this configuration $Th1$ increases ($dTh1/dt > 0$) and, thus, helps the differentiation of $Ts2$ (i.e., $dTs2/dt > 0$) which then depletes $Th2$ (i.e., $dTh2/dt < 0$), which, in turn, disables differentiation and expression of the $Ts1$ population (i.e., $dTs1/dt < 0$), which, if present and active, would be capable of attacking $Th1$. The stable "suppression" configuration corresponds to a ($Ts1$, $Th1$, $Ts2$, $Th2$) state vector of (1, 0, 0, 1) or $dTs1/dt > 0$, $dTh1/dt < 0$, $dTs2/dt < 0$, and $dTh2/dt > 0$ leading analogously to an increase in $Th2$ and a decrease in $Th1$.

The shifting of the CRC between the help and suppression configurations can be accomplished by addition of an ARC. The simplest ARC's consists of a single strong vertex, corresponding to the B cells and their interactions with the Th cells through soluble products. The single strong ARC vertex can be linked to the CRC in two ways to preserve the bipartite nature of the entire network (Figure 4, middle and bottom). The first of these networks, B_4+C_4, (Figure 4, middle) corresponds to the "suppressive ARC" Herzenberg model[4] and consists of two circuits of lengths 4 of opposite signs with two edges in common and a single turbulent vertex, namely $Th2$. This turbulent vertex is assigned the OR relationship since the interactions of B and $Ts2$ with $Th2$, as indicated by the directed edges, are separate processes. The truth table for this $B_4 + C_4$ network (Table 2) illustrates the process of first generating local truth tables for each circuit and then combining the local truth tables into a global truth table by using the logical relationships at turbulent vertices.[3] From the global truth table in Table 2, only one of the 32 possible ($Ts1$, $Th2$, $Ts2$, $Th2$, B) state vectors, namely (1, 0, 0, 1, 0), is seen to lead to a stable configuration. This stable configuration has $dTs1/dt > 0$, $dTh1/dt < 0$, $dTs2/dt < 0$, and $dTh2/dt > 0$ corresponding to the stable "suppression" configuration of the isolated CRC (Table 1). This simple example, which corresponds to one of the Herzenberg networks, shows how addition of an ARC to a CRC can establish the CRC into one of its two stable configurations.

The other possible way of adding a single strong vertex (B) to the Herzenberg CRC while preserving the bipartite nature of the network (Figure 4, bottom) leads to a five-vertex $B_4 + B_2$ network consisting of positive circuits of length 4 and 2 sharing a common vertex. This shared vertex ($Th1$) is, thus, turbulent and is assigned the OR relationship as in the $B_4 + C_4$ case above. The truth table for this $B_4 + B_2$ network (Table 3) indicates a single stable configuration corresponding to the stable "help" configuration of the isolated CRC (Table 1) with $dTs1/dt < 0$, $dTh1/dt > 0$, $dTs2/dt > 0$, and $dTh2/dt < 0$ opposite to the stable "suppression" configuration of the $B_4 + C_4$ network discussed above.

This analysis, thus, indicates simple ways of attaching a fifth vertex (B) to a given vertex ($Th1$) of the simple four-vertex Herzenberg CRC to lock it in either the "suppression" or "help" configurations. The help-stimulating Herzenberg network[4] is a sum of the B_4+C_4 and B_4+B_2 networks in Figure 4 (middle and bottom) with the $B_4 + B_2$ behavior obviously dominating (i.e., a greater weight of the $B \rightarrow Th1$

edge relative to the $B \rightarrow Th2$ edge). The allotype-suppressive Herzenberg network[4] is still more complicated consisting of a seven-vertex $B_4 + C_4 + C_4 + B_2$ network having a triply turbulent $Th1$ vertex common to all four feedback circuits as well as a (doubly) turbulent $Th2$ vertex.

5. SUMMARY

This paper shows that flow topological methods developed for the analysis of the dynamics of chemical reaction networks[1–3,12–17] can also be used for the analysis of immunological control networks with appropriate modifications to account for the differences between the two types of networks. Thus, the chemical reaction networks of interest have included systems exhibiting periodic oscillations or aperiodic chaos whereas the immunological control networks represent nonoscillatory systems having one or more stable configurations (e.g., "help" and/or "suppression"). In addition, the complementary idiotype/anti-idiotype interactions in the immunological control networks lead to bipartite graphs, a feature not generally found in chemical reaction networks.

The Herzenberg models[4] for immunological control networks are well suited for the flow topological methods outlined in this paper and lead to the locking in stable "help" and/or "suppression" configurations observed experimentally. More generally, this paper demonstrates the application of flow topology for the analysis of immunological control networks to reproduce some of the basic features of such networks in simple systems using as few as five internal independent variables.

ACKNOWLEDGMENTS

I am indebted to the Office of Naval Research for partial support of this work. I would also like to acknowledge helpful discussions with Prof. Jerome Eisenfeld (Department of Mathematics, University of Texas at Arlington) at the Conference on "Perspectives in Biological Dynamics and Theoretical Medicine" in Bethesda, Maryland, April, 1986.

REFERENCES

1. R. B. King(1980), "Chemical Applications of Group Theory and Topology. 8. Topological Aspects of Oscillating Chemical Reactions," *Theor. Chim. Acta (Berlin)*, **56**, 269–296.
2. R. B. King (1982), "The Flow Topology of Chemical Reaction Networks," *J. Theor. Biol.*, **98**, 347–368.
3. R. B. King (1983), "Chemical Applications of Topology and Group Theory. 14. Topological Aspects of Chaotic Chemical Reactions," *Theor. Chim. Acta (Berlin)*, **63**, 323–338.
4. L. A. Herzenberg, S. J. Black, and L. A. Herzenberg (1980), "Regulatory Circuits and Antibody Responses," *Eur. J. Immunol.*, **10**, 1–11.
5. M. Kaufman, J. Urbain, and R. Thomas (1985), "Towards a Logical Analysis of the Immune Response," *J. Theor. Biol.*, **114**, 527–561.
6. R. Thomas (1979), *Kinetic Logic: A Boolean Approach to the Analysis of Complex Regulatory Systems* [Lecture Notes in Biomath., 29], (Berlin: Springer).
7. J. Eisenfeld and C. DeLisi (1985), "On Conditions for Qualitative Instability of Regulatory Circuits with Applications to Immunological Control Loops," *Mathematics and Computers in Biomedical Applications*, Ed. J. Eisenfeld and C. DeLisi (Amsterdam: Elsevier).
8. J. Eisenfeld, "Compatibility Tests on Immunological Control Loops," *Ann. N.Y. Acad. Sci.* **504**, 132–150.
9. N. K. Jerne (1974), "Towards a Network Theory of the Immune System," *Ann. Immunol. (Paris)*, **125C**, 373–389.
10. S. H. Caldwell(1967), *Switching Circuits and Logical Design* (New York: Wiley).
11. S.-T. Hu (1968), *Mathematical Theory of Switching Circuits and Automata* (Berkeley and Los Angeles, California: California Press).
12. L. Glass and S. A. Kauffman (1973), "The Logical Analysis of Continuous Non-linear Biochemical Control Networks," *J. Theor. Biol.*, **39**, 103–129.
13. L. Glass (1975), "Classification of Biological Networks by Their Qualitative Dynamics," *J. Theor. Biol.*, **54**, 85–107.
14. L. Glass (1975), "Combinatorial and Topological Methods in Nonlinear Chemical Kinetics," *J. Chem. Phys.*, **63**, 1325–1335.
15. L. Glass (1977), "Combinatorial Aspects of Dynamics in Biological Systems," *Statistical Mechanics and Statistical Methods in Theory and Application*, Ed. U. Landman (New York: Plenum).
16. L. Glass (1977), "Global Analysis of Nonlinear Chemical Kinetics," *Statistical Mechanics*, Part B, Ed. B. J. Berne (New York: Plenum).
17. L. Glass and J. S. Pasternack (1978), "Stable Oscillations in Mathematical Models of Biological Control Systems," *J. Math. Biol.*, **6**, 207–223.
18. R. Thomas (1984), "Logical Descriptions, Analysis, and Synthesis of Biological and Other Networks Comprising Feedback Loops," *Adv. Chem. Phys.* **55**, 247.

HANS B. SIEBURG
Developmental Biology Laboratory, The Salk Institute for Biological Studies,
P.O.Box 85800, La Jolla, California 92138-9216

A Logical Dynamic Systems Approach to the Regulation of Antigen-Driven Lymphocyte Stimulation

INTRODUCTION

The immune system is an anticipatory system serving as part of the host's recognition and defense mechanisms.[6] It is *hierarchically organized* and consists of two major functionally overlapping compartments, termed the innate and the adaptive immune system.[21] Both systems are integrated mechanisms cooperating by virtue of a large number of cellular and humoral components of varying specialization. Important cellular components are the phagocytes, e.g., macrophages, which form part of the innate immune system and the lymphocytes, e.g., B and T cells, which mediate adaptive immunity. Intruding agents, generically defined as antigen, are specifically recognized as non-self by receptors on the surface of B and T cells. T cells detect antigen in the context of cell surface structures (self) expressed on macrophages and B cells. Upon contact with the macrophage-antigen complex (altered self), T cells are activated and can secrete a variety of activating and/or proliferation-inducing factors. These factors can be effective in a feedback stimulation to induce the proliferation of T cells themselves. Other factors can act on B cells. When B cells come in contact with the specific antigen in the presence of T cell-derived factors, they will proliferate and differentiate into antibody-secreting

Furthermore, upon interaction with the antigen-macrophage complex, memory B and T cells are generated. The ability of the adaptive immune system to preserve information is the basis of successful vaccination.

Mathematically, the system presents itself as large, heterogeneous, and dynamically complex: properties, which largely prevent a unified mathematical treatment by traditional approaches, such as differential equations or stochastic processes (also see [2]). Fortunately, the immune system is also discrete. This suggests that discrete mathematical approaches can be employed for its analysis. Two major lines of thought have been followed in this direction; one based on kinetic logic (Kaufman and Eisenfeld, this volume; Prueitt, this volume; also see [15,25]) and the other based on automaton theory.[22] Both approaches are characterized by technical simplicity, thus are comprehensible for the non-specialist, allow direct accessment of—verbal—immunological knowledge, and are distinguished by their capacity to derive "ante facto" predictions. These properties constitute, in our mind, three "must elements" of successful mathematical modeling in the field of immunology. However, when separately applied to the modeling of an entire immune system, kinetic logic and automaton theory experience the "complexity melt-down"* typical for traditional methodologies.

We here introduce a novel "object-oriented" approach to an antigen-driven mathematical model of the immune system, called a Cellular Device Machine.[22] This approach, which combines kinetic logic and automaton theory with principles from artificial intelligence, in particular knowledge engineering, offers several advantages. First, it remains simple. Second, by defining the immune system in terms of "objects" and "communication" between them, the problem of complexity melt-down does not occur. Third, due to our choice of mathematical techniques, the Cellular Device Machine (CDM) can be directly implemented on a personal computer and be easily extended and modified to allow simulation of both the normal and diseased immune system. Fourth, multiple time scales can be introduced at the "object" level. Consequently, CDM behavior can closely match the intrinsic parallelism of immune system behavior.

The concept of Cellular Device Machines emphasizes as a principle that "communication" in large biological systems is based on strategies of dynamic information sharing among a large number of "objects". Each "object" is capable of a small number of discrete states and is defined using a recursive mathematical structure, called nested automaton.[22] These "automata within automata" are applied to compress dynamically complex data-structures, e.g. the development of immune system cells *and* their interactions, into a set of decision rules. The elements of this set are given by small systems of simple Boolean equations. In contrast to network approaches, the information flow between the components of a CDM is

*Complexity melt-down: Loss of control over a model resulting from unrestricted accumulation of continuously refined building blocks. Widely used building blocks are sentences or equations.

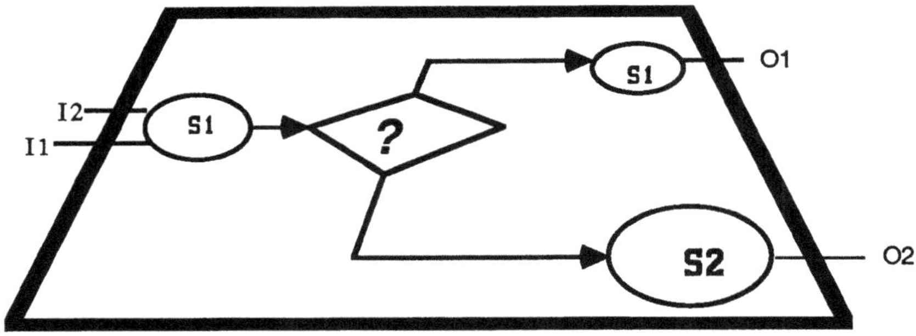

FIGURE 1 A two-state, decision-making automaton, the states denoted by S1 and S2, respectively. A new state is calculated by a Boolean function on the input symbols I1, I2, i.e., $S1 = f(I1, I2)$, $S2 = g(I1, I2)$. The outputs O1, O2 are given as Boolean functions of the states, i.e., $O1 = h(S1)$, $O2 = k(S2)$. Biologically, such a decision-making automaton can be used to represent a well-defined stage in the development of a cell. The inputs are given in this case by free or cell-surface bound molecules, the outputs indicating, e.g., the secretion of a factor or the expression of a particular cell-surface receptor.

event driven rather than part of the model definition. This means that, at various points during a simulation, two or more components can establish or interrupt communication linkages dynamically, i.e., as functions of their states.

CELL DEVICES

The immune system is a discrete system whose components arise from pluripotent hematopoietic stem cells residing in the bone marrow. Through a series of as-yet-unidentified differentiation steps, stem cells continuously replenish a tree of cell classes of varying specialization (for review see [20]). We conjectured that each such class is characterized by self-similarity in design. This implies that a cell class can be fully fathomed by a functional description of a single representative.

The number of developmental changes a representative is able to undergo during its lifetime is small, and biologically characterizable by parameters such as cell size or the expression of particular surface receptors. Mathematically, we interpreted each developmental step as a simple decision-making automaton with very few inputs and outputs

Then we considered the decision-making automata as "states" of a finite automaton with output (FAO); the processes leading from one developmental step to another defining its "transition graph." In reference to their construction from immune system cells, we called these FAO cell devices.

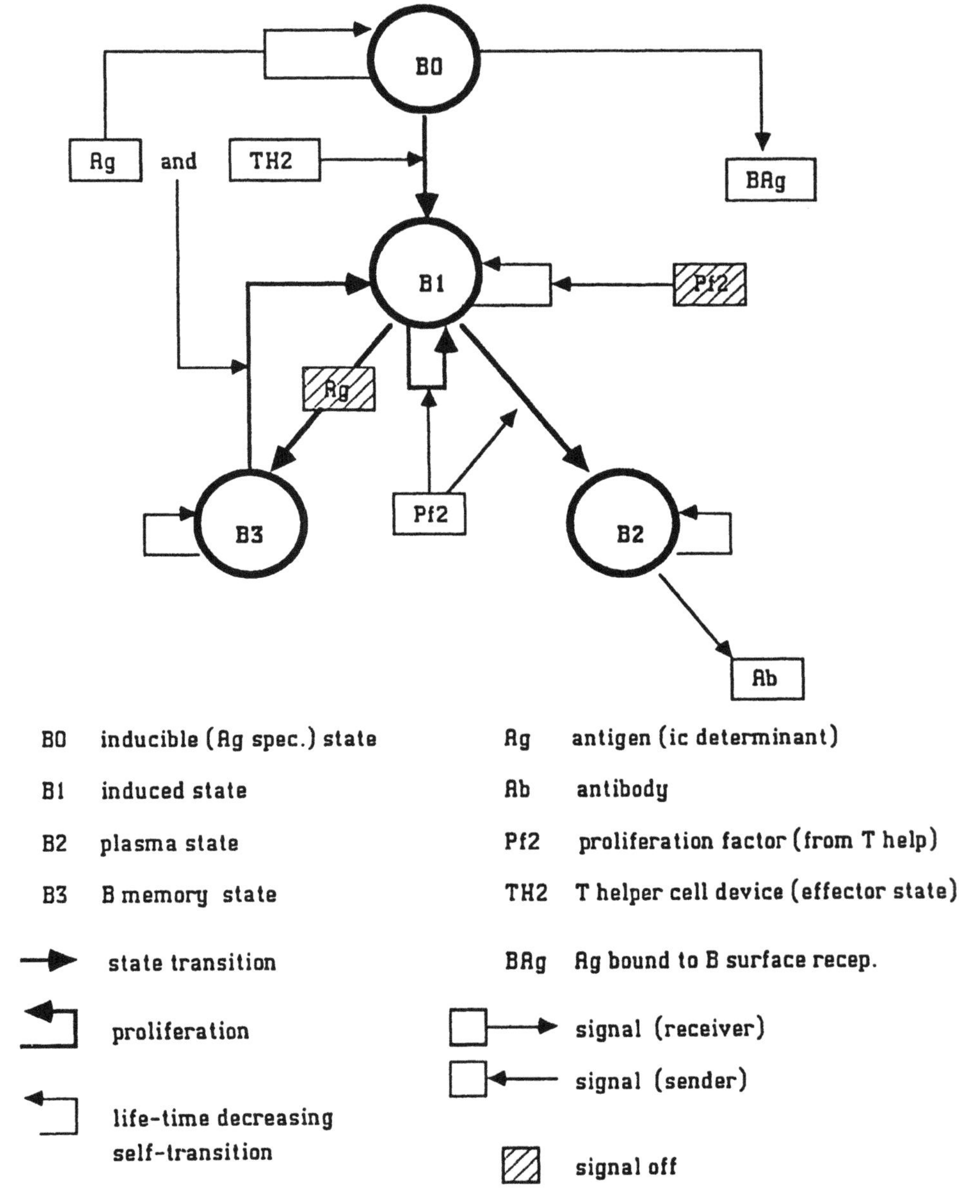

FIGURE 2 Transition graph of a B-cell device (four-state, antigen-driven automaton).

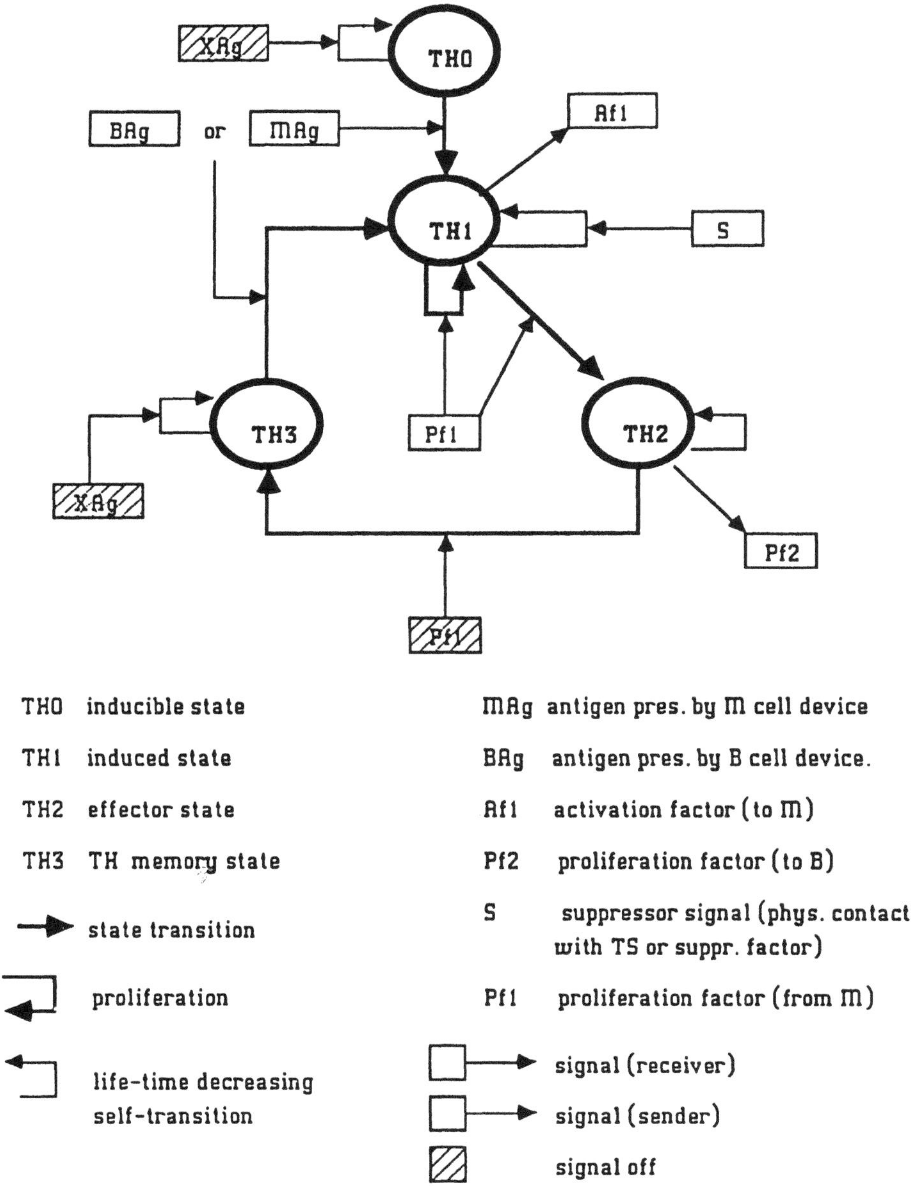

FIGURE 3 Transition graph of a T-helper cell device (four-state, antigen-driven automaton).

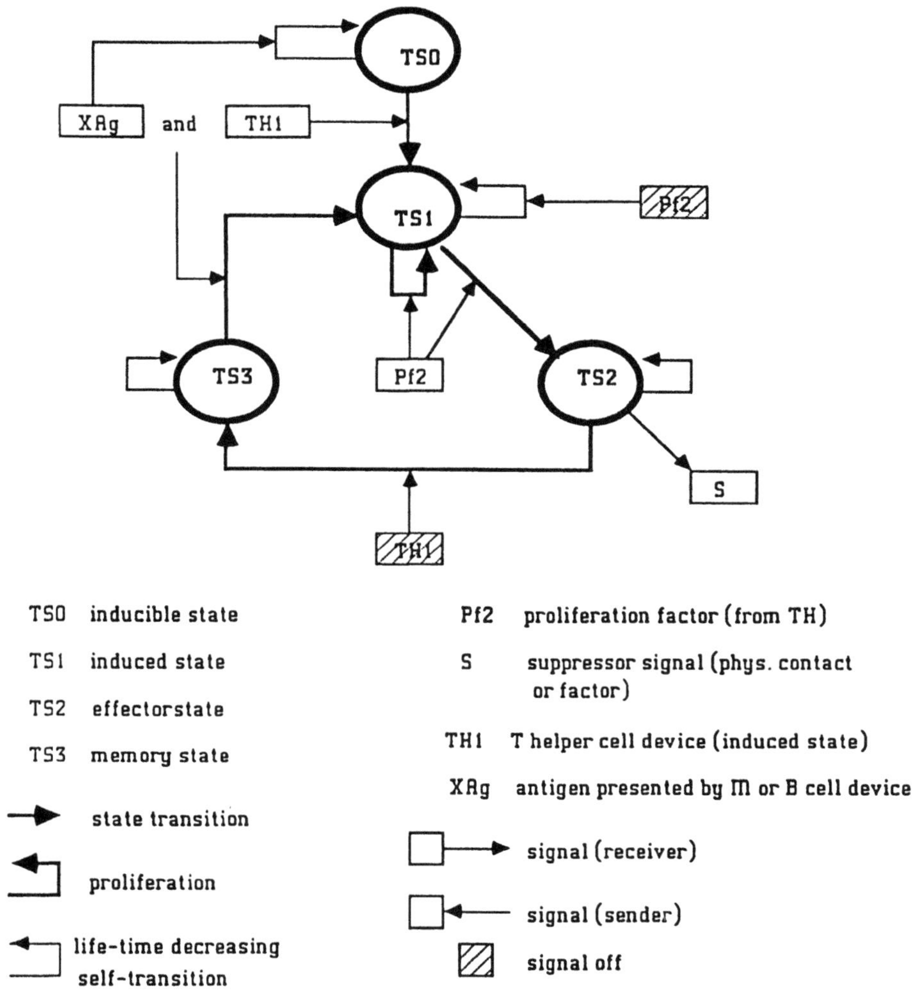

FIGURE 4 Transition graph of a T-suppressor cell device (four-state, antigen-driven automaton).

For purposes of introduction, we will limit our presentation to committed and mature immune system cells, specifically to B-, T-helper, T-suppressor lymphocytes and macrophages. Under these constraints, we found that four state automata, containing two induction states, one effector state and a memory state, constitute minimal configurations realistically reflecting known lymphocyte behavior [1,17]

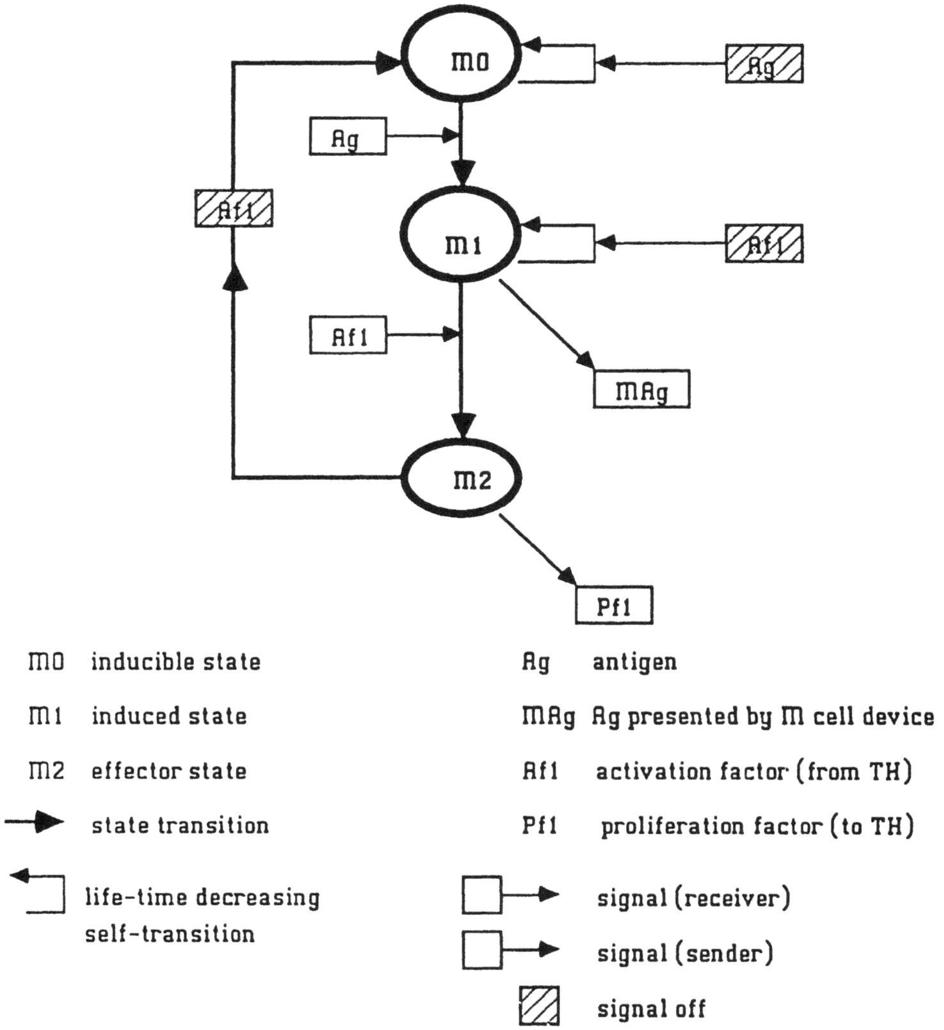

FIGURE 5 Transition graph of a Macrophage-cell device (three-state, antigen-driven automaton).

To explain briefly how these transition-graphs are applied, we consider the situation where a B-cell device (Figure 2) B in state BO specifically recognizes antigen Ag. In this case a complex BAg is formed which can stiumlate specific T-helper-cell devices TH in state TH0 (Figure 3). Given the availability of appropriate signals, such TH can undergo two proliferation and differentiation steps to arrive at state TH2. In turn, given the cell-cell contact between BAg and TH2, B0 can transition to differentiation state B1. In summary, we require two signals (Ag and TH2) for this transition, not one (Ag). At this stage, three possibilities arise. On one hand, given sufficient quantitites of proliferation factor PF2 produced by TH2 in its neighborhood, B1 will transition to a plasma cell state or B2, which can take up specific antibody production. On the other hand, in case PF2 is not sufficiently available, but Ag is still present, B1 will remain in this state until its life-time is exceeded (further details regarding life-times are provided later in this section). Alternatively, if neither Ag nor PF2 are available, B1 will become a B3 or memory-cell device. Logically, B3 and B0 are equivalent states. However, with regard to their history, B-cell devices in state B3 and B0 exhibit distinct behaviors when stimulated by antigen Ag.

A three-state automaton, containing two induction states and one effector state, suffices to describe the macrophage compartment.[10]

As mentioned in the introduction, the capacity to use many different designs of cell devices, is one major advantage of our approach. Therefore, the designs presented in Figures 2–5, although engineered to exhibit established experimental knowledge, may be considered as first approximations. One can then choose to apply an approach based on evolution, e.g., the one suggested in [14], to arrive at optimal designs for cell devices.

The signals a cell device is able to receive define its input set and those it is able to send define its output set. In this context, "receiving" and "sending" substitute for the biological terms "binding" and "secreting," respectively. The term signal generally defines the presence or absence of molecular entities, such as antibody molecules, interleukins or other factors, antigen or cell-cell contact occurring during the interaction of T-helper lymphocytes and antigen-presenting cells. This definition *a fortiori* accommodates the sender-receiver duality observed in immunological signals.

With respect to state transition, we apply the following rule-of-thumb derived from immunological knowledge: being at a certain stage in its development, a cell device may remain in its present state or may proceed to a different state ("differentiation") or may produce progeny ("proliferation") upon signal reception. Decision-making usually depends on more than one signal. Antigen is assumed to play a crucial role during the induction steps. Therefore, cell devices and all configurations derived from them, are antigen driven. Differences between cell types and among the individual cells of one cell type are represented at the level of their input sets. For example, in order to generate the capacity for specific and discriminative recognition (in B-cell, T-helper and T-suppressor cell devices) certain signals, such as antigen-specific antibodies and other recognitive surface proteins, are encoded as bit patterns (a similar technique was used in [8] in the context of idiotype networks). The set of all bit-encoded signals defines the BTM-Machine repertoire. The repertoire is considered as changing in time reflecting the (also) bit-encoded "antigen world" in terms of virtual images. The time dependency of the repertoire implies that a virtual image may be empty ("repertoire holes"). Other signals (e.g., activation or proliferation factors) remain of general type, i.e.,are not bit encoded. Also, the input set of one cell device may contain different signals than the input set of another.

Since cell devices "are" transition graphs of nested automata, their computer implementation is exceedingly simple; they can be directly "compiled" into "subroutines." Technically, this is accomplished by using a graphics terminal to engineer the design, and a graph-logic compiler for accommodation into the larger set-up we will discuss in the Cellular Device Machine section. The effectiveness of this procedure is such that a CDM containing a start-up population of 10^3 cell devices can be easily implemented on a Macintosh Plus personal computer.

CELL MODULES

From the experimentally known or theoretically postulated interactions between B lymphocytes, T lymphocytes and macrophages, we are able to deduce a principle of similarity in interaction. More specifically, we consider the basic mechanisms of interaction between an arbitrarily chosen representative of one cell type and any representative of another, to be the same.

Using this principle, the interactions between different cell types can be mathematically represented as the transition graph of an FAO. The state set of the FAO consists of cell devices, each representing a particular cell type. With respect to our special consideration of B-, T-helper, T-suppressor lymphocytes and macrophages and in accord with current computer science terminology, we have called the nested automaton implementing these cell types *and* their interactions, a BTM-cell module. Figure 6 above depicts the module derived from the designs presented in the previous section.

CELLULAR DEVICE MACHINES

Without indulging into a general mathematical discussion, we simply define a Cellular Device Machine as a (potentially infinite) two-dimensional cellular automaton, whose active sites contain cell modules (for details concerning 2-d cellular automata see [28,29]). This cellular automaton constitutes a mathematical embedding structure, which biologically represents a host's body. Alternative embedding structures can be considered. In the context of cellular automata, the name Cellular Device Machine relates to the automaton classes fundamental for its configuration, i.e., cellular automata and cell devices, and the logical definition of the term "machine."[13]

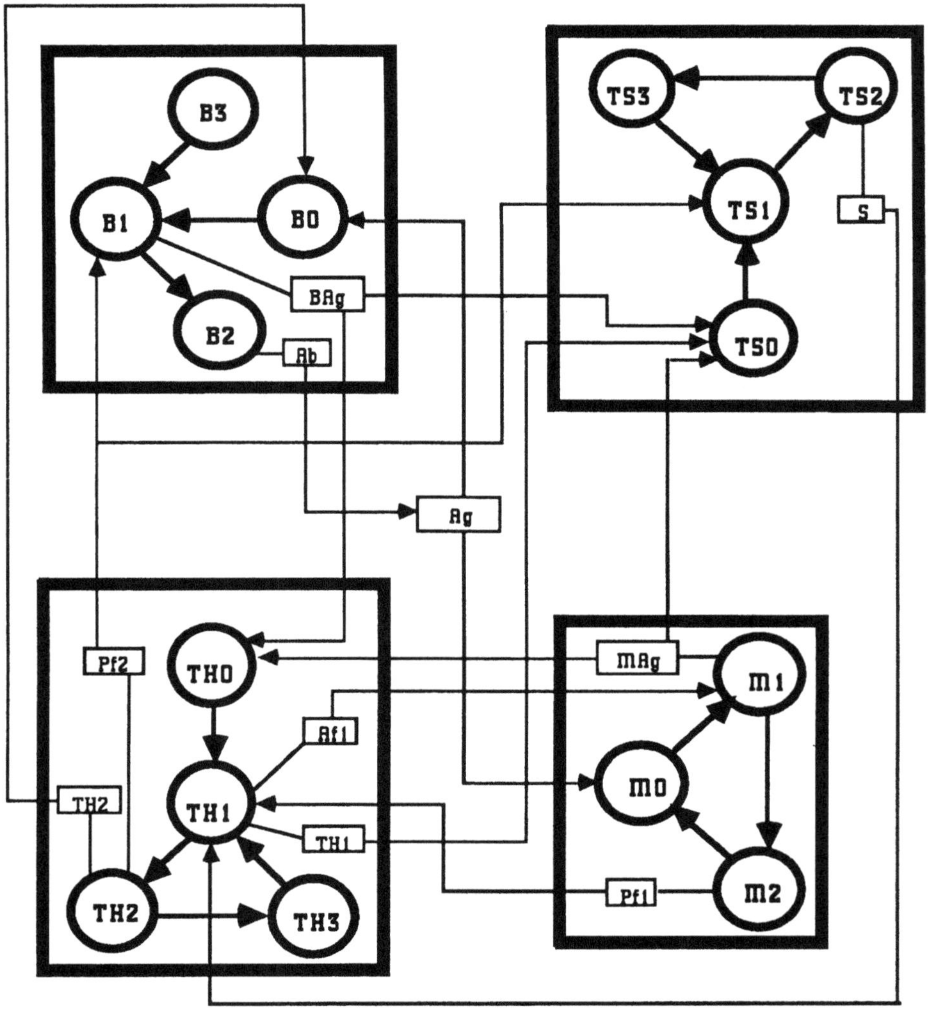

FIGURE 6 Transition graph of the BTM cell module. Using the notation introduced in Figures 2–5, this figure indicates the "virtual" connectivity between module components through input and output "messages." When the BTM-Machine recognizes antigen within the scope of its repertoire, message linkages are established at various points during the response, while others are interrupted. In other words, the connectivity becomes partially "real." The situation returns to "virtual," when antigen disappears from the system.

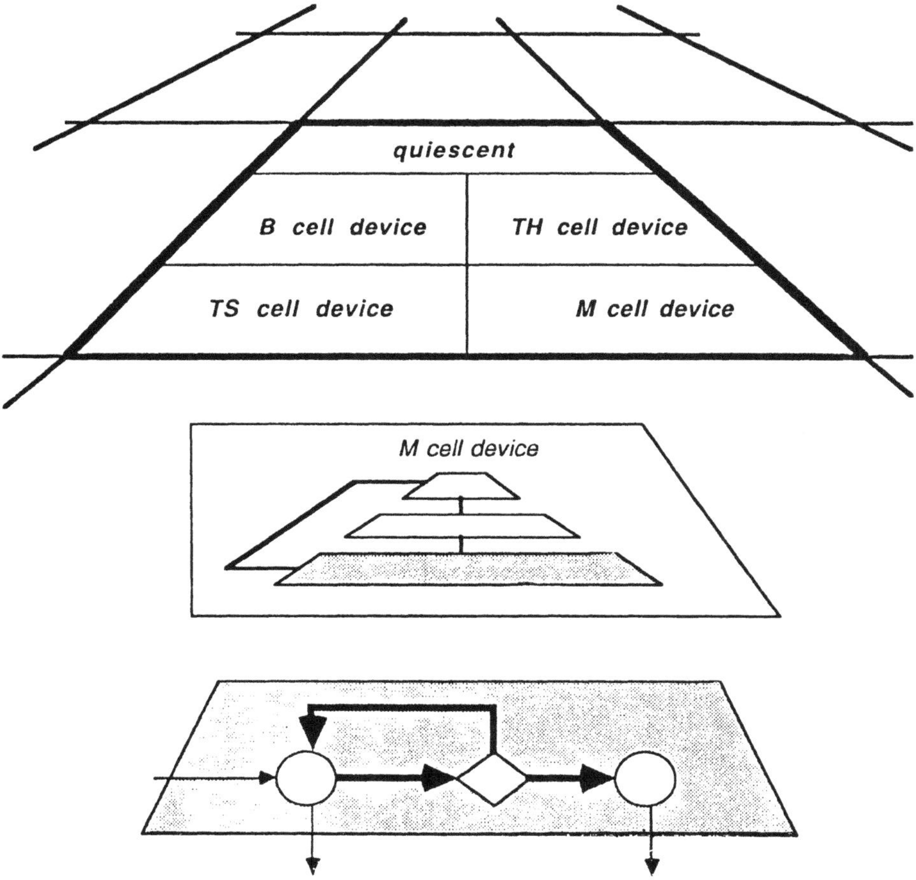

FIGURE 7 The three-level nested automaton used in the BTM-Machine. The top figure depicts a BTM-module embedded into the structure of a 2-d cellular automaton. In biological terms, the latter (or any other alternatively chosen embedding structure) simulates a host's body. The middle figure shows a particular cell device. Biologically, a cell device represents the developmental capacity of a cell type in relation to its environment. The bottom figure depicts a decision-making automaton. Such automata are used to simulate particular stages in the development of an immune system cell.

In order to constitute a special purpose Cellular Device Machine, one defines each site of the cellular automaton to be in exactly one of two states, i.e., either in a quiescent state or a state representing a cell device which is part of a cell module. Figure 7 depicts the CDM constructed from BTM-modules.

Considering other cell types and differing numbers of cells, we arrive at different Cellular Device Machines. Therefore, it is possible to model the immune system on different levels of complexity. The highest level of complexity, achieved by employing all the cell types found in the immune system, deserves the term Artificial Immune System (this notion was, as far as we know, first employed by Klaus Rajewsky to describe an *in vitro* immune system).

The biology represented by a Cellular Device Machine in the context of cellular automata is given by a (Boolean) transition function, which operates on the sites, depends on the signals as variables and is locally effective in small neighborhoods around sites

In order to be able to account for threshold dependency, half-life and strength of each signal, we introduced the notion of signal-energy. The energy of a signal is defined by a weighted sum over (discrete) time of the number of its active sources. A similar notion of energy was used to provide each cell device with its own internal clock. When a cell device first appears in the machine, either derived from an artificial bone marrow or as progeny during proliferation, an initial value, which is specific for its kind and state, is assigned as the energy of this particular cell device. This value is decreased with each machine cycle by a fixed amount.

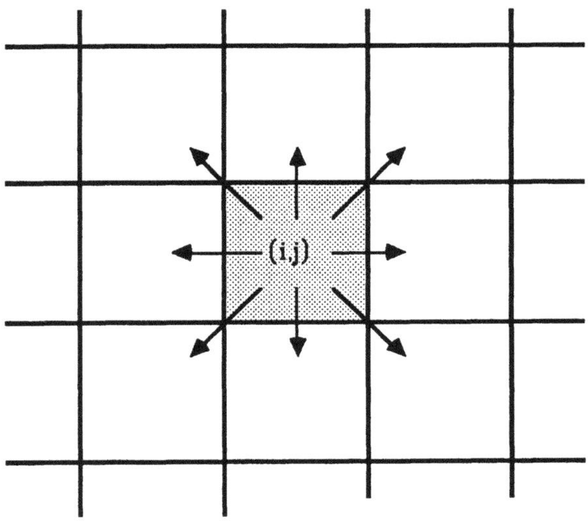

FIGURE 8 Transmission of signals from (→) an arbitrary site (i.j) of the BTM-Cellular Device Machine into its neighboring sites. In the context of the 2-d rectangular topology used to represent the embedding cellular automaton, each site has nine neighbors (itself included).

Therefore, if a cell device remains unstimulated over a period of cycles, its site will become quiescent, thus indicating its death. Devices stimulated to differentiate to another state, however, will gain a certain quantity of energy with each state transition. Therefore, the more evolved a cell becomes, the longer it will remain in the system.

In summary, we have the following means of characterizing signals: first, by biological function (e.g., antigen, antibody, factor, cell-cell contact), second, by "purpose" (i.e., general or special, indicated by a bit pattern of length less than or equal to 1 or greater than 1, respectively) and, third, by threshold values and half-lives (given by energy). Our means of characterizing cell devices summarize to: first, by biological function (cell type), second, by state, third, by lifetime (energy), and fourth, with respect to the input set (e.g., specificity).

					TH			B		TH	TH	M	B	M	B		TS	B	
	TS	TH	TH	TS	TS	B	TS	B		B		TH	TH						
		TH	TS		TH		B	TS	B			B	TH		TH	TH	TS	TH	TS
TH	TS		B	TH	TH		TH	TH	TH				TH	TH	TH		M		TH
			TS		TH			TH	B	TS	B					B	B		
B				TH	TH	TH		B		TH	B	TH				TS	TH		
		TH	B		TH	TS		TS	B	TS	B			B	M		B	TH	TS
M			M	TS	TH				B		TH	TH	TS			B	TH		B
	TH	TH	TH	B	TH	B	TH	TS	B			TS				TH	TH		TH
TH	TH			TH	TH			TH	TH			TH			B		B		
	TS		B		TH				B			TS	TH				B	TH	
	B	TH	B	M		TH		TH			B	TS	TS		TH	M	TS	TH	
B	TS		TS				TH	TH		TH			TH	B			TH	TH	TS
B		TH	TH		TH		TH				B		TS	TH	M		B	TH	
TS	B	TS				TH			TH	TH		TH	TH	TS		TS	TH	TS	
TS		M		TH		TS		TH		TS	TH		TH	TH	TH	TS	B		
M	TH					TS	TH	B	TH			TH	TH	M	TS	TH	B	TH	TS
		TH		TH	TH			TH		TH	B	B	TS	TS	TH	TS	TS		
B	B	TH		TS		TS		TH	B		TH	B	TH	M		TH	TS	TH	TH
TS	M	B		TH	TH			TH		B	M		TH		TH				B

FIGURE 9 Random section of the BTM-Cellular Device Machine before addition of antigen. The section shows the grid outline of the cellular automaton hosting the BTM-modules (dotted lines). The letters B, TH, TS, M denote the respective cell devices which were found active when the snapshot was taken. An empty site indicates that a particular module is in it's quiescent state. The B:(TH/TS):M ratio considered here was 25:65:10.

LYMPHOCYTE STIMULATION AND REGULATION

This section summarizes some of our experiences concerning the stimulatory and regulatory behavior of the BTM-Machine. The start-up populations we considered in our simulations so far, ranged from 10^3 to 10^6 randomly distributed B-, T-helper, T-suppressor and macrophage cell devices.

Since our system is antigen driven, the introduction of antigen (from the outside or the inside), once recognized as non-self with respect to the repertoire, initiates the organization of cell devices into cell clusters or local cellular networks. The elapse time between the introduction of antigen and the initial formation of these networks can be measured and is, since recognition is specific, characteristic for this particular antigen and the state of the BTM-Machine (immune system) repertoire at this time. With this, we are able to define the term stimulation as a dynamic observable characterized by the organization of interacting cell devices into local networks. Figures 9 and 10 show the onset of organization starting from a random distribution of cell devices.

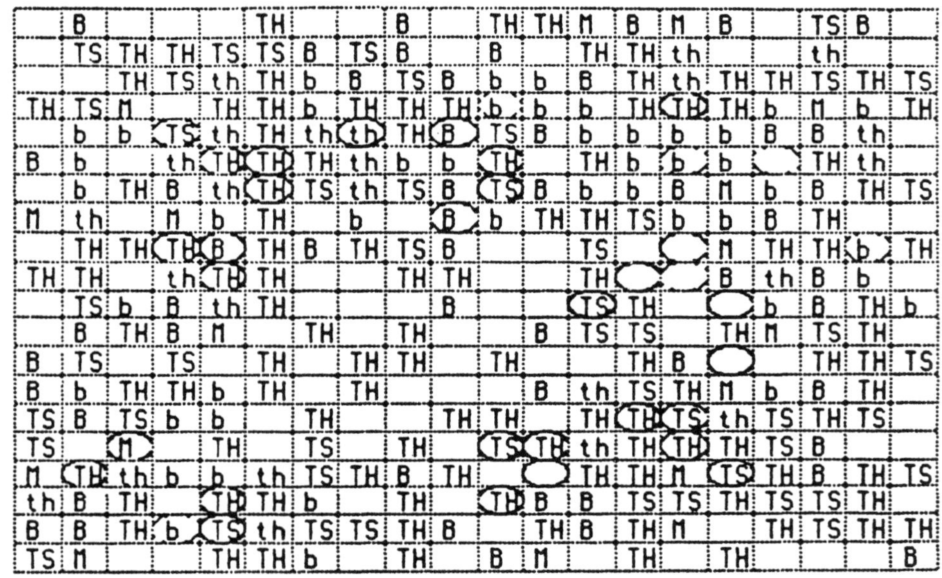

FIGURE 10 Onset of cell-cluster formation. The circles within particular sites indicate that the antigen concentration in these sites exceeds a certain threshold level. The system does not respond to antigen concentrations below this threshold. Small letters b and th denote progeny from proliferating B and TH cell devices.

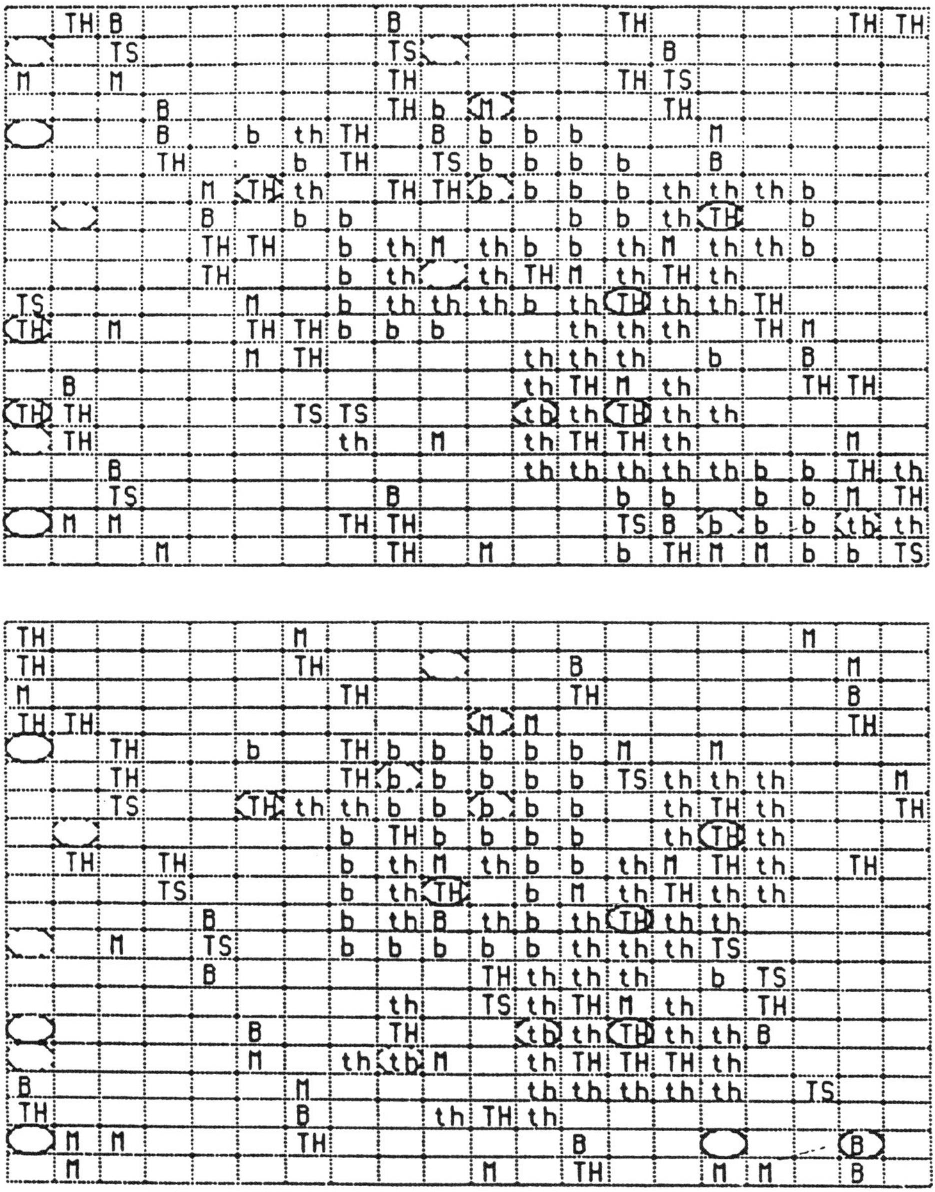

FIGURE 11 The self-regulatory capacity of a BTM-Machine is revealed through spatial patterns (cell device clusters), which remain limited throughout a response. Both snapshots were taken at different times during the specific response to an antigen introduced to the system earlier.

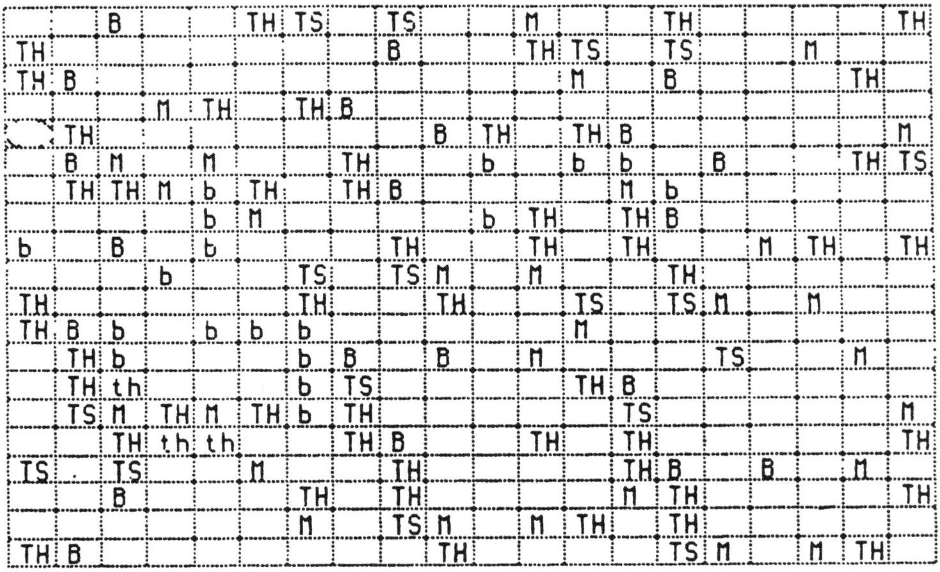

FIGURE 12 A specific response has led to the disappearance of an antigen from the system. The aftermath of this response is reflected by decomposing cell device clusters. These clusters contain b and th cell devices, which remain in the system as memory cells. The circle in the left-most column of sites indicates a sufficient threshold concentration of an antigen, which may not yet have been recognized by the cell devices in its neighborhood.

Once initiated, a cluster is maintained, though altering in form and composition (e.g., due to proliferation, or caused by "percolating" signals, which weakly attract devices in its close environment), as long as antigen is present in the system. This is indicated by the snapshots provided in Figure 11.

When antigen disappears from the system (which can mean "removal" or "hiding"), a cluster gradually decomposes into smaller ones, until only single devices remain.

These devices remain in the system over varying and sometimes considerable, but always finite, lengths of time. Recurrence of the same antigens reveals the capacity of the cell devices to generate memory as the onset of cluster formation and the processes leading to antigen "removal" are considerably faster than in an "unprimed" system.

The observation that a cell cluster or local network is spatially limited at all times, indicates that a BTM-Machine (and any CDM in general) has the property of self-regulation. More strongly, we can define regulation as a dynamic observable characterized by the alteration and final decomposition of spatially confined networks. The elapse time between the onset and offset of organization can be measured, thus providing a measure for the duration of a particular response.

An interesting observation concerns the behavior of purely numerical parameters, such as numbers of cells or concentrations of factors, which we are able to measure prior to, during and after a response. For example, when the BTM-Machine does not recognize any antigen, these parameters exhibit oscillatory behavior over time,[9] defining the "antigen-free" "state." This state seems to have a basin of attraction which changes location and form over time. We have also found strange attractors, especially in models of infectious diseases of the immune system,[22] such as the Acquired Immune Deficiency Syndrom (AIDS) caused by the Human Immune Deficiency Virus (HIV).

DISCUSSION

In this paper, we have introduced a novel mathematical structure, called a Cellular Device Machine. This approach is distinguished by several features. The concepts of nested automata and self-similarity compress the vast data derived from complex dynamical systems into manageable structures. These structures can be easily

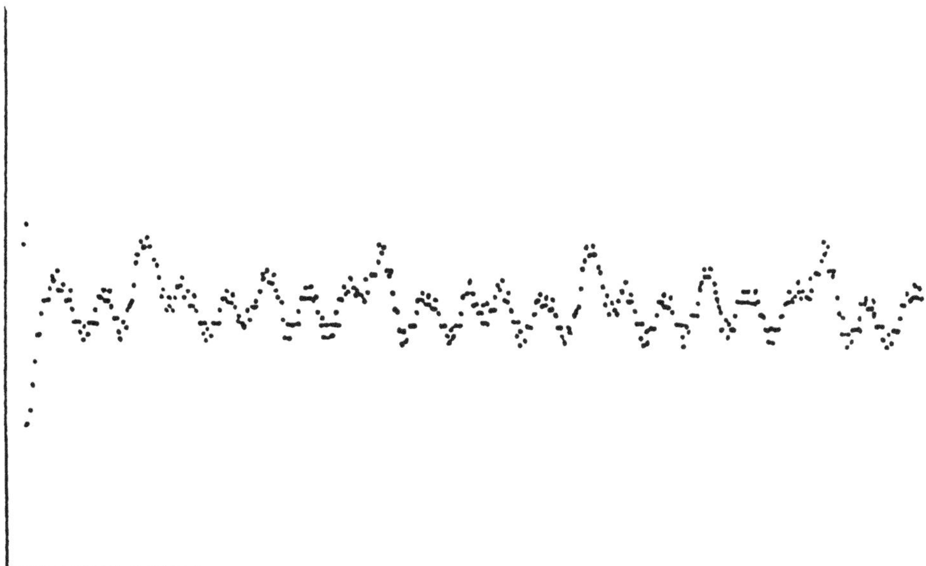

FIGURE 13 T-helper cell device kinetics in the "antigen-free" BTM-Machine. The curve depicts the number of T-helper cell devices (vertical axis) versus time (horizontal axis).

implemented on a computer. Interactions among them are event driven. Self-regulation is obtained through multiple time scales. Model behavior is not part of the model definition. Due to these properties, Cellular Device Machines can be applied to realistically simulate very large and heterogeneous biological systems.

In the present paper, we constructed an antigen-driven Cellular Device Machine, called the BTM-Machine. We presented computer-screen snapshots showing how cellular clusters "form" in a self-organized fashion in response to antigen. This very process was used to define the term stimulation. Further pictures indicate continuing interaction until antigen disappears from the scope of a cluster's repertoire. Though changing in size and composition during a response, cellular clusters are spatially limited at all times, therefore can be termed local networks. Locality reflects the BTM-machine's capacity for (self-)regulation, which interestingly does not depend on the presence of actively suppressive elements. Specific and discriminating (pattern) recognition with respect to a time-variable repertoire always precedes the initiation of a response. Its aftermath is reflected by dynamic memory, meaning the conservation over long (but finite) periods of time of adapted sections of the repertoire carried by freely ranging machine components previously involved in this response .

As far as we know, this application of automata to immune system modeling and simulation is completely new. Further, the emergence of form and its application to describe immune response phenomenology, establish for the first time a linkage between immune system modeling and modern "shape mathematics" (a detailed discussion of this aspect will be provided in [23]). However, ideas related to the fundamentals of our Cellular Device Machine concept have been suggested, or put into practice, in other fields, in particular in physics, molecular biology, biochemistry and computer science (see [3–5,10–13,16,24,29]).

Although aware of the simulator's value in supporting basic and clinical immunological research (see also [21]), we have barely started to access its full capacity. From the data provided by the BTM-Machine so far, we chose only a small sample serving to introduce the ideas and concepts underlying its construction. With this purpose in mind, we consciously neglected its rigorous deduction from more general mathematical objects, which we called kinetic sets, as well as any detailed consideration of quantitative aspects (some of those were discussed during the actual talk). These and other specialized topics, such as the application of fractal methods,[14] which we found to be convenient tools in the mathematical analysis of local networks, computational complexity, and planning in the presence of unreliable and/or sparse knowledge, will receive a separate discussion elsewhere.

The simulator is presently implemented on the VAX 11/785 system in the C programming language. Smaller, but equally functional, versions are available written in the BASIC and assembler programming languages for the Apple Macintosh Plus. In order to study long-term immune system behavior, we have developed plans for its "permanent" installation.

ACKNOWLEDGMENTS

I would like to express my deepest thanks to Dr. Melvin Cohn, who made this adventure of a mathematician into a biological research laboratory possible. Without his encouragement, and intellectual stimulation and support, the project could not have been started. Also, I am glad to acknowledge the invaluable help and the many patiently suggested clarifications I received from Drs. Christa E. Muller-Sieburg and C. David Pauza in numerous discussions about the immune system and its diseases. Finally, the technical help of Francis X. Chapman, who implemented the VAX version of the BTM-Machine, is acknowledged .

The work reported in this paper was supported under NIH Grant 2 RO1 AI 05875 to M. Cohn.

REFERENCES

1. Alcover, A. et al. (1987), "Functional and Molecular Aspects of Human T Lymphocyte Activation via T3-Ti and T11 Pathways," *Immunol. Rev.* **95**, 5.
2. Bruni, C. et al. (ed.) (1979), "Systems Theory in Immunology," *Lecture Notes in Biomathematics* (Berlin-Heidelberg: Springer Verlag), vol. 32.
3. Carter, F. L. (1984), "The Molecular Device Computer: Point of Departure for Large-scale Cellular Automata," *Physica* **10D**, 175–194.
4. Cocho, G., R. Perez-Pascual, and J. L. Rius (1987), "Discrete Systems, Cell-Cell Interactions and Color Patterns of Animals, I. Conflicting Dynamics and Pattern Formation," *J. Theor. Biol.* **125**, 419–435.
5. Cocho, G., R. Perez-Pascual, and J. L. Rius (1987), "Discrete Systems, Cell-Cell Interactions and Color Patterns of Animals, II. Clonal Theory and Cellular Automata," *J. Theor. Biol.* **125**, 437–447.
6. Cohn, M. (1970), "Anticipatory Mechanisms of Individuals," *Control Processes in Multicellular Organisms* (a CIBA Fnd. Symp.), Eds. G. E. W. Wolstenholme and J. Knight (London: J. A. Churchill).
7. Farmer, D., T. Toffoli, and S. Wolfram (eds.) (1984), *Cellular Automata* (Amsterdam: North-Holland) (reprinted from *Physica* **10D**, vol. 1&2).
8. Farmer, J. D., N. H. Packard, and A. S. Perelson (1986), "The Immune System, Adaptation, and Machine Learning," *Physica* **22D**, 187–204.
9. Guerry, D., D. C. Dale, M. Omine, S. Perry, and S. M. Wolff (1973), "Periodic Hematopoiesis in Human Cyclic Neutropenia," *J. Clini. Invest.* **52**, 3220–3230.
10. Holland, J. H. (1966), "Universal Spaces: A Basis for Studies of Adaptation," *Automata Theory*, Ed. E. R. Caianiello (New York-London: Academic Press).
11. Holland, J. H.(1970), "Hierarchical Descriptions, Universal Spaces, and Adaptive Systems," *Essays on Cellular Automata*, Ed. A. W. Burks (Urbana: University of Illinois Press).
12. Holland, J. H. (1970), "Outline of a Logical Theory of Adaptive Systems," *Essays on Cellular Automata*, Ed. A. W. Burks, (Urbana: University of Illinois Press).
13. Hopcroft, J. E., and J. D. Ullman (1979), *Introduction to Automata Theory, Languages and Computation* (Reading, MA: Addison-Wesley).
14. Kauffman, S. A., and R. G. Smith (1986), "Adaptive Automata Based on Darwinian Selection," *Physica* **22D**, 68–82.
15. Kaufman, M. (1985), "Logical Analysis of Lymphocyte Interactions during an Immune Response," *Dynamical Systems and Cellular Automata*, Eds. J. Demongeot, E. Goles and M. Tchuente (London: Academic Press).
16. Langton, C. G. (1986), "Studying Artificial Life with Cellular Automata," *Physica* **22D**, 120–149.
17. Linch, D. C., D. L. Wallace, and K. O'Flynn (1987), "Signal Transduction in Human T Lymphocytes," *Immunol. Rev.* **95** 137.

18. Lumb, J. R. (1983), "The Value of Theoretical Models in Immunological Research," *Immunol. Today* **4**, 209.
19. Mandelbrot, B. B. (1983), *The Fractal Geometry of Nature* (New York: W.H. Freeman and Company), revised edition.
20. Müller-Sieburg, C. E., G. F. Tidmarsh, I. L. Weissman, and G. J. Spangrude, "Maturation of Hematopoietic Cells that Express Thy-1," to appear in *Thy-1: Immunology, Neurology, and Therapeutic Applications*, Eds. A. E. Reif and M. Schlesinger (New York: Marcel Dekker).
21. Roitt, I. M., J. Brostoff, and D. K. Male (1985), *Immunology* (London-New York: Gower Medical Publishing).
22. Sieburg, H. B., and C. E. Muller-Sieburg (1987), "Cellular Device Machines: A Knowledge-engineered Dynamical Systems Approach to Mathematical Modeling of Infectious Diseases," abstract to appear in *Proc. of the Int. Conf. on the Physics of Chaos and Systems far from Equilibrium, Monterey 1987.*
23. Sieburg, H. B., "Form and Self-similarity in the Immune System," manuscript in preparation.
24. Stanley, H. E. and N. Ostrowsky (eds.) (1986), *On Growth and Form* (Boston-Dordrecht-Lancaster: Martinus Nijhoff).
25. Thomas, R. (ed.), "Kinetic Logic: A Boolean Approach to the Analysis of Complex Regulatory Systems," *Lecture Notes in Biomathematics* (Berlin-Heidelberg-New York: Springer Verlag), vol. 29.
26. Toffoli, T. (1984), "CAM: A High-performance Cellular-Automaton Machine," *Physica* **10D**, 195–204.
27. Unanue, E. R., and P. M. Allen (1987), "The Immunoregulatory Role of the Macrophage," *Hospital Practice*, April 15 (1987), 87.
28. Wolfram, S., and N. H. Packard (1985), "Two-Dimensional Cellular Automata," *Journal of Statistical Physics* **38**, 901. Reprinted in: S. Wolfram, (1987), *Theory and Applications of Cellular Automata* (Singapore: World Scientific.).
29. Wolfram, S. (1987), *Theory and Applications of Cellular Automata* (Singapore: World Scientific).

PETER A BRETSCHER
Department of Microbiology, University of Saskatchewan, Saskatoon, Saskatchewan S7N 0W0 Canada

Decision Criteria And Coherence In The Regulation Of The Immune Response

INTRODUCTION

Recent studies on the regulation of the immune response have revealed a complexity undreamt of a decade ago. Richard Gershon, in probably his last review,[1] describes the cellular interactions he believed to be involved in specifically inhibiting the humoral response. Interactions between 11 subclasses of regulatory T cell are outlined. It is clear that this depiction of cellular regulation is more of a provisional description than a final and complete one. Even so, the complex state of affairs suggested by these studies must make one wonder whether significant features of immune regulation can be elucidated without obtaining a full characterization of all the cells and all their interactions.

I believe that a partial but very useful and valid description of cellular regulation is possible. This belief stems from two considerations. Firstly, we would try to understand the main characteristics of the cellular interactions, even if we had in our possession a full description of them, by focussing on the decision points in the network, e.g., what determines whether antigen induces the first T cell in a series, or whether antigen preferentially activates helper or suppressor T cells. A similar approach may be possible even in the face of our current ignorance. We must pay particular attention to the decision criteria the immune system employs in our attempt to understand cellular organization. Secondly, there have been experimental studies examining how the quantitative magnitude of some variables, such as the concentration of antigen, determines the kind of response induced. These studies reveal that the immune response is very coherently regulated. This coherence, to be defined in detail later, implies that there are universal decision criteria applying to the response to all the epitopes of an antigen molecule. Coherence must impose restraints on possible models of cellular regulation. I shall describe here some examples of this coherence, and their implication for cellular regulation, as well as some ideas on the "decision criteria" embedded in the cellular circuits of the immune system. I shall also outline attempts to experimentally test the validity of these proposed decision criteria. This will give rise to a partial description of cellular interactions that is in conflict with decision criteria proposed by others. I therefore feel it appropriate to discuss why I feel these latter decision criteria are difficult to reconcile with some general features of immune regulation.

AN EXAMPLE OF COHERENCE

Mitchison carried out several years ago a very extensive series of experiments on the immunogenicity of the antigen bovine serum albumin in adult mice.[2] It was known at the time that very high as opposed to moderate doses of some antigens were not only unable to induce an antibody response, but made the animals specifically unresponsive to a subsequent challenge with a moderate dose of antigen. Mitchison examined how the administration of different amounts of antigen to mice affected their subsequent antibody response to a challenge of the antigen that induced antibody in an untreated animal. He found that an amount of antigen resulting in a concentration in the mouse's body fluids of 10^{-7} M sensitized the animal to mount a larger humoral response than in an untreated animal, whereas administering amounts resulting in concentrations of 10^{-8} M or 10^{-5} M resulted in a reduced response to the subsequent challenge. He thereby defined "low-zone" and "high-zone" paralysis. These studies were made at a time before cell interactions in the regulation of immune responses were either known or envisaged. A natural hypothesis to explain these observations, discussed at the time, can be referred to as the "counting" model: a low degree of saturation of the B cell's receptors was envisaged to result in the inability of the B cell to be subsequently induced (low-zone paralysis), a medium degree of saturation to result in sensitization, and a high degree of saturation to again result in inactivation of the B cell (high-zone paralysis). This counting model accounts for the progression from low-zone paralysis to sensitization and high-zone paralysis as the concentration of antigen is increased.

There is a difficulty with this model. It was recognized even at the time of Mitchison's experiments that there is normally a great heterogeneity in the binding constants of the antibody induced, and therefore of the receptors on the B cells activated. A conservative estimate of the B cell receptor's affinity heterogeneity is that it varies over a thousand-fold. This heterogeneity, within the context of the counting model, would not allow the B cell population as a whole to respond sharply to variations in the antigen concentration. A simple precise rule, if applied to a heterogeneous population, would result in a smeared out dose-response curve for the population as a whole. Mitchison's observations seem to suggest that B cells with heterogeneous receptors can respond to changes in antigen concentration in a more "coherent" manner than expected on simple models of induction. I shall refer to any phenomenon, in which a heterogeneous population of cells behaves more coherently than expected on simple models of regulation, as revealing coherence. I shall discuss later another surprising example of coherence, and how its existence might tell us something about the cellular network.

THE BIOLOGICAL MEANING OF SOME ADAPTABLE STATES

The biological significance of low- and high-zone paralysis was unclear at the time of their discovery, and is still apparently a controversial issue. I shall outline findings that provide, in my view, strong grounds for believing that low- and high-zone paralysis reflect regulation whose normal biological function is to control the class of response an antigen induces. The most compelling observations are:

a. The induction of different classes of immunity by antigen is exclusive. Different conditions of immunization are, thus, required to induce different classes. This exclusiveness was originally observed for the induction of IgG antibody and delayed-type hypersensitivity (DTH), a major subclass of cell-mediated immunity (known to be important in controlling the spread of several pathogens). Exclusiveness is even observed when the antigen induces both DTH and antibody; DTH is induced first, and declines as IgG antibody is produced. The existence of exclusion must be accounted for by any comprehensive description of how the immune response is regulated (reviewed in 3).

b. The induction of an antibody response to an antigen generally renders the animal unresponsive for the induction of DTH to this antigen .[3–6] This phenomenon, termed immune deviation when discovered about twenty years ago, suggests either that precursor cells for DTH are directly inactivated during a humoral response, or that antigen-specific regulatory elements (molecules or cells) are produced that inhibit the induction of DTH.

c. Parish, in studies modelled on those of Mitchison, found that the conditions leading to both low- and high-zone paralysis were just those under which DTH is induced.[7] These findings suggested that, when DTH is induced, precursor cells involved in the humoral response are either directly inactivated, or specific cells and/or molecules are produced that inhibit their induction. Parish's observations appear to reflect a phenomenon that is the complement of immune deviation: the induction of DTH is associated with unresponsiveness for a humoral response, whereas in immune deviation the induction of antibody is associated with unresponsiveness for the induction of DTH. It seems highly likely that the regulation responsible for these phenomena serves the biological purpose of ensuring that the induction of strong DTH and IgG responses is exclusive.

d. This latter proposal is strongly supported by further studies. My colleagues and I showed that mice mounting a strong humoral response contain $Ly1^+ Ly2^- Ia^-$ T cells able to inhibit the DTH response, and mice mounting a strong DTH response contain $Ly1^- Ly2^+ Ia^+$ T cells able to inhibit the antibody response. These two subclasses of suppressor T cell, referred to respectively as TsDTH and TsAb, differ in three characteristics, namely the conditions under which they are induced, the surface markers they bear, and the class of immunity whose induction they inhibit.[8–10] Their existence was predicted by a theory whose purpose was to describe how antigen interacts with lymphocytes to determine the class (DTH, IgM and IgG antibody) of response induced.[3,11,12] These findings, and Parish's observations that low- and high-zone paralysis for the antibody response are associated with a state of DTH, imply that low- and high-zone paralysis should be associated with the generation of TsAb. TsAb are present after giving high amounts of antigen leading to high-zone paralysis.[13] The existence of TsAb in mice given doses of antigen subimmunogenic for an antibody response, and leading to low-zone paralysis, was demonstrated later.[14]

A DECISION CRITERION

Exclusivity between DTH and IgG responses may be accounted for by the induction of TsAb and TsDTH, as I favour, but this proposal, even if true, only highlights the question of how the induction of these regulatory cells themselves is regulated. The exclusivity between the induction of TsDTH and TsAb is easily understood if they each directly or indirectly inhibit each other's induction; in this case the subclass that is initially slightly more efficiently induced will inhibit the induction of its adversary, with the consequence that the inhibition of its own induction is relieved and it becomes dominant. This proposal raises the further question of what conditions allow one class of cell to be initially more efficiently induced than the other. The same question is more often posed from the stand point of the regulation of the humoral response, namely why does antigen sometimes preferentially induce helper T cells (ThAb) and under other circumstances suppressor T cells (TsAb)? A most important clue comes from another example of coherence.

An immunologist, faced with the problem of wanting to raise IgG antibody to a hapten, will couple it to a carrier able to induce such a response, and will immunize with the conjugate under conditions that favour an IgG response to the carrier. A favoured carrier with these properties is kehole limpet hemocyanin (KLH). Similarly, if the immunologist wants predominantly IgM or IgE anti-hapten antibody he couples the hapten to carriers able to induce such responses, for example to bacterial polysaccharides for IgM responses and extracts of Ascaris for IgE responses. Similarly, if he wishes to induce DTH to the hapten he will couple it to a carrier able to induce such a response. This strategy reflects a generalization so well known that its significance is in danger of being unappreciated, namely that there is a strong tendency for the class of response to all the different determinants or epitopes on an antigen to be coordinately, or coherently, regulated. This must mean that there is at least one property of the carrier that is "recognized" by the immune system and this "recognition" employed in the process that determines the class of immunity an antigen induces.

It is an experimental generalization that antigens with few foreign epitopes, either because they are small in size or because most of their epitopes are self, cannot induce IgG antibody but can induce DTH .[6,11,12,15] Such antigens only induce IgG antibody when coupled to an entity with many foreign sites, thereby becoming part of such an entity. Examples of antigens with few foreign sites able to induce only DTH are minor histocompatibility antigens and slightly denatured self antigens and small peptides.[6,7,11,15] Conversely. antigens with many foreign sites are potent inducers of IgG antibody. For example, antigens from animals closely related phylogenetically to mice are less immunogenic in mice for the induction of antibody than the homologous antigen from a more distantly related species. (A single intravenous shot of 500 million rat red blood cells induces a substantial DTH and IgM response in mice, but undetectable IgG antibody, whereas a similar dose of more foreign red blood cells will induce substantial IgG antibody [my unpublished observations]).

These experimental generalizations must mean that the immune system has a mechanism for assessing the foreignness of an antigen. This assessment requires there to be specific cells, or molecules derived from them, that recognize foreign as opposed to self epitopes. It is generally accepted that there are insignificant numbers of helper T cells specific for self epitopes, and that helper T cells specific for foreign epitopes on an antigen are required for the antigen to induce cell-mediated[16] and humoral immunity.[17] It is appealing, in view of this, to suggest that the generation of a few helper T cell-dependent signals (by an antigen with few foreign epitopes for which there are consequently few helper T cells) results in the induction of DTH and TsAb, and the generation of more of these signals (by antigens with many foreign sites for which there are many helper T cells) in the induction of IgG and TsDTH.[11] Furthermore, I have already discussed observations that impose considerable restraints on how these helper T cells might act. The class of response mounted against a haptenic epitope is in general the same as that mounted against the carrier epitopes. This requires the "foreignness" of the carrier to be assessed when an anti-hapten precursor cell is induced, a requirement satisfied if the helper T cells act via the linked recognition of antigen. We can express the essence of this proposal by stating that the induction of different classes of immunity requires the formation of different numbers of complexes between the antigen-specific receptor on the relevant precursor cell, antigen and antigen-specific factors on, or derived from, helper T cells; the formation of a few (on the surface of, for example, a precursor for TsAb) results in the induction of TsAb and DTH, and the formation of many (on the surface of, for example, a precursor for TsDTH) in the induction of TsDTH and IgG antibody. This is the decision criterion I wish to discuss and whose validity I have attempted to test. I shall refer to it as the *Decision Hypothesis*. I have fully described elsewhere the evidence on which it is based.[6,11,12] I shall only point out here one of its features that I find appealing. The number of complexes formed between receptors, antigen and helper T cell factors will depend on the concentration of antigen present. Consider an antigen with many foreign sites present at an optimal concentration for the formation of inductive complexes, and, hence, present under those conditions leading to the induction of antibody and TsDTH. Increasing or deceasing the antigen concentration will result in the formation of fewer complexes and consequently the induction of DTH and TsAb. (Higher concentrations of antigen result in the increasing saturation of both the specific receptors on precursor cells and helper T cell-derived specific factors, and, hence, the formation of fewer complexes.) The Decision Hypothesis, thus, accounts for "low-zone" and "high-zone" paralysis. The hypothesis is quantitative and it accounts for all the quantitative observations on conditions favouring the induction of DTH and IgG responses.[11,12]

AN EXPERIMENTAL TEST OF THE DECISION CRITERION

Consider an antigen with many foreign sites impinging upon the immune system at a concentration optimal for the formation of inductive complexes and, thus, leading to the induction of antibody. The Decision Hypothesis accounts for the facts that increasing or decreasing the antigen concentration results in the induction of TsAb and DTH, as discussed above. The Decision Hypothesis predicts another way in which the response can be switched from antibody to DTH. Consider what is expected to happen as the number of lymphocytes is steadily decreased in the presence of the same optimal concentration of antigen able to induce an antibody response in a normal animal. The number of helper T cells will diminish as the number of lymphocytes decreases. A point will be reached when few rather than many inductive complexes will form, and TsAb and DTH should therefore be induced instead of TsDTH and antibody. Furthermore, the addition of specific helper T cells to such a reduced population of cells should result in the specific and coordinate suppression of DTH and the mounting of an IgG response. This "thought experiment" tests the legitimacy of an equivalence relationship predicted by the Decision Hypothesis: *an optimal concentration of an antigen with many foreign sites in the presence of a few unprimed lymphocytes will induce the same DTH response as an optimal concentration of an antigen with few foreign sites in the presence of many lymphocytes.* I have attempted to achieve an approximation to this idealized experiment.

I chose to examine the immune response of lethally irradiated mice which had been reconstituted with different numbers of normal spleen cells and a constant amount of antigen in an attempt to test the Decision Hypothesis. The lymphocytes of the irradiated mice cannot respond to the antigen as a result of irradiation, and the immune response observed is due to the donated normal spleen cells. All mice were given the same dose of antigen that induces a strong and brisk IgG response in a normal mouse. The number of spleen cells given to reconstitute the irradiated mice was varied and the size of the DTH and antibody responses measured six or seven days after reconstitution and antigen challenge. (The experimental details have been fully described elsewhere.[18]) The observations were used to calculate the relative efficiency with which DTH, IgM and IgG precursor cells are induced as a function of the number of normal spleen cells used to reconstitute the irradiated mice. Three modes of response are found: i) the efficiency of induction of all responses decreases to undetectable levels when few (<3 million) spleen cells are employed for reconstitution, ii) only a DTH and small IgM response when a medium number (5-7 million) of cells are used, and iii) substantial IgM and IgG responses and negligible DTH reactivity are induced with 35 million spleen cells. Three features of this pattern are noteworthy: a) the efficiency of induction of all responses initially increases with increasing cell number, a phenomenon most likely reflecting the requirement for helper T cells in the induction of all these responses (see later), b) the induction of DTH and IgG responses are exclusive and c) a smaller innoculum of normal spleen cells is induced by antigen to express DTH than is required to obtain an IgG response.[18] This is consistent with the Decision Hypothesis that more helper T cell-dependent signals are required to induce IgG than DTH responses. Further observations are consistent with more specific predictions of the hypothesis. In this system, 5 million normal spleen cells allow antigen to induce only a DTH response, and this response declines and an IgG response mounted as the size of the innoculum is increased. An experiment was designed to determine whether this decline in DTH reactivity and increase in the antibody response was due to the increase in number of T cells as predicted by the Decision Hypothesis. The response was assessed of irradiated mice reconstituted with a number of normal spleen cells that by themselves induce DTH (5 million) and a further number of T cell-depleted spleen cells that would have supported an IgG response if undepleted. These mice mount only a DTH response, demonstrating that the T cells present in larger innocula of normal spleen cells are responsible for both the decline in DTH reactivity and the production of IgG antibody. Furthermore, it was shown that the provision of increasing numbers of antigen-specific T cells to 5 million normal spleen cells, that by themselves support only the induction of a DTH response, results in the increasing suppression of this DTH response and the increasing production of IgG antibody. These cells act in an antigen-specific manner; they cause a switch from a DTH to an antibody response only when specific for the challenging antigen, and bear markers characteristic of helper T cells ($Ly1^+Ly2^-$).[18] It is important to stress that they coordinately suppress the DTH and help the antibody response, i.e., they have a see-saw effect. This observation provides very strong evidence that these T cells are responsible for determining whether antigen induces a DTH or an antibody response, i.e., control this decision process. It provides direct support for the Decision Hypothesis. The coordinate expression of the ability to suppress DTH and help IgG antibody responses is consistent with but does not prove the possibility that both are the consequence of the same activity of the same subclass of T cell.

The Decision Hypothesis described above incorporates more than one decision criterion. It implies a requirement for helper T cells in the induction of both DTH and antibody responses, and it only has meaning if no response occurs in the absence of helper T cells. There has been evidence for several years that IgG responses require helper T cells,[17] but no evidence until relatively recently for a similar requirement in the induction of DTH. The requirement for such cells can be observed in an experimental system similar to the one described above. In brief, the density at which normal spleen cells are cultured determines the class of response an antigen induces in vitro in a manner similar to the way in which the number of spleen cells, used to reconstitute irradiated mice, determines the class of response induced by antigen in vivo. No response is induced at a relative density of 1 unit, only DTH is induced at a relative density of 3 units and only IgM antibody is induced at a relative density of 9 units.[19] The cells in unit density cultures mount DTH responses if provided with specific helper T cells, and these T cells act most efficiently when specific for epitopes that are physically linked to the antigen whose DTH response is being helped, i.e., these helper T cells act, in the parlance of immunologists, by the linked recognition of antigen.[16] Secondly, the DTH response mounted by cells at a relative density of 3 units can be specifically suppressed and the IgM response specifically helped by providing specific helper T cells.[10] Helper T cells, thus, appear to determine whether any response is made, and the number of helper T cell-dependent signals determines whether antigen induces a DTH or an antibody response.

COHERENCE AND ITS IMPLICATIONS FOR CELLULAR REGULATION

The regulation responsible for the coherence seen in the responses induced by antigen must serve some biological function. I do not have the space to discuss here why it is advantageous for different kinds of antigen to induce different classes of response, and why the same class should be induced to the different epitopes of an antigen molecule under a given set of conditions of immunization, even though these are central aspects of the theory.[6,11,12] I shall assume that the mere existence of coherent regulation of the immune response to different epitopes on the same antigen molecule is sufficient evidence of its biological importance.

I gave an example of coherence at the beginning of this article, and explained why it was difficult to explain within the context of a simple counting model of induction. I wish to explore here the idea that, given the existence and therefore the importance of coherence, some of the complexity in the cellular regulation of the immune response might be required to achieve such coherence. Consider the findings I described above on how the density at which normal spleen cells are cultured determines the kind of response antigen induces: unit density cultures are not induced to mount any response, 3 unit density cultures mount only a DTH response, and 9 unit density cultures only mount an IgM response. I shall assume for the purposes of this discussion that this density dependence reflects the decision criterion that the number of helper T cell-dependent signals required for the induction of the DTH and IgM response is different. Consider the simplest of models consistent with this idea. We suppose that there are no other regulatory cells besides the helper T cell, and that the T cell precursors for DTH, pDTH, and the IgM B cell precursor, pIgM, have the same number of antigen-specific receptors (a trivial assumption for the sake of convenience and simplicity) with the same range of binding constants. We then try to explain the cell density dependence of the induction of different classes by postulating a difference in the number of inductive complexes required for activating pDTH and pIgM cells. We also assume that the formation of these complexes depends on the value of the receptor's binding constant. This is really a variation of the counting model we discussed earlier. In this case it is difficult to explain how the class of immunity induced by antigen can depend so sharply on cell density, particularly in view of the fact that complex antigens (xenogeneic red cells with a multiplicity of different epitopes) were employed in these studies. A partial solution to this quandary is to admit the model is too simple, and to point out that we know cellular regulation is more complex; TsAb cells are induced under conditions favourable to the induction of DTH, and TsDTH cells are generated under conditions favourable to the induction of antibody. This solution raises the further question of how the induction of these different subclasses of T cell can themselves be so coherently regulated that they are able to impose coherence on the activation of pDTH and pIgM cells. It seems to me that heterogeneous populations of cells can be made to respond coherently in response to external variables if the cellular circuits have two components: i) a series of decision criteria which determine what is required (e.g., induction of helper T cells for DTH and TsAb) and ii) highly cooperative interactions that then allow an appropriate but heterogeneous subclass of lymphocyte to be induced or suppressed in a coherent fashion.[19] A likely example of such highly cooperative interactions has become apparent recently. Some helper T cells possess the autocrine property of not only producing IL-2 upon appropriate antigenic stimulation but requiring IL-2 and antigen to grow (see Kevrekidis, Zecha and Perelson, this volume and [20]). Are such regulatory loops essential for the immune system to achieve coherence? It is possible to imagine, for example, that once a decision criterion is satisfied antigen is "allowed" to stimulate high affinity "helper" T cells to produce IL-2. The autocrine property of this subclass should then allow the stimulation of other cells belonging to the same subclass but with lower affinity for the antigen. In this way it seems possible to imagine how heterogeneous cells can be coherently regulated. It seems to me that coherence cannot be explained by simple models and that coherence can therefore be used to explain the need for complexity.

CONFLICTING DECISION CRITERIA

A number of other ideas are current concerning the factors that control whether antigen induces helper or suppressor T cells and whether antigen induces antibody or cell mediated immunity. These ideas are difficult to reconcile with the Decision Hypothesis I have described. It is therefore appropriate to discuss them here.

Several different explanations rely on the idea that the response of different classes or subclasses of lymphocyte are restricted to different epitopes, and, hence, different antigens bearing different epitopes induce different subclasses. Probably the earliest example of such an idea was that haptens can only induce B cells. This idea arose from studies demonstrating that a vigorous secondary response to a hapten requires preimmunization with both the hapten and the carrier even when the challenging conjugate contains several hapten molecules.[17] This requirement suggested to many that B-T collaboration could occur through hapten-carrier epitope recognition but not through hapten-hapten recognition, and this inference could in turn be explained if haptens were unable to induce helper T cells. I think this conclusion unnecessary as the original observations can be naturally accounted for on quantitative grounds. The idea that helper T cells cannot recognize haptens is in conflict with the generalization that the immune system has the capacity to recognize virtually anything, and no general property of haptens has been suggested that would form a basis for their lack of recognition by helper T cells. The idea that foreign haptens cannot in general induce a subclass of lymphocyte also conflicts with the Decision Hypothesis. It is now known that haptens can induce helper T cells, e.g., [21]. Another example of the idea that different classes of lymphocyte are restricted to recognizing different kinds of epitopes is reflected in the frequent statement that B cells "see" conformational determinants and T cells "see" sequential determinants. This statement is sometimes used as an explanation for why some antigens induce cell-mediated and others antibody responses. It is alright if meant as a rough (and not universally true) experimental generalization, but not if meant as a basic decision criterion. In the latter case it is in conflict with the Decision Hypothesis. Two examples show that the idea of innate restriction of B cells to conformational epitopes and T cells to sequential epitopes is false. Firstly, B cells can respond to haptens, and anti-hapten antibody will bind the hapten of a hapten-antigen conjugate independently of the nature of the antigen, and to the appropriate fragments obtained by digesting the hapten-carrier conjugate with proteolytic enzymes. In other words, haptens are perfect examples of sequential determinants, and are "seen" by B cells. Indeed, it is possible to make antibodies to small "haptenic" peptides expressing only sequential epitopes by immunizing with them when coupled to an immunogenic carrier such as kehole limpet hemocyanin. Again, it is the carrier which determines in large measure whether a particular response is mounted. The small peptide is not immunogenic (for the induction of antibody) in contrast to antigenic because it contains (according to the Decision Hypothesis) too few foreign sites. Many small peptides have the three properties of being too small to have conformational epitopes and of being immunogenic for

T but not for B cells. These properties form in part the basis for the idea that T cells see only sequential determinants. It is known, however, that DTH can be induced to slightly denatured self antigens,[11,15] as revealed in contact sensitivity to metal ions; it is inconceivable in this case that the T cells recognize the metal ion directly, but rather self antigens that have bound the metal ion and consequently changed their three dimensional structure. It seems that the epitopes recognized on a protein would have to be conformational as the normal and distorted self protein would have the same sequence. The third example I wish to discuss is the proposal that the specialization of different subclasses of T cell to the recognition of different epitopes is the basis on which the immune system decides whether a particular antigen should induce helper or suppressor T cells. An antigen with relatively many "helper epitopes" would preferentially induce help, and one with relatively many "suppressor epitopes" would preferentially induce suppression.[22] (This hypothesis is incomplete and awkward if the induction of helper and/or suppressor T cells requires the existence, as is likely, of further epitopes through which helper or inducer T cells can act.) According to this proposal, the intrinsic nature of the epitopes is the crucial factor in determining whether helper or suppressor T cells are preferentially induced. It often takes the form that the decision between the preferential induction of helper or suppressor T cells depends on the ratio at which these two subclasses of cell are present. This is inconsistent with the Decision Hypothesis, and is difficult to reconcile with the finding that the same antigen can preferentially induce helper or suppressor T cells depending on the antigen concentration. Decreasing the antigen concentration will not systematically change the effective ratio of specific helper to specific suppressor T cells and cannot therefore account for the induction of TsAb in low-zone paralysis. The experiments outlined above and carried out to test the Decision Hypothesis are also inconsistent with the idea that the ratio of suppressor to helper T cells determines whether DTH or antibody in induced; in this case the same population of cells is induced by antigen to express DTH or antibody depending on the number of cells used to reconstitute irradiated mice. The ratio of helper to suppressor T cells in mice reconstituted with 5 million cells, and allowing antigen to induce DTH, is the same as the ratio in mice reconstituted with 35 million of the same cells, allowing the mice to mount an antibody response.

Finally, an almost universally held idea is that the limiting step in the induction of an immune response is the association of antigen, or its processed fragments, with class II molecules. I shall refer to this as the antigen presentation model. The strength of this association is regarded as representing the decision criterion of whether to respond or not. There is strong evidence to support this view. The recent measurements of the strength of the interaction between antigenic fragments and the appropriate class II antigens would appear to provide quantitative observations in its favour.[23] I find it most difficult to accept the conclusions drawn from these studies as generally valid for two reasons. Firstly, according to the Decision Hypothesis, the decision criterion of whether to respond or not depends on whether there are a sufficient number of helper T cells available, a parameter that would depend for the most part on the number of foreign epitopes an antigen bears.

This property is unrelated to the antigen's intrinsic ability to interact with class II molecules and, hence, the antigen presentation model appears to be in conflict with the Decision Hypothesis. It is also in conflict with the general finding that the more similar an antigen is to a self antigen the more difficult (within certain limits) it is to raise an immune response against it. It is next to impossible to raise immune responses against legitimate self antigens (without employing tricks(!) such as adjuvants), and yet there is nothing about such antigens that prevents them from interacting with self class II molecules as well as foreign antigens or peptides. In other words, what we know about the properties of an antigen that limits the induction of an immune response at the whole animal level does not correlate in any way with the ability of the antigen to interact with class II molecules. Secondly, given the kind of interaction postulated, it is very difficult to understand why the site or sites on class II molecules are not saturated with self antigens or their fragments and, thus, unable to bind the foreign antigen/peptide. Self and foreign antigens must interact differently with their respective precursor helper T cells if there is to be self non-self discrimination. This need led to the postulate that specific helper T cells are required to induce helper precursor cells,[24] so that helper T cells specific for self are not induced in the absence of effector helper cells. The postulate that the primary induction of help requires T-T cooperation, mediated by the linked recognition of antigen, has been recently tested.[25]

PROSPECTS

I am optimistic that we shall be able to delineate decision criteria without characterizing every detail of every cell interaction, and that the validity of some decision criteria will be convincingly demonstrated as we manage to use them to specifically modulate the immune response (as already achieved in part). It is very important, in attempts to test the validity of proposed decision criteria, to be clear of what is essential to a proposal and what is merely a matter of simplicity and assumed as a convenience in discussion. I am aware that I have simplified what I wish to say in the hope that I could make it more comprehensible. There is always a danger of oversimplifying to such an extent that one's meaning is incomprehensible and/or misunderstood. I should state, in order to avoid confusion, that the Decision Hypothesis I have outlined does not require that the same subclass of helper T cell be involved in the induction of all precursor cells, and that one of the non-trivial simplifications I have used is to employ a static picture of cell interactions. It is clear that the formation of inductive complexes is not instantaneous, and will change with time as helper T cells are induced. The Decision Hypothesis I have described will have to be reformulated in a more elaborate form, bringing in such parameters as the time it takes cells to divide and differentiate to express their regulatory activity. The dynamic properties of the regulation have been ignored in my discussion. I should like to end by describing some of our recent observations that both illustrate this dynamic property and provide a basis for the hope that the immune response can be modulated in certain diseases. It is known that some people make an ineffective response upon natural infection by microorganisms. People making only antibody to the causative microorganism of leprosy are far more ill than those expressing DTH. It might be thought that such people harbour TsDTH cells and that such cells present an irreversible barrier against modulating their response to achieve the expression of DTH. A dynamic view of regulation would suggest, however, that such patients are continually induced to synthesize antibody, and that the TsDTH cells, having a finite half life, must be continually regenerated. The establishment of conditions unfavourable to this regeneration might lead to their decay and the development of DTH. We believe we have achieved this. Cells humorally immune to xenogeneic red cells, expressing no DTH and containing TsDTH, can upon appropriate culture give rise to cells mediating DTH to this antigen.[26] These findings illustrate the reversibility of a decision process, a phenomenon most easily understood within the context of a dynamic view of cellular regulation. I believe they provide strong grounds for optimism; procedures for modulating on-going immune responses of an inappropriate class to a more effective one can and will be developed.

REFERENCES

1. Green, D. R., P. M. Flood, and R. K. Gershon (1983), "Immunoregulatory T Cell Pathways," *Ann. Rev. Immunol.* **1**, 439–463.
2. Mitchison, N. A. (1969), in *Regulation and the Antibody Response*, Ed. B. Cinader (Illinois: Thomas Springfield).
3. Bretscher, P. A. (1981), "Significance and Mechanisms of Cellular Regulation of the Immune Response," *Fed. Proc.* **40**, 1473–1478.
4. Asherson, C. R., and S. H. Stone (1962), "Selective and Specific Inhibition of 24-Hour Skin Reactions in the Guinea Pig. I. Immune Deviation: Description of the Phenomenon and the Effect of Splenectomy," *Immunol.* **9**, 217.
5. Battisto, J. R., and J. Miller, "Immunological Unresponsiveness Produced in Adult Guinea Pigs by Parenteral Introduction of Minute Quantities of Hapten or Protein Antigen," *Proc. Soc. Exp. Biol. Med.* **111**, 111–115.
6. Crowle, A. J., and C. C. Hu (1966), "Split Tolerance Affecting Delayed Hypersensitivity Induced in Mice by Pre-Immunisation with Protein Antigens in Solution," *Clin. Exp. Immunol.* **1**, 323–335.
7. Parish, C. R. (1972), "The Relationship Between Cell-Mediated and Humoral Immunity," *Transplant. Rev.* **13**, 35–66.
8. Ramshaw, I. A., P. A. Bretscher and C. R. Parish (1976), "Regulation of the Immune Response I. Suppression of Delayed-Type Hypersensitivity by T Cells from Mice Expressing Humoral Immunity," *Eur. J. Immunol.* **6**, 674–679.
9. Ramshaw, I. A., P. A. Bretscher and C. R. Parish (1976), "Regulation of the Immune Response II. Repressor T Cells in Cyclophosphamide-Induced Tolerant Mice," *Eur. J. Immunol.* **6**, 674–679.
10. Ramshaw, I. A., I. F. C. McKenzie, P. A. Bretscher and C. R. Parish (1977), "Discrimination of Suppressor T Cells of Humoral and Cell-Mediated Immunity by Anti-Ly and Anti-Ia Sera," *Cell. Immunol.* **31**, 364–369.
11. Bretscher, P. A. (1974), "Hypothesis: On the Control Between Cell-Mediated and Humoral Immunity," *Cell. Immunol.* **13**, 171–195.
12. Bretscher, P.A. (1977), in *B and T Cells in Immune Recognition* (New York: Academic Press), pp. 457–485.
13. Basten, A. (1974), in *Immunological Tolerance* (New York: Academic Press), pp. 107–119.
14. Weber, G., and E. Kolsch (1975), "Transfer of Low-Zone Tolerance to Normal Syngeneic Mice by Theta Positive Cells," *Eur. J. Immunol.* **3**, 767–772.
15. Pearson, M. N., and S. Raffel (1971), "Macrophage-Digested Antigen as Inducer of Delayed Hypersensitivity," *J. Exp. Med.*, 494–505.
16. Tucker, M. J., and P. A. Bretscher (1982), "T Cells Cooperating in the Induction of Delayed-Type Hypersensitivity Act via the Linked Recognition of Antigenic Determinants," *J. Exp. Med.* **155**, 1037–1049.

17. Cheers, C., and J. F. A. P. Miller (1972), "Cell-to-Cell Interaction in the Immune Response IX. Regulation of Hapten-Specific Antibody Class by Carrier Priming," *J. Exp. Med.* **136**, 1661–1665.
18. Bretscher, P. A. (1983), "Regulation of the Immune Response by Antigen I. Specific T Cells Switch the In Vivo Response From a Cell-Mediated to a Humoral Mode," *Cell Immunol.* **81**, 345–352.
19. Bretscher, P. A. (1983), "In Vitro Analysis of the Cellular Interactions between Unprimed Lymphocytes Responsible for Determining the Class of Response an Antigen Induces: Specific T Cells Switch a Cell-Mediated Response to a Humoral Response," *J. Immunol.* **131**, 1103–1107.
20. Cantrell, D. A., and K. A. Smith (1984), "The Interleukin-2 T Cell System: A New Cell Growth Model," *Science* **224**, 1312–1316.
21. Bretscher, P. A. (1984), "Hapten-Ficoll Conjugates Induce T Cell-Dependent IgM Anti-Hapten Responses and T Cells Mediating Hapten-Specific Delayed-Type Hypersensitivity," *Cell. Immunol.* **85**, 396–405.
22. Goodman, J. W., and E. E. Sercarz (1983), "The Complexity of Structures in T-Cell Activation," *Ann. Rev. Immunol.* **1**, 465–498.
23. Buus,S., A. Sette, S. M. Colon, C. Miles and H. M. Grey (1987), "The Relation Between Major Histocompatibility Complex (MHC) Restriction and the Capacity of Ia to Bind Immunogenic Peptides," *Science* **235**, 1353–1358.
24. Bretscher, P. A., and M. Cohn (1971), "A Theory of Self-Nonself Discrimination," *Science* **169**, 1042–1048.
25. Bretscher, P. A. (1986), "A Cascade of T-T Interactions, Mediated by the Linked Recognition of Antigen, in the Induction of T Cells Able to Help Delayed-Type Hypersensitivity Responses," *J. Immunol.* **137**, 3726–3733.
26. LeClercq, S., and P. A. Bretscher, "Immune Deviation is Reversible," *Eur. J. Immunol*, in press.

PETER J. HINGLEY
Kingston Polytechnic, Penrhyn Road, Kingston upon Thames, Surrey, England, KT1 2EE

Antibody Affinity Distribution-Based Models for Vaccine Potency Assessment

ABSTRACT

Extensions of conventional methods for measurement of the potency of foot and mouth disease vaccine batches are considered. A timecourse model is described; based on a single differential equation for antibody of affinity K, with such equations collected together over all clones. The model exhibits useful properties, the persistence of mean antibody affinity over total antibody concentration giving a simple mathematical expression for immune memory. The model is related to information obtainable from immunoassays, specifically the neutralization test and the ELISA test. A logistic model is shown to be applicable to ELISA data, and areas for further experimentation are identified.

INTRODUCTION

This paper will consider new approaches to the potency assessment of vaccines against foot and mouth disease (FMD), giving rise to methodology that also might possibly be applicable to other vaccines. A timecourse model will be described that has useful properties, and the practical problems of applying such a model to data will be discussed.

Conventional methods for the potency assessment of FMD vaccines do not utilize all the available information on the immune responses provoked in the tested animals. Methods vary from country to country, but in the UK a total of twenty four cattle are used in three groups of eight, with each group receiving a different vaccine dilution. At twenty-one days after vaccination, live virus is inoculated into the tongues of the animals, and protection to disease is defined in terms of inhibition of the subsequent development of characteristic lesions on the animals feet. Potency is estimated as the dilution that provides 50% probability of protection against such lesions (Hingley and Pay, 1987). It can be shown that these potency measures have wide confidence intervals (Figure 1).

Antibody levels are obtained, as a routine adjunct to the challenge test, in the form of neutralization test titres. Since the immune response is a time dependent process, it should also be possible to increase information on potency by making repeat measurements on animals over time. Methods have been described for determining potency from such measurements, but they are also subject to large errors. It may be practical to apply other immunoassays, such as ELISA, to serum samples; and this might be preferable if such an assay could be shown to reflect biochemical facets of the underlying immune process more precisely.

A TIMECOURSE MODEL

A model for antibody levels has been defined which is based on the ideas of clonal selection and, it is hoped, will prove simple enough to fit to immunoassay data (Hingley, 1985, 1988). The essential feature of the model is, that if antigen is introduced at a time t_0, the rate of change with time of concentrations of antibody of specific clonal affinity K is governed by the following equation.

$$\frac{\delta y(t,K)}{\delta t} = Ay(t_0,K) + cKx(t)y(t,K) - Ay(t,K) \qquad c>0,\ A>0,\ K>0. \tag{1}$$

where:

$y(t,K)$ is the concentration of antibody of affinity K at time t,

$x(t)$ is the concentration of antigen at time t, assumed to decline through a process of exponential decay,

A is the excretion rate, and also the background production rate, for antibody from the system,

c is a constant of proportionality.

In order to ensure that the baseline of antibody can be maintained before t_0 and also that, after t_0, $y(t,k)$ returns to baseline as antigen becomes depleted,

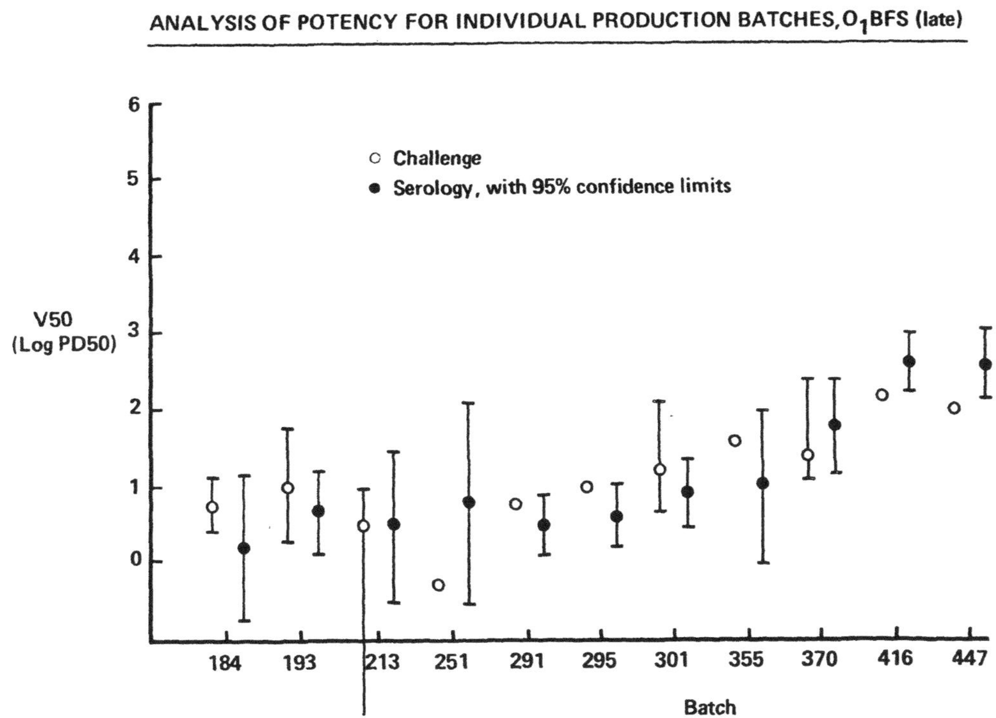

FIGURE 1 Analysis of potency for individual production batches of FMD vaccine, type O_1BFS. V50 indicates estimated log vaccine dilution to provide 50% probability of protection to challenge. ○ Challenge (with 95% fiducial limits where calculable, in other cases limits can be considered to be infinite). ● Serology (with 95% fiducial limits).

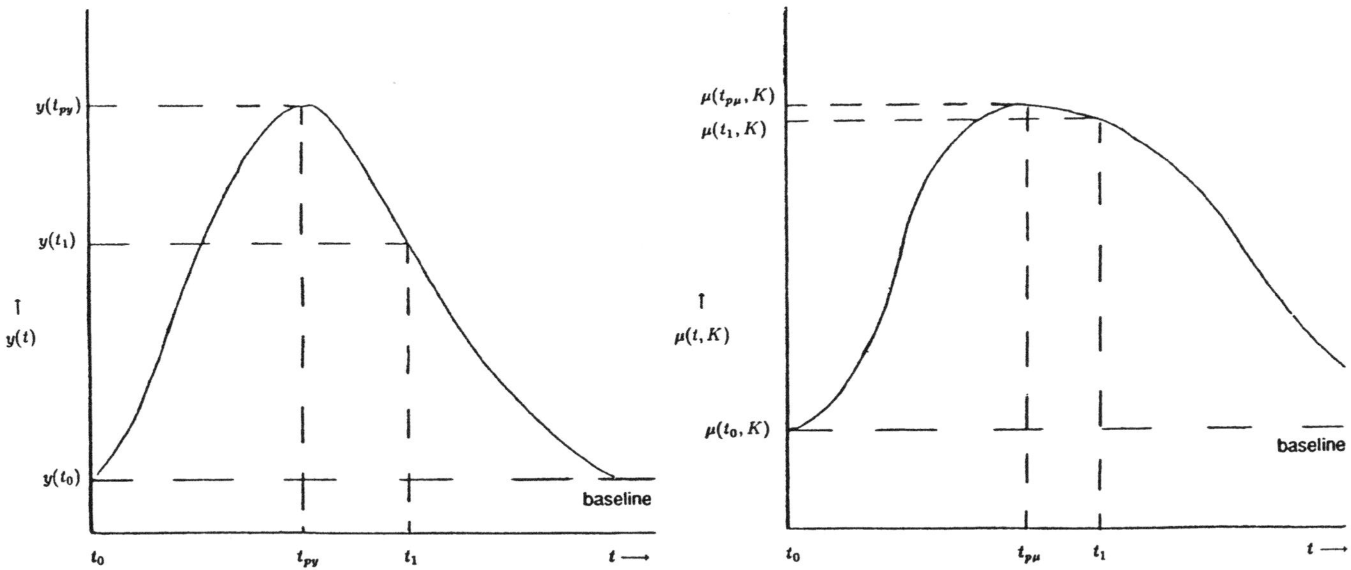

$$raty(t_1) = \frac{y(t_1) - y(t_0)}{y(t_{py} - y(t_0)}$$

$$rat\mu(t_1) = \frac{\mu(t_1, K) - \mu(t_0, K)}{\mu(t_{p\mu}, K)_\mu(t_0, K)}$$

FIGURE 2 Schematic timecourse response profiles of total antibody concentration $(y(t))$, and mean antibody affinity $(\mu(t, K))$. It can be shown that the ratio $rat\mu(t)$ persists at higher levels than the ratio $raty(t)$, providing a description of immune memory.

a constant production term, $Ay(t, K)$, and a negative exponential excretion term, $Ay(t, K)$, have been incorporated into Eq. 1. A is assumed to be a positive constant, and can be considered both as a turnover rate of background antibody and as the rate of excretion of antibody from the animal. A partial differential equation is used to indicate a derivative with respect to t, holding K constant. The equation is replicated over all clones, covering all allowable values of K. Since there are many clones, a continuous approximation, if necessary, can be made to the distribution of K among all antibodies. It is possible to solve Eq. 1 and to compute $y(t, K)$ for all desired values of t and K by a numerical method.

The total antibody at time t can be obtained by integration.

$$y(t) = \int_0^{\infty} y(t, K)dK \tag{2}$$

A probability density can be defined for antibodies of affinity K at time t, by normalization of $y(t, K)$.

$$P(t, K) = \frac{y(t, K)}{y(t)} \tag{3}$$

P(t,K) will be termed the "affinity density" at time t. It is possible to use Eqs. (1)–(3) to obtain the following differential equations for the rate of change of total antibody $y(t)$, and mean antibody affinity $\mu(t, K)$, with time.

$$\frac{dy(t)}{dt} = cx(t)\mu(t, K)y(t) + A\,[y(t_0) - y(t)] \tag{4}$$

$$\frac{d\mu(t, K)}{dt} = cx(t)Var[t, K] + \frac{Ay(t_0)}{y(t)}[\mu(t_0, K) - \mu(t, K)] \tag{5}$$

Using these equations it can be shown that when $y(t)$ and $\mu(t, K)$ have peaked and are declining back towards baseline, $\mu(t, K)$ persists at higher levels than $y(t)$ (Figure 2). It is proposed that, if $y(t)$ represents measurable total antibody concentration, $\mu(t, K)$ can be considered to represent the propensity of the system to react to further antigenic stimulation. This feature of the model could quantify the phenomenon of immune memory usefully for some types of data, even though the model does not differentiate between memory cells, lymphocytes, humoral antibody etc.

The model has been used to simulate immune profiles of antibody responses to FMD vaccine, using appropriate parameter values. To demonstrate the model, a functional form must be assumed for the affinity density. In the absence of specific evidence, a simple two-point distribution will be used.

$$P(t, K) = \begin{cases} p(t), & K = K_1 \\ 1 - p(t), & K = K_2 \end{cases}$$

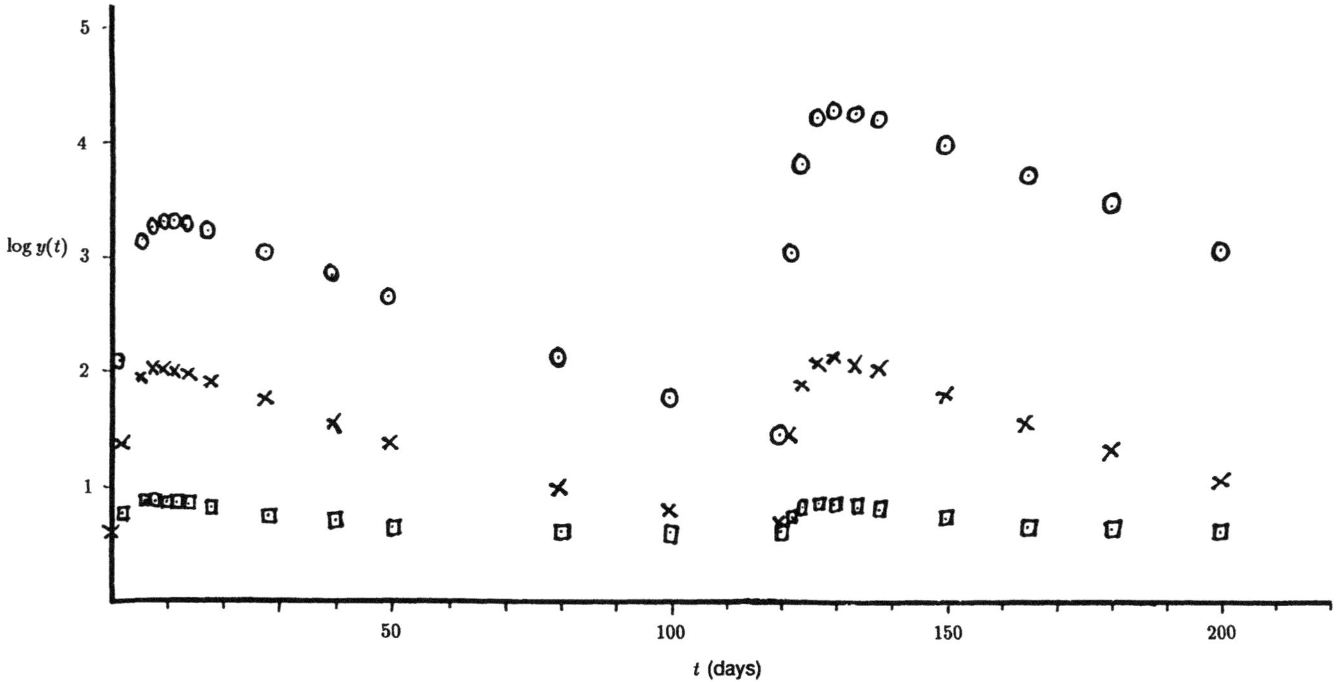

FIGURE 3 Timecourse model with a two-point affinity density, showing effects of dose on profiles of total antibody concentration $y(t)$ vs. time t. Vaccination occurs at $t_0 = 0$ and revaccination at $t = 120$. $y(t_0) = 4$, $A = 0.042$, $c = 0.01$, $K_1 = 0.001$, $K_2 = 10$, antigen excretion rate $= 0.4$, ⊡ $x(t_0) = 5$, X $x(t_0) = 17.5$, ⊙ $x(t_0) = 30$ (with same doses at revaccination). Time units are days.

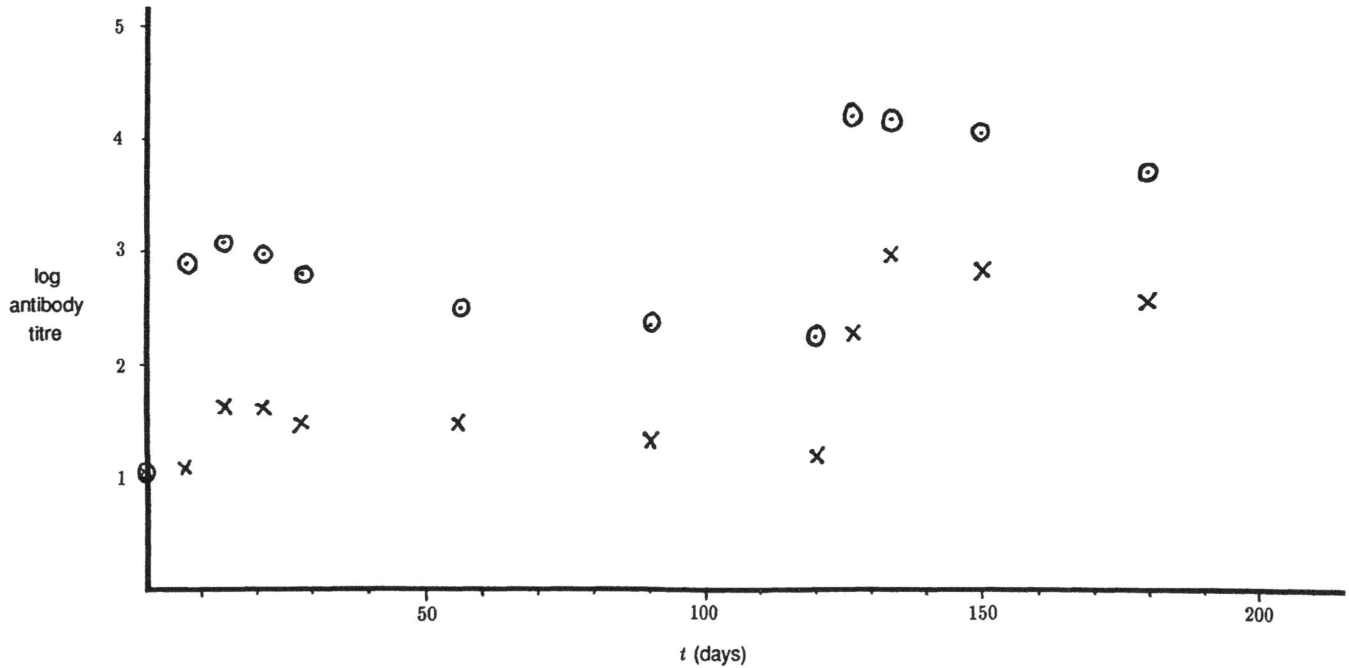

FIGURE 4 FMD neutralization test timecourse data. Profiles for two initial antigen doses (with same doses at revaccination). X $x(t_0) = 10^{1-6}$ ng antigen per dose, $\odot$ $x(t_0) = 10^{5-6}$ ng antigen per dose. Primary vaccination, means of twelve animals. Secondary vaccination, means of four animals (Rweyemamu et al., 1984).

Initially, $p(t_0) = 0.5$.

Simulated curves (Figure 3) show some agreement with measured antibody response profiles (e.g., Figure 4), though problems remain with regard to the dose dependence of the magnitude of the antibody response. These problems might be reduced by a better choice of affinity density, or possibly by replacement of $x(t)$ in Eq. (1) by some more appropriate function of $x(t)$. The paucity of data suggests that it might be more advisable to carry out further biological experiments before performing any such modifications to the model.

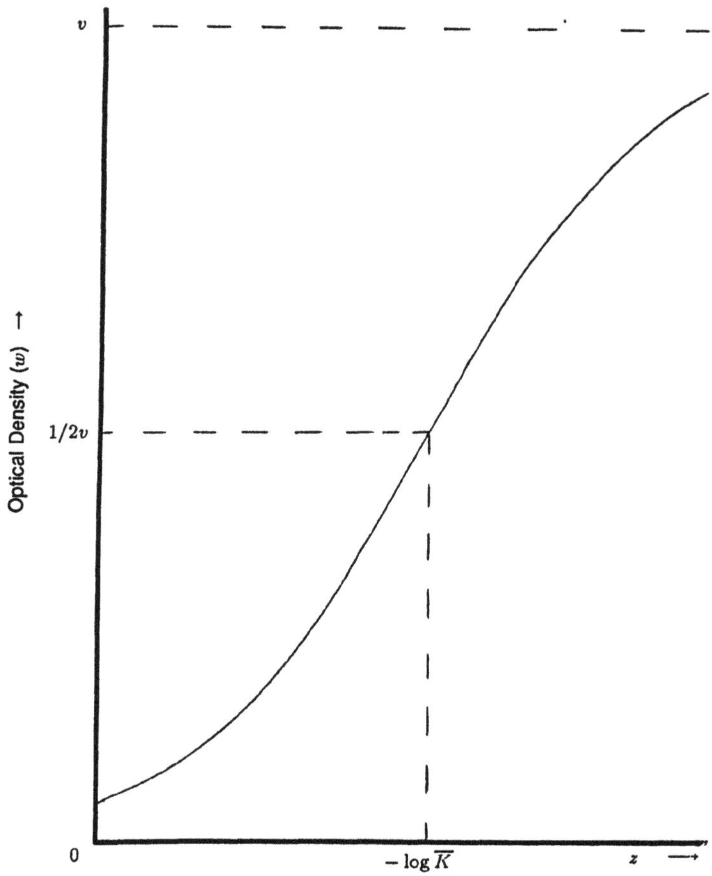

FIGURE 5 Logistic model for ELISA saturation curve data.

MEASURING PARAMETERS OF THE AFFINITY DISTRIBUTION

The timecourse model is heavily dependent upon the concept of the affinity density. A big problem with the resolution of models of this type is that data are collected by immunoassays whose results must be related to parameters of underlying immune processes. For example, the mean log antibody titres of Figure 4 are obtained as 50% endpoints from titrations of serum against standard virus preparations. It is not necessarily true that this variable should be equivalent to the total antibody concentration in any particular timecourse model. It is important to search for immunoassays that correspond more directly to parameters of the underlying affinity distribution, or alternatively to search for models that correspond more exactly to available immunoassays.

The Enzyme Linked ImmunoSorbent Assay (ELISA) seems to be a system with potential. In common with equilibrium dialysis and radioimmunoassay, the extents of reaction of various concentrations of test virus with a serum sample can be measured quantitatively. A characteristic sigmoid "saturation curve" is produced (Figure 5), which it has been found possible to model by using a logistic function (Rodbard, 1974; Hingley and Ouldridge, 1985). The parameters of the curve can be estimated using iterative nonlinear least squares, assuming a homoscedastic error distribution.

$$w = \frac{v(\overline{K}e^z)^a}{1+(\overline{K}e^z)^a} \tag{6}$$

where:

w = observed optical density,
z = log test virus concentration,
$\overline{K}$ = location parameter,
a = heterogeneity index (steepness parameter of logistic function),
v = plateau value of optical density at full saturation.

How are we to relate the estimated parameters of the fitted curve to the parameters of the underlying affinity distribution of antibody? Following Sips (1948), and Bruni et al. (1976), if the simplistic assumption is made of a monovalent antibody-antigen reaction, a saturation curve can be obtained from the affinity density $P(t, K)$ as follows:

$$w = v\int_0^\infty \frac{Ke^z}{1+Ke^z}P(t,K)dK = vE_t\left[\frac{Ke^z}{1+Ke^z}\right] \tag{7}$$

where $E_t[\ldots]$ indicates the expected value at time t.

It is interesting to take $Q = \log K$, and to express the affinity density as a function, $P(t, Q)$, in the metric of Q.

$$w = v\int_{-\infty}^{\infty} \frac{e^{Q+z}}{1+e^{Q+z}} P(t,Q)dQ = vE_t\left[\frac{e^{Q+z}}{1+e^{Q+z}}\right] \tag{8}$$

Working in this metric, it can be shown that the logistic saturation curve (Eq. (6)) corresponds to a specific symmetric density, known as the Sips density, with mean $E_t[Q] = \log K$. The variance is a monotonically decreasing function of a, given by Bruni et al. (1976)

$$Var_t[Q] = \frac{\pi^2(1-a^2)}{3a^2}.$$

In this formulation a is constrained in the interval $0 \leq a \leq 1$, corresponding to all possible levels of $Var_t[Q]$. However, in estimating a from experimental FMD data, it is often found that values greater than 1 are obtained—which can probably be explained by the complexity of the immune reaction between antibody and antigen.

AN INVESTIGATION

Under the assumptions of the particular timecourse model suggested above, *any* sensible density function $P(t_0, K)$ can be modelled, which will in general provide a continuous sequence of densities of changing functional form $P(t, K)$ as the immune response progresses. The exact form of the affinity density during a real immune response at any one time is likely to be difficult to elucidate. Since it seems possible empirically to use a logistic model in the case of FMD data, the most sensible approach may be to assume that the density is of Sips form and investigate both the time dependence of the estimated parameters, and the possible evidence of any significant lack of fit to the logistic model.

Dr. C. Hamblin has provided some data from which a preliminary analysis of this sort can be made. Saturation data were generated using a liquid phase blocking sandwich ELISA. In this assay the serum under test is used to inhibit the reaction of virus with standard antibody, so that optical density *increases* as the serum dilution increases. Another difference is that a fixed test virus concentration is reacted with varying log dilutions of the serum under test. The data do not, therefore, fit Eq. (1), which was formulated in terms of fixed serum and variable antigen concentrations, but Hingley and Ouldridge (1985) have shown that logistic models appear to fit well to data with either component as the variable quantity. The interpretation of a in terms of the affinity density is uncertain in this case, but analysis of the data using this model is informative (Two dimensional ELISA experiments are, in fact, possible, in which both components are allowed to vary; but in this case the logistic model would be inadequate and some other approach would be required).

TABLE 1 Individual Liquid Phase-Blocking Sandwich ELISA Experiment[1]

Log serum dilution	Optical density readings	
1.81	0.157	0.144
2.11	0.303	0.229
2.41	0.538	0.410
2.71	0.620	0.599
3.01	0.863	0.817
3.31	1.049	1.039
3.61	1.147	1.158
3.91	1.184	1.274

[1] Table includes data for an individual liquid phase-blocking sandwich ELISA experiment, using fixed antigen concentration (100 ng. FMD 140S antigen per cell, type O_1BFS), and variable serum dilutions. (Serum from Animal 7, 14 days after vaccination).
Estimates: $\overline{K} = 0.00176$, (s.e. = 0.0002); $v = 1.3406$, (s.e. = 0.0501); and $a = 0.9207$, (s.e.= 0.07). Residual sum of squares = 0.0222 (13 degrees of freedom). Lack of fit sum of squares = 0.0057 (5 degrees of freedom). Pure error sum of squares = 0.0165 (8 degrees of freedom).

ELISA data were generated for twelve cattle, with sera collected at 0, 7, 14, and 21 days after vaccination. The logistic model was successfully fitted to 46 of the 48 resulting saturation curves. Table 1 shows an example of the analysis applied to one particular set of optical density readings. In Figure 6, mean parameter estimates are shown with 95% confidence limits calculated at each time point. Mean affinity ($\log\overline{K}$) increases from t_0, while a decreases, suggesting an increase in the variance of the affinity density as the immune response heads towards its peak. v also approaches a plateau. Unfortunately data were only available for the early stages of the immune responses, but the limited results are consistent with predictions that could have been made by the timecourse model.

The goodness of fit of the logistic model will now be assessed. In Table 2, the pooled residual sum of squares across times for each animal is split into a lack of fit sum of squares and a pure error sum of squares term. An approximate F test to

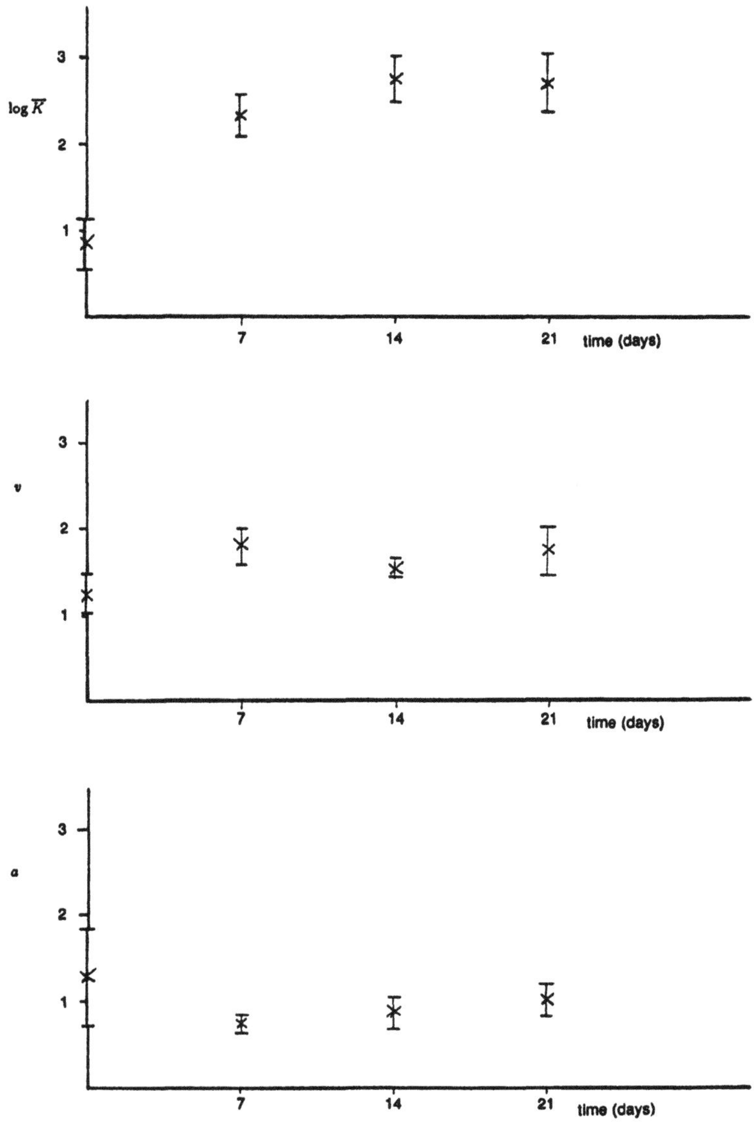

FIGURE 6 Analysis of liquid phase blocking sandwich ELISA optical density results by three parameter logistic model. Twelve animals were vaccinated with O_1BFS FMD vaccine, and sera were extracted at 0, 7, 14, and 21 days after vaccination. Dilution series of individual animal sera were reacted with fixed quantities of O_1BFS test virus in an ELISA system. Points show mean animal responses with 95% confidence limits based on standard deviations between animals.

TABLE 2 Lack of Fit F Tests for Logistic Model (applied to each animal separately)

Animal	Lack of fit sum of squares	Degrees of freedom	Pure error sum of squares	Degrees of freedom	F	P
1	0.2980	16	0.3750	28	1.39	>0.05
2	0.0634	11	0.0492	20	2.34	<0.05
3	0.1460	16	0.5468	28	0.47	>0.05
4	0.0432	16	0.0662	28	1.14	>0.05
5	0.1133	16	0.1155	28	1.72	>0.05
6	0.1295	16	0.1237	28	1.83	>0.05
7	0.1199	16	0.1428	28	1.47	>0.05
8	0.0934	16	0.1399	28	1.17	>0.05
9	0.2420	16	0.4715	28	0.90	>0.05
10	0.1076	15	0.0417	24	4.13	<0.01
11	0.1503	16	0.3227	28	0.82	>0.05
12	0.1332	16	0.2796	28	0.83	>0.05

compare these sums of squares allows the significance of lack of fit of the logistic model to be carried out (Draper and Smith, 1981). For only two animals out of twelve is a significant lack of fit established at the 95% significance level. In Table 3, the sums of squares are tested across animals for each time point, and here it emerges that a significant lack of fit to the logistic model exists only on the day of vaccination, before the immune system has been stimulated by antigen.

TABLE 3 Lack of Fit F Tests for Logistic Model (applied to each time point separately)

Time after vaccination (days)	Lack of fit sum of squares	Degrees of freedom	Pure error sum of squares	Degrees of freedom	F	P
0	0.2326	11	0.3249	44	2.86	<0.01
7	0.3431	60	0.7016	96	0.78	>0.05
14	0.7089	55	1.1336	88	1.00	>0.05
21	0.3546	60	0.5145	96	1.10	>0.05

The logistic model seems adequate for most parts of the data set, but this does not, of course, imply that this is necessarily the "correct" model. Even in those cases for which no lack of fit was found, another form of model may be more appropriate. A problem of robustness is posed when an empirical function is fitted to a set of data generated by a distinct correct model. The problem is further compounded when the empirical model is nonlinear with respect to the parameters to be estimated. Both simulation studies and analytical results suggest that strong estimation bias can occur, particularly when the functions are asymmetric or not well centered in the experimental design region (Hingley, 1985; Gallant, 1987).

CONCLUSIONS

The results reported above are consistent with the applicability of a simple timecourse model, though it would be advisable to generate more data to confirm this. A simple two-point affinity distribution was used for illustrative purposes, but clearly an immunological response to antigen involves a broad range of antibody types and might be better represented by a continuous distribution. It would be interesting to apply the model to data from other animal systems. Immunoassays should also be found that estimate parameters from the model, such as the mean and variance of the underlying affinity density, as accurately as possible. The investigations reported here suggest that the ELISA may be such an assay.

As formulated in this paper, the timecourse model is applicable only to killed vaccines, since antigen levels are strictly decreasing. The term $y(t, K)$ represents antibody entities of all types with clonal affinity K, but even so the model seems to exhibit the useful property of an immunological memory. For FMD vaccine, six monthly revaccinations of livestock are usually considered necessary, and the limited extent of the immunological memory allows the model to provide adequate representations of the secondary response profiles. The model may not be so successful when applied to vaccines giving more permanent immunity, and in these cases it might be necessary to distinguish between cellular and humoral antibody moieties in order to obtain a successful representation.

REFERENCES

1. Bruni, C., A. Germani, G. Koch, and R. Strom (1976), "Derivation of Antibody Distribution from Experimental Binding Data," *J. Theor. Biol.* **61**, 143.
2. Draper, N. R., and H. Smith (1981), *Applied Regression Analysis* (New York: John Wiley), 2nd edition.
3. Gallant, A. R. (1987), *Nonlinear Statistical Models* (New York: John Wiley).
4. Hingley, P. J. (1985), "Problems in Modelling Responses of Animals to Foot and Mouth Disease Vaccine," Ph.D. thesis (University of Reading).
5. Hingley, P. J. (1988), "A Clonal Selection Based Timecourse Model for Antibody Responses of Animals to Killed Vaccine," paper to be resubmitted to *Biometrics*.
6. Hingley, P. J., and E. J. Ouldridge (1985)., "The Use of a Logistic Model for the Quantitative Interpretation of Indirect Sandwich Enzyme Labelled Immunosorbent Assays (ELISA) for Antibodies and Antigens in Foot and Mouth Disease," *Computers in Biology and Medicine* **15**, 137.
7. Hingley, P. J., and T. W. F. Pay (1987), "Sources of Variability in Foot and Mouth Disease Vaccine Potency Estimates Based on Serum Neutralizing Antibody Assay," *J. Biol. Standardization* **15**, 127.
8. Rodbard, D. (1974), "Statistical Quality Control and Routine Data Processing for Radioimmunoassays and Immunoradiometric Assays," *Clinical Chemistry* **20**, 1255.
9. Rweyemamu, M. M., L. Black, A. Boge, A. C. Thorne, and G. M. Terry (1984), "The Relationship between the 140S Antigen Dose in Aqueous Foot and Mouth Disease Vaccines and the Serum Antibody Responses of Cattle," *J. Biol. Standardization* **12**, 111.
10. Sips, R. (1948), "On the Structure of a Catalyst Surface," *J. Chem. Phys.* **16**, 490.

R. R. MOHLER and Z. H. FAROOQI
Department of Electrical and Computer Engineering, Oregon State University, Corvallis, Oregon 97331

On Immune Process Stochastic Structure

ABSTRACT

A simple building-block synthesis of the immune system is presented here. It is shown that, mathematically, this complex system consists of numerous cascades of bilinear processes which are themselves coupled together by nonlinear gain elements. These coupling elements include stimulation and inhibition terms which are a focus of much immunological research. The mathematical structure is further divided for convenience into cellular and molecular components with mainly molecular terms affecting nonlinear coupling. Then, the derived finite-state estimator shows the significance of antigen information which is the single key nonlinear feedback in the humoral process.

Compartmental models are presented, and, after modeling experimental data for lymphocyte circulation in rats, a discrete, stochastic bilinear, time-series approximation is statistically analyzed. This lymphocyte model should be considered as one approximation and not the final answer. It also shows the need for more complete data.

INTRODUCTION

Interest in mathematical immunology has been growing. This is reflected in the several monographs that have been published and the conferences that have been held lately. Examples of recent work are available in Bell, Perelson, and Pimbley (1978); Merrill (1980); Mohler, Bruni, and Gandolfi (1980); DeLisi (1983); Marchuk (1983); Marchuk and Belykh (1983); Marchuk, Belykh, and Zuev (1985); and Mohler (1987). The mathematical study of events that are involved at the cellular level in transmission of information seems to be generally missing in the literature, the reference here being to the study of recirculation of lymphocytes. These events functionally interconnect many of the parts of the immune response by physically and biochemically conveying information and providing defense. This paper is concerned with a convenient structural decomposition and a particular approximation for the distribution of recirculating lymphocytes. Besides the healthy state of an organism, such distribution is also of significance for disease states. Any maldistribution could be a symptom of a disease, for example in humans, Hodgkin's disease, hepatitis B, psoriasis, and chronic lymphocytic leukemia. Thus, quantification of the norm of distribution pattern, statistical analysis of the deviation from the norm and/or maldistribution, and modeling and identification of the models can be useful in diagnosis and estimation of disease severity, and in the classification and differential diagnosis of abnormalities in migratory patterns of lymphocytes.

DECOMPOSITION

Obviously, the immune process is extremely complex and it would be impossible to develop an encompassing mathematical model to address all questions. However, immunological theory and experimentation has developed to the point that mathematical models and system studies can be useful for specific applications and as a basis for a general mathematical structure for future research. This approach may lead eventually to a systematic understanding of body immune defense and its stimulation for effective health care.

Experimentation shows that decay rate of bacteria is proportional to the product of its concentration with that of antibody and that of complement (prior to saturation), i.e., parametric control. If the effect of antibody and complement were linearly additive, as for conventional linear systems, the bacterial decay would not be so rapid since superposition would apply. Such multiplicative effects (rather than superposition) seem very prevalent throughout immunology.

IMMUNO-MATHEMATICAL DECOMPOSITION

Roughly, the immune system is a communication command and control system to defend the body from alien intrusion and infection. Various immune subsystems, including complement, humoral system, and cell-mediated systems, are activated and deactivated according to antigen level and chemical structure in concert with other substances, such as antibody, suppressors, helpers, lymphokines, etc.

CELLULAR-MOLECULAR STRUCTURE

Cellular and molecular kinetics are the basis of the entire immune process. These processes are quite well defined from conservation equations and chemical mass-action principles. In general, the *cellular population* (or concentration), x_i, of ith class may be described by

$$\frac{dx_i}{dt} = \text{source rate} - \text{death rate} + \text{division rate} + \text{rate differentiation to} \\ - \text{rate differentiation from},$$

or

$$\frac{dx_i}{dt} = v_i(t) - \frac{x_i}{\tau_i} + p_i(\cdot)x_i + \sum_{j \neq i} 2p_j(\cdot)p_{ji}(\cdot)x_j - \sum_{k \neq i} 2p_k(\cdot)p_{ik}(\cdot)x_k, \qquad (1)$$

where $v_i(t)$ is the source term (from bone marrow via blood); τ_i is death time constant; $p_i(\cdot)$, $p_{ji}(\cdot)$, and $p_{ik}(\cdot)$ are appropriate growth coefficients (including probabilities of stimulation and differentiation from one class to the other). These coefficients or probabilities represent parametric feedback control in the immune system of a very complex nature. Indeed, it is these terms upon which much of immunological research is currently focused, i.e., what manner is cell production activated and controlled by mainly molecular substances. Consequently, $p_i(\cdot)$, $p_{ji}(\cdot)$, and $p_{ik}(\cdot)$ are functions of primarily molecular concentrations. They may be deterministic functions or random processes depending on the approximation used.

Here i refers to different cell types; e.g., this might refer to B, T, and macrophage cell lineages, such as resting cells, excited cells, cytotoxic cells, suppressors, helpers, memory cells, and plasma cells (which generate Ab). Also, other killer cells and mast cells could be included.

An mth class of molecular concentrations, y_m, may be described by

$$\frac{dy_m}{dt} = \text{molecular source rate} - \text{death rate} + \text{dissociation rate of appropriate} \\ \text{complexes} - \text{association rate of appropriate complexes}.$$

This is usually approximated by

$$\frac{dy_m}{dt} = w_m(t) + \sum_i \beta_{im} x_i + \sum_{n \neq m} c_{mn} y_n - \sum_{l \neq m} c_{lm} y_l y_m - y_m/\tau_m, \tag{2}$$

or more accurately described by

$$\frac{dy_m}{dt} = w_m(t) + \sum_i \beta_{im}(\cdot) x_i + \sum_{n \neq m} c_{mn}(\cdot) y_n - \sum_{l \neq m} c_{lm}(\cdot) y_l y_m - y_m/\tau_m. \tag{3}$$

Here, $w_m(t)$ is an external source rate; τ_m is a lifetime; β_{im} is an ith-cell source rate for generating y_m; and c_{mn}, c_{lm} are appropriate coefficients of dissociation and association, respectively. In Eq. (3), these coefficients would be functions of the appropriate $y_m(t)$. They may be assumed to be deterministic or stochastic, depending again on the approximation desired or the information available. m refers to the mth class of molecules, such as Ab, Ag, appropriate cell receptor, appropriate lymphokine, IFN (interferon), or other molecular substance.

COMPARTMENTAL STRUCTURE

If x_i^s, y_m^s refer to a particular *compartment*, s, or organ with migration between compartments, then Eqs. (1) to (3) must include a net migration term such that

$$\begin{aligned}\frac{dx_i^s}{dt} =& v_i^s(t) - \frac{x_i^s}{\tau_i^s} + p_i^s(\cdot) x_i^s + \sum_{j \neq i} 2p_j^s(\cdot) p_{ji}^s(\cdot) x_j^s - \sum_{k \neq i} 2p_k^s(\cdot) p_{ik}^s(\cdot) x_k^s + \\ & \sum_{u \neq s} \delta_{i,s,u}(\cdot) x_i^s - \sum_{r \neq s} \delta_{i,r,s}(\cdot) x_i^r.\end{aligned} \tag{4}$$

Here, superscript s shows possible dependence on the compartment or organ. (For example, an inflamed spleen should cause more stimulation of appropriate cells.) In general, the migration coefficients, $\delta_{i,s,u}(\cdot)$ and $\delta_{i,r,s}(\cdot)$, could be deterministic or stochastic functions of appropriate $y_m(t)$ since certain lymphokines (e.g., macrophage migration inhibition factor, MIF) manipulate migration coefficients. A similar molecular version of Eq. (4) may be developed.

TUMOR MODEL

Similar to cellular model (1), a conservation equation would yield the following for tumor-cell population $z(t)$:

$$\frac{dz}{dt} = v_z(t) + p_z(\cdot)z - \frac{z}{\tau_z} + \sum_j 2p_j(\cdot)x_j - u_z(\cdot)z, \tag{5}$$

where $v_z(t)$ is an external source of malignant cells (such as from a metastasized tumor); $p_z(\cdot)$ and $p_j(\cdot)$ are appropriate growth coefficients (including probability of stimulation for division); $p_{jz}(\cdot)$ is a differentiation coefficient (probability). All three coefficients may be functions of z itself, certain lymphokines, and interferon (IFN). However, they commonly may be assumed constant. $P_{jz}(\cdot)$ considers the possibility of a normal lymphocyte becoming malignant. $u_z(\cdot)$, the parametric control, is a function of concentrations (populations) of T_c, M_a (angry macrophages), and possibly Ab, N_k (natural killers), C (complement), IFN, and z itself. For example, $u_z(\cdot) = [k_z(x_M + x_T)]/(z + L_z)$ would seem appropriate if any control limiting constraint is handled by the denominator term with k_z and L_z appropriate constants, and $x_M(t)$, $x_T(t)$ concentrations of M_a and cytotoxic T cells (T_c), respectively, assuming these are the major killers in question. τ_z is the natural death time constant of the tumor. Migration of malignant cells from the tumor could be included in the net source term $v_z(t)$ or similar to above compartmental model.

Tumor debris of concentration x_d might be generated to enhance tumor recognition by the immune mechanisms. This would be modeled by the latter term in Eq. (5) with a natural death or removal time constant τ_d so that

$$\frac{dx_d}{dt} = u_z(\cdot)z - \frac{x_d}{\tau_d}. \tag{6}$$

Consequently, the tumor antigen concentration $\rho_a(\cdot)$ as presented for stimulation by the macrophages M_ϕ and M_a of concentrations x_ϕ and x_a, respectively, might be approximated by

$$\rho_a = \alpha_d(\cdot)\,(x_\phi + x_a) \tag{7}$$

where α_d might be limited according to

$$\alpha_d = \frac{c_d x_d}{L_d + x_c}$$

with c_d and L_d appropriate constants which follows the results of DeBoer, et al. (1985). Again stochastic parameters may be utilized depending on the accuracy desired and the information available (through theory and data). More commonly, $\alpha_d \doteq c_d x_d$.

Several complications to tumor control need to be studied more thoroughly to be modeled accurately through parametric "control" terms as above, but with added equations in some cases. These consist of tumor blocking, tumor escape (metastases), relevant lymphokines, and conversion of healthy cells to tumor cells. Of course, there are appreciable random effects associated with these processes.

COUPLED-BILINEAR STRUCTURE AND ESTIMATION

In general, it is seen that the *immune system structure* consists of numerous multiplicatively (parametrically) coupled systems. Or, they are structured from basic bilinear system (BLS) components which, in some cases, are nonlinearly coupled by limiters or saturation elements. Fortunately, considerable theory has been derived for the analysis of BLS. Some of this may be useful in terms of understanding individual BLS immune elements, such as the multiplicative effects of controlling bacteria. Indeed, the drastic controllability requirement of the immune system suggests bilinear control of stable and unstable modes as borne out by the conditions for completely controllable BLS (see Mohler, 1973). If the appropriate outputs are measured (i.e., those which formulate the multiplicative controls or inputs), the above coupled BLS become conditionally linear systems (CLS). Even with nonlinear parametric coupling, the CLS structure is retained if those states formulating the nonlinear parameters are measured. State and measurement noise may be added to include certain random phenomena in these processes.

In general, CLS with state and measurement noise are described by the following form:

$$\frac{dx}{dt} = f(t,y) + F(t,y) + G(t,y)w_t, \tag{8}$$

$$\frac{dy}{dt} = h(t,y) + H(t,y)x + v_t, \tag{9}$$

where $x_t \epsilon R^n, y_t \epsilon R^m$, and w_t, v_t are appropriate noise vectors (if the models are assumed to be stochastic); $f(\cdot)$, $F(\cdot)$, $G(\cdot)$, $h(\cdot)$, and $H(\cdot)$ are vectors or matrices of appropriate dimension. (Note that here x_t and y_t do not necessarily refer to cells and molecules as above.) Again, Eq. (8) may be envisioned as a BLS with nonlinear output (measurement) feedback control. The structure, with the appropriate definition of state variables and output measurements, is assumed to be dynamical. Or this stochastic CLS may be considered as a nonlinear system with all "nonlinear states" (and possibly some "linear states") measured directly but with certain "linear states" not measured.

The *immune system models* are *CLS*, Eqs. (8) and (9), in all cases that the coupling (feedback control) parameters are bilinear in cellular and molecular populations. When nonlinear (limiting) parametric feedback is considered in the stimulation, inhibition, differentiation, and migration terms, then the immune model is a CLS only with appropriate measurements (observations). In the latter case, it would be necessary to make certain appropriate molecular counts such as antibody (Ab), antigen (Ag), and interferon (IFN). While the above models are time invariant (in terms of parameters), Eqs. (8) and (9) do allow time-variant parameters such as might arise, for example, if certain lymphokine secretions or other molecular terms are not monitored but are approximated by given functions of time.

This structure seems to be significant here, since it allows mathematical (computer) solution of the state-estimation and parameter-estimation problems in many cases. Also, it leads to a separation principle which allows solution of the optimal feedback policy of the measurements as shown by Kolodziej and Mohler (1986). The latter has potential significance for *immunotherapy*, stimulation of the appropriate mechanisms of the immune system to better use its natural defenses against alien threats such as carcinogens and viruses.

Since data collection and analysis normally proceeds in discrete fashion, Eqs. (8) and (9) are replaced by

$$
\begin{aligned}
x(t+1) &= f(t,y) + F(t,y)x_t + G(t,y)w_t, &(10)\\
y(t+1) &= h(t,y) + H(t,y)x_t + v_t, &(11)
\end{aligned}
$$

where a unit sampling period is assumed for convenience, and the probability distribution $P[x_{ito} \mid y_{to}] \leq a_i$, $i = 1, 2, \ldots, n$. As usual, v_t, w_t are assumed to be independent white Gaussian processes for convenience. However, in general, x_t, y_t are not Gaussian processes. If x_{to} is conditionally Gaussian with respect to y_{to}, x_t/y_t will be conditionally Gaussian for all time. Certain identification and control solutions in this case are given by Lipster and Shiryayev (1977, 1978). For the more general case, the following minimum mean-square error state estimator, $\hat{x}(t)$, can be derived from Kolodziej and Mohler (1986) (1.42) to (1.49):

$$
\begin{aligned}
\hat{x}(t+1) &= f + F\hat{x}(t) + P(t)H\nu(t+1),\\
\nu(t+1) &= y(t+1) - \big(h + H\hat{x}(t)\big),\\
P(t) &= \overline{P}(t) + \phi\big(I_t(2) - I_t^2(1)\big)\phi^T(t),\\
\overline{P}(t) &= \big(F\overline{P}(t) + \overline{P}(t)F^T + GG^T - \overline{P}(t)HH^T\overline{P}(t)\big),\\
\overline{P}(0) &= 0,\\
I_t(k) &= \frac{\sum_i a_i^k \exp\big(a_i^2 F_1(t) + a_i F_2(t)\big)p_i}{\sum_i \exp\big(a_i^2 F_1(t) + a_i F_2(t)\big)P_i}, \quad k = 1, 2, &(12)\\
F_1(t+1) &= -\frac{1}{2}\big(H\phi(t)\big) + \big(H\phi(t)\big)^T, \quad F_1(0) = 0,\\
F_2(t+1) &= \phi(t+1)H\big(\nu(t+1)\big) + \phi(t)HI_t(1), \quad F_2(0) = 0,\\
\phi(t+1) &= \big(F - H^T\overline{P}(t)H\big)\phi(t), \quad \phi(0) = I.
\end{aligned}
$$

Here $P(t)$ is the conditional covariance matrix; $\nu(t)$ is an innovation process; $\phi(t)$ is the conditional state-transition matrix, and the other intermediate variables are defined for conveniences as given. Obviously, the solution resembles a Kalman-Bucy filter, but is more complicated.

A steady-state solution might be used as an approximation or if an on-line solution is not needed. However, it might be desired to perform on-line estimation in many cases such as in running a very expensive experiment only until satisfactory results are obtained or in clinical application.

CIRCULATORY LYMPHOCYTE ANALYSIS

MODEL DEVELOPMENT

While Eq. (4) provides a very general model for immune response in specific organs with migration considered, data available to the investigators at the present time is constrained to circulating-lymphocyte trace experiments for rats by Smith and Ford (1983). A specific circulatory model for a general class of lymphocytes is now developed which is consistent with the limited data available. That is, immune dynamics (stimulation, death, etc.) is assumed to be slow (and neglected) relative to the fast circulatory terms. Feedback of various terms which affect migration coefficients are modeled as parametric white-noise sequences for lack of appropriate data. A discrete-time model is used to be more consistent with limited temporal data.

The most important factors that influence lymphocyte migration are (i) hemodynamic and (ii) physico-chemical interaction between cells and vascular endothelium (Ford, Smith, and Andrew, 1978).

The whole blood behaves as a non-Newtonian fluid with the blood viscosity varying with the hematocrit (the percent of blood volume occupied by cells). Blood flow in large vessels (i.e., those with diameter much larger than cell diameter) may be considered as homogeneous and obeying Hagen-Poiseuille's law, while microcirculation is characterized by low Reynold's number and nonhomogeneous flow with two phases (one plasma and the other cells). In the capillaries, there is a continuous variation in blood flow velocity and the hematocrit is unsteady (Fung, 1984).

While the erythrocytes tend to flow along the tubal axis, the leukocytes (including lymphocytes) prefer to stick to the vascular endothelium and roll along it (DeSousa, 1981). This causes a shear on the leukocyte which can be expressed as a nonlinear function of the diameter of white blood cells, WBC, considered as a sphere, diameter of blood vessel, maximum velocity of undisturbed flow in the vessel, linear velocity of centroid of the WBC, hematocrit, Reynold's number, plasma density, and plasma coefficient of viscosity (Fung, 1984).

This shear is strongly dependent on hematocrit which is stochastic *in vivo* and, consequently, the shear fluctuates in time.

A leukocyte adhering to the endothelium is subjected to shear due to blood flow and also to an effective frictional drag. Thus, accelerations and cell concentrations are functions of distance along the vessel axis, thus there results a concentration gradient. Then by Fick's first law of diffusion, a flow exists which is proportional to concentration differences across the interface. The coefficient of multiplication depends on the shear and drag is thus stochastic in nature.

For the purposes of constructing the stochastic model, blood, lungs, spleen, and bone marrow are separate compartments. All the other organs are lumped into three compartments, assuming homogeneity of structure and function within a compartment: (i) tissues like gut, liver, Peyer's patches and miscellaneous nonlymphoid tissue as "other tissue," (ii) the lymph nodes (LN) draining the other tissues as "LN-a," and (iii) all the other LN as "LN-b."

Because of the presence of inherent stochasticity and the lack of precise knowledge of various variables, as noted above, the model in discrete-time vector form becomes from Fick's law and cell conservation:

$$x(n+1) = A\,x(n) + \beta(n)\,x(n) + \epsilon(n+1), \tag{13}$$

where

$$x(n) = \left[x_1(n)\ \ldots\ x_7(n)\right]^T$$

$$\epsilon(n) = \left[\epsilon_1(n)\ \ldots\ \epsilon_7(n)\right]^T, \qquad T \text{ stands for transposition}$$

$$A = \begin{bmatrix} 1-\alpha_1 & & & & & & a_{17} \\ & 1-\alpha_2 & & & \phi & & a_{27} \\ & & 1-\alpha_3 & & & & a_{37} \\ & & \alpha_3 & 1-\alpha_4 & & & a_{47} \\ \phi & & & & 1-\alpha_5 & & a_{57} \\ & & & & & 1-\alpha_6 & a_{67} \\ \alpha_1 & \alpha_2 & 0 & \alpha_4 & \alpha_5 & \alpha_6 & 1-\alpha_7 \end{bmatrix} \tag{14}$$

$$\alpha_7 = \sum_{i=1}^{6} a_{i7} \tag{15}$$

$$\beta(n) = \begin{bmatrix} -\beta_1(n) & & & & b_{17}(n) \\ & \cdot & \phi & & \cdot \\ & & \cdot & & \cdot \\ & & & \cdot & \cdot \\ \phi & & & \cdot & \cdot \\ & & & \cdot & b_{67}(n) \\ \beta_1(n) & \cdot\ \cdot\ \cdot\ \cdot & & \beta_6(n) & -\beta_7(n) \end{bmatrix} \tag{16}$$

$$b_{i7}(n) = \frac{a_{i7}}{\alpha_7}\beta_7(n) \quad \text{such that} \quad \beta_7(n) = \sum_{i=1}^{6} b_{i7}(n). \tag{17}$$

$$\epsilon_7(n+1) = \sum_{i=1}^{6} \epsilon_i(n+1) \tag{18}$$

where:

$x_i(n)$ = state of compartment i at time instant n, being the number of lymphocytes in compartment i expressed as a percentage of the total number in the system as measured by the presence of radioactive label.

α_i = deterministic part of transfer parameters (leaving i), a_{i7} is the portion of α_7 going from 7 to i. α_i (and a_{i7}) scales the % activity in compartment i (or 7) at instant n (dimensionless).

$\beta_i(n)$ = stochastic part of transfer parameters (leaving i), $b_{i7}(n)$ is the portion of $\beta_7(n)$ going from 7 to i. $\{\beta_i(n)\}$ are i.i.d. and constitute multiplicative noise (dimensionless).

$\epsilon_i(n)$ = additive noise, added at the input and/or output of the compartment, $\{\epsilon_i(n)\}$ are i.i.d. (dimensions of % activity).

$\{\beta_i(n)\}$ and $\{\epsilon_i(n)\}$ are independent. ϕ designates zero elements except where indicated.

The subscripts are as follows:

1 = bone marrow
2 = spleen
3 = other tissue
4 = LN-a
5 = LN-b
6 = lungs
7 = blood

It is assumed that there are no births and deaths, and no loss of label. Experimentally, the number of cells in any compartment is measured by the amount of radiolabel activity. The 7×7 matrix $\beta(n)$ is compartmental and is interpreted as the random coefficient of flow. $\epsilon(n)$ is a random 7-vector whose presence may be justified by changes in the same variables as for $\beta(n)$, but are due to intercompartmental stochasticity rather than intracompartmental randomness (as for $\beta(n)$). Also, $\epsilon(n)$ may mimic the presence of different types of immunities and other physiological factors, and the random variation in different organisms within the same species.

Eq. (13) can be viewed as a system of difference equations with both slope and intercept random or as a Markov chain generated by two stochastic processes (i.e., a Markov chain in a random environment), but such processes will not be discussed here. The second term on the right-hand side of Eq. (13) causes the structure to be that of a bilinear time series.

Literature is available on bilinear time series with one and two inputs (Nicholls and Quinn, 1982; Subba Rao and Gabr, 1984). A bilinear time series with two inputs has sometimes been called random coefficient autoregression. It is generally assumed that, in Eq. (13), A is a constant matrix, $\beta(n)$ is a matrix such that $E[\beta(n)] = 0$ and $E[\beta(n) \circ \beta(n)] = C_\beta$, and $\epsilon(n)$ is such that $E[\epsilon(n)] = 0$ and $E[\epsilon(n)\epsilon^T(n)] = G$.

$\{\beta_i(n)\}$ and $\{\epsilon_i(n)\}$ are independent. $\circ$, the Kronecker product, is defined such that the ij-th block of $A \circ B$ is $[A \circ B]_{ij} = a_{ij}B$. Eq. (13) may be used to estimate the mean and variance of the state. Results on second-order stability, second-order stationarity, their relationship, strict stationarity, and asymptotic properties of least squares and maximum likelihood estimators are given in Nicholls and Quinn (1982). Their biological interpretation, that is relevant here, follows.

If the variation in the number of recirculating lymphocytes in any compartment reaches an equilibrium which does not depend on the number of cells present initially in the compartment, the mean number of cells also reaches an equilibrium. This means that at equilibrium the distribution of these cells in that compartment has attained a unique constant mean and standard deviation. When the random perturbations in the recirculating lymphocyte pool are independent and identically distributed, the necessary and sufficient condition for the lymphocyte distribution in any compartment to have a constant mean and standard deviation at equilibrium is that the variation in the number of lymphocytes there be stable. Under these conditions the distribution is also asymptotically time-invariant. This implies that the number of recirculating lymphocytes can be characterized in terms of their population distributions in different compartments.

Smith and (late) Ford (1983) studied the recirculating lymphocytes under conditions as close as possble to the physiological conditions. In the paper, they report in effect three experiments: (i) a survey of distribution of lymphocytes; (ii) tempo of recirculation from blood to thoracic duct lymph, and (iii) estimation of time taken for lymphocytes to cross high endothelial venules into LN. Only the one that is relevant to distribution is summarized here.

AO rats, adult male donors and adult female recipients were used. *In vitro* labeling of thoracic duct TDL was done with sodium-[^{51}Cr]-chromate, then passaged from blood to lymph in an intermediate rat to ensure using recirculating live lymphocytes. The results of the experiment are given in Table 1. Five rats were sacrificed at most of the 13 sample time points. After removal of the 13 relevant organs (vis., blood, lungs, spleen, liver, right and left popliteal LN, coeliac LN, superifical cervical LN, deep cervical LN, mesenteric LN, Peyer's patches, gut and bone marrow), the results of scintillation counting were calculated as percentages of injected dose per organ and per gram of tissue. The results tabulated in Table 1 are the means and standard deviations of activity per organ expressed as a percentage of the injected dose.

Soon after injection (1 minute), 40% of the injected dose is in the lung, 40% in the blood, and 13% in the liver and very little elsewhere. After 2 minutes, the radioactivity in the lungs follows that in blood. Until about 1/2 hour, it falls sharply in both and substantially in the liver. In the spleen, and all the LN and Peyer's patches, the label steadily increases. Cell localization in gut and bone marrow seems flat from 1/2 hour on. At least in the spleen, LN, and Peyer's patches, the rate of cell entry is apparently directly proportional to their concentration in the blood as suggested by experiments on isolated organs (Smith and Ford, 1983).

TABLE 1 Original Data from Smith & Ford (1983)
(mean + standard deviation); % radioactive label (relative to injected dose) per organ

Time (min):	0	1	2	5	10	30	60
superficial cervical LN	0.00	0.18 +.06	0.24 +.11	1.02 +.68	2.20 +1.01	5.15 +1.18	7.46 +.76
deep cervical LN	0.00	0.06 +.02	0.09 +.04	0.49 +.30	1.11 +.45	2.17 +.47	3.55 +.40
coeliac LN	0.00	ns	ns	0.09 +.07	0.15 +.06	0.41 +.15	0.60 +.13
mesenteric LN	0.00	0.14 +.05	0.17 +.07	0.86 +.65	1.66 +.65	4.91 +1.40	8.18 +1.43
Peyer's patches	0.00		0.19	0.85 +.70	1.58 +.69	3.97 +1.07	6.68 +.96
small intestine	0.00	0.90 +.21	0.87 +.18	1.29 +.69	1.35 +.39	1.20 +.45	0.81 +.78
spleen	0.00	2.29 +1.79	5.31 +1.68	10.56 +2.64	19.51 +6.04	38.92 +5.45	39.41 +5.58
lung	0.00	35.50 +6.14	40.08 +6.13	27.63 +6.73	22.97 +4.87	8.46 +1.25	2.50 +.60
liver	0.00	12.84 +2.76	12.13 +2.82	16.24 +6.16	11.24 +2.70	6.22 +.83	3.92 +.83
bone/tibia	0.00	1.15 +.21	1.15 +.21	1.36 +.49	1.61 +.69	1.81 +.69	2.26 +.50
blood/10ml	100.00	41.47 +4.94	36.59 +1.80	38.91 +4.80	29.58 +5.34	11.64 +2.93	3.34 +1.15
R. popliteal LN	0.00				0.24 +.07	0.63 +.16	0.73 +.09
L. popliteal LN	0.00				0.06 +.02	0.16 +.06	0.21 +.04

TABLE 1
(continued)

150	360	540	720	900	1080	1440
7.71 +1.91	8.43 +1.72	8.97 +1.72	9.40 +1.13	8.88 +.43	8.75 +.99	7.83 +1.53
3.63 +.95	4.35 +.40	4.32 +.64	4.99 +.74	4.62 +.55	4.29 +.28	3.97 +.54
0.60 +.13	0.84 +.17	0.71 +.18	0.94 +.16	0.84 +.14	0.68 +.17	0.81 +.13
9.32 +1.40	10.79 +1.32	12.65 +3.80	15.97 +3.25	16.34 +1.55	17.27 +2.48	15.62 +1.71
5.73 +.50	6.16 +1.06	6.09 +.90	7.46 +1.53	6.68 +1.23	6.44 +.99	6.39 +1.27
1.69 +.66	2.05 +.99	2.65 +1.35	1.99 +1.23	2.56 +1.47	1.54 +.51	1.72 +1.02
32.31 +3.42	21.32 +2.97	17.90 +3.72	19.14 +3.33	14.86 +3.22	15.48 +1.92	15.48 +.78
2.73 +.51	2.33 +.20	1.85 +.26	1.66 +.33	2.02 +.36	1.86 +.41	2.03 +.38
3.66 +.39	3.66 +.58	4.04 +.39	4.11 +.45	4.56 +1.09	4.24 +.32	5.26 +.51
2.23 +1.29	1.33 +.38	1.24 +.55	2.17 +1.22	0.72 +.34	1.01 +.50	0.73 +.15
4.06 +.88	3.25 +.34	3.06 +.52	2.41 +.56	3.13 +.92	2.46 +.58	3.04 +.65
0.72 +.13	0.77 +.10	0.78 +.20	0.78 +.06	0.84 +.17	0.73 +.04	0.75 +.14
0.23 +.06	0.22 +.06	0.28 +.08	0.33 +.06	0.28 +.05	0.25 +.02	0.23 +.03

STATISTICAL ANALYSIS

Meaningful data interpretation is available in the form of state *means* and *standard deviations* which are obtained from Eq. (13) for parameter estimation. Estimates for the elements of A are obtained using the equation for means and those for the variances of the random variables from

$$P_d(n+1) = A_d P_d(n) + C_d P_e(n+1) \tag{19}$$

where

$P_d(n)$ = a 7-vector diagonal of the variance-covariance matrix,

$$A_d = \begin{bmatrix} (1-\alpha_1)^2 + \beta_1^2 & & & & a_{17}^2 + b_{17}^2 \\ & \cdot & \phi & & \cdot \\ & & \cdot & & \cdot \\ & & \cdot \ \cdot & & \cdot \\ \phi & & & \cdot & \\ & & & \cdot & a_{67}^2 + b_{67}^2 \\ \alpha_1^2 + \beta_1^2 & \cdot \ \cdot \ \cdot & \cdot & \alpha_6^2 + \beta_6^2 & (1-\alpha_7)^2 + \beta_7^2 \end{bmatrix}$$

$$C_d = \text{diag } [\sigma_{\epsilon i}], \quad i = 1, \ldots, 7$$

and

$$P_e(n) = \text{diag } \{E[\epsilon(n)\epsilon'(n)]\}$$

Here, $\sigma_{\epsilon i}$ refers to standard deviation, and ϕ denotes all zero elements except as indicated. Some important covariances, like those between blood and the other compartments, are not available in the experimental data, and Eq. (19) does not take them into account.

The weighted least-squares cost function

$$Q\chi^2(\cdot) = \frac{q}{nk} \sum_{i=1}^{k} \sum_{j=i}^{n} \frac{\left(x_{io}(j) - \hat{x}_i(j)\right)^2}{\sigma_{io}^2(j)}$$

is used. It is a function of the sample variance of the residual errors, so that effectively this is what is minimized to obtain the "best" parameter estimates. Being divided by variance from the data as weights, $Q.(.)$ is a dimensionless ratio. For the first moments min $\chi^2 = 5.73$, while for the second moments min $Q\chi^2(C_\beta, G) = 116428.22$. The estimated parameter values are tabulated in Table 2. The minimized χ^2-value is significant, which indicates that the residual errors are quite large. In other words, this preliminary model does not fit the data very well. However, according to their definitions given above with Eq. (13), α_i and a_{i7} provide, respectively, relative transfer rates of lymphocytes leaving

TABLE 2 Estimated Parameters Values for the Discrete Model Weighted Least-Squares Estimation

Deterministic Parts of Multiplicative Parameters					
α_1	.11948E−01	a_{37}	.35198E−01	α_6	.38460E+00
α_{17}	.48594E−01	α_4	.33771E−02	a_{67}	.39131E+00
α_2	.78494E−02	a_{47}	.34741E−02	α_7	.49617E+00
a_{27}	.65323E−01	α_5	.41243E−03		
α_3	.15659E−02	a_{57}	.36640E−02		
Standard Deviations of the Multiplicative Noises					
$\sigma_{\beta 1}$	.10245E−02	$\sigma_{\beta 4}$	.15762E−03	$\sigma_{\beta 6}$	.93364E+00
$\sigma_{\beta 2}$	.99294E+00	$\sigma_{\beta 5}$	.10000E+01	$\sigma_{\beta 7}$	.29113E+00
$\sigma_{\beta 3}$	.99673E+00				
Standard Deviations of the Additive Noise					
$\sigma_{\epsilon 1}$	.23895E−01	$\sigma_{\epsilon 4}$	.56450E−05	$\sigma_{\epsilon 6}$	.37201E−06
$\sigma_{\epsilon 2}$	.19190E−05	$\sigma_{\epsilon 5}$	.74743E−01	$\sigma_{\epsilon 7}$	.23510E+01
$\sigma_{\epsilon 3}$	.22762E+01				

the ith compartment and that portion of α_7 (leaving blood) which is associated with entering the ith compartment. While their accuracy is questioned here, their relative values have physical significance.

Here, A is an irreducible matrix. It is column stochastic and, thus, its spectral radius is 1. The seven compartments are not linearly independent because blood is the sum of the other compartments and, overall, it is a closed system.

With regards to the second moments, the matrix $(A \circ A + C_\beta)$ is a 49×49 sparse matrix, and the usual method for finding eigenvalues is not efficient. Ranges of eigenvalues were estimated using Gershgorin's Circle Theorem (Lancaster, 1969) and the fact that the spectral radius of a matrix is always greater than or equal to the largest diagonal element. The spectral radius is greater than 1. (The eigenvalues lie in the interval $[-6.27, 8.5]$ and the spectral radius > 1.9992).

The implication is of second-order instability and second-order non-stationarity (Nicholls and Quinn, 1982). This being so and the minimized χ^2 being significant at $\alpha = 0.01$, further analysis was done to check into the goodness of fit of the compartments of the model. Since one of the conditions for existence and convergence of moments, $\rho(A \circ A + C_\beta) < 1$, is the same as the condition for stability which is violated, the second moment diverges (Feigin and Tweedie, 1985). There is a possibility of this happening because the variances are estimated by using only the diagonal of the $E[x(n)x^T(n)]$ matrix, covariances not being available in the data.

TABLE 3 Statistical Analyses: Discrete-Time Model

First Moments Weighted Least-Squares Estimation								
I(a): Friedman Test			I(b): Wilcoxon and Wilcox Text From tables: $D_{WW;\alpha} = \begin{cases} 32.5 & \alpha=0.05 \\ 38.0 & \alpha=0.01 \end{cases}$					
Compartment #	Sum of Ranks		5 (63)	7 (62)	1 (59)	4 (58)	6 (34)	2 (24)
1	59	3 (64)	1	2	5	6	30	40**
2	24	5 (63)		1	4	5	29	39**
3	64	7 (62)			3	4	28	38**
4	58	1 (59)				1	25	35*
5	63	4 (58)					24	34*
6	34	6 (34)						10
7	62							
	$\hat{\chi}^2_R = 25.68$**							

Second Moments Weighted Least-Squares Estimation								
Compartment #	Sum of Ranks		3 (72)	6 (58)	1 (50)	6 (44)	7 (41)	4 (17)
1	50	3 (82)	10	24	32	38**	41**	65**
2	72	2 (72)		14	22	28	31	55**
3	82	5 (58)			8	14	17	41**
4	17	1 (50)				6	9	33*
5	58	6 (44)					3	27
6	44	7 (41)						24
7	41							
	$\hat{\chi}^2_R = 45.33$**							

* indicates significance at $\alpha = 0.05$.
** indicates significance at $\alpha = 0.01$.

By doing so, the number of equations used in the estimation of the parameters is reduced and, in turn, the number of constraints on the parameter is also reduced, thus leading to "misestimation."

Since the exact distributions are not known, the following null and alternative hypotheses were tested on the residual errors between the model output and the experimental data:

$$\begin{aligned} &H_o : F_1 = F_2 = \ldots = F_7 \\ &H_1 : \text{Not all } F_1 \text{ are equal} \end{aligned} \tag{20}$$

Friedman's two-way analysis of variance based on ranks was done. The null hypothesis concerns distributions (not means). For the first moments, the test statistic $\hat{\chi}^2_R = 25.68$ was obtained, which is significant ($\chi^2 = 16.81$, $k = 7$, $\alpha = 0.01$). Thus, the errors in all the compartments do not have the same distributions. Wilcoxon and Wilcox multiple-comparison procedure is used to do all pair-wise comparisons. The distribution of errors in compartment 2 is significantly different from the distributions in compartments 1, 3, 4, 5, and 7 as seen from Table 3.

For the second moments also Friedman's analysis showed differences in distributions $\hat{\chi}^2_R = 44.60 > \chi^2 = 16.81$, $k = 7$, $\alpha = 0.01$). Wilcoxon and Wilcox test gave three subsets of compartments with similar distributions:

$$\{(1,2,3,5),(1,2,5,6,7),(4)\}$$

It was shown above that, for the estimated parameter values, the model mean is marginally stable. Stability of the second moment is a function of the squares of the deterministic parts of the parameters and the variances of the parametric noise. Because of the parametric noise, the model is a variable structure system (actually structurally unstable system) with random variation in the structure. As the estimated variances of the noises have large values in the model, the distribution of recirculating lymphocytes is unstable. Comparing lymphocyte populations in any compartment, there is a big difference in the estimated number of cells and the experimental quantity. For the mean, compartment 2 (spleen) appears to be the main source of error, while for the variance (standard deviation) compartments 3 (LT) and 4 (LN-a) may be the main cause.

Apparently, the model can be significantly improved by appropriate covariance terms, appropriately correlated parametric noise sequences and/or feedback of appropriate states to generate the diffusion coefficients—all of which would require further data.

ACKNOWLEDGEMENTS

This research was supported by NSF Grant No. 8215724.

REFERENCES

1. Bell, G.I., A. S. Perelson, and G. H. Pimbley, Jr. (1978), "Theoretical Immunology," *Immunology Series* (New York: Marcel Dekker), Vol. 8.
2. DeLisi, C. (1983), "Mathematical Modeling in Immunology," *Ann. Rev. Biophys. Bioeng.* **12**, 117–138.
3. DeBoer, R.J., P. Hogeweg, et al. (1985), "Macrophage T Lymphocyte Interactions in the Anti-Tumor Immune Response," *J. Immunology* **134**, 1–11.
4. DeSousa, M. (1981), *Lymphocyte Circulation: Experimental & Clinical Aspects* (New York: John Wiley & Sons).
5. Feigin, P.D., and R. L. Tweedie (1985), "Random Coefficient Autoregressive Processes: A Markov Chain Analysis of Stationarity and Finiteness of Moments," *J. Time Series Anal.* **6(1)**, 1–14.
6. Fung, Y.C. (1984), *Biodynamics: Circulation* (New York: Springer-Verlag).
7. Lancaster, P. (1969), *Theory of Matrices* (New York: Academic Press).
8. Lipster, R.S., and A. N. Shiryayev (1977), *Statistics Processes I—General Theory* (New York: Springer-Verlag).
9. Lipster, R.S., and A. N. Shiryayev (1977), *Statistics Processes II—General Theory* (New York: Springer-Verlag).
10. Marchuk, G.I. (1983), *Mathematical Models in Immunology* Translation, Series in Mathematics and Engineering (New York: Optimization Software).
11. Marchuk, G.I., and L. N. Belykh (eds.) (1983), "Mathematical Modeling in Immunology and Medicine," *Proceedings of the IFIP-TC 7 Working Conference on Mathematical Modeling in Immunology & Medicine, Moscow, USSR, 5-11 July, 1982* (Amsterdam: North-Holland Publishing Co.).
12. Marchuk, G.I., L. N. Belykh, and S. M. Zuev (1985), "Mathematical Modelling of Infectious Diseases," *Problems of Computational Mathematics and Mathematical Modelling, Advances in Science and Technology in U.S.S.R. (Mathematics and Mechanics Series)*, Eds. G. I. Marchuk and V. P. Dymnikov (Moscow: Mir Publishers), 223–240.
13. Merrill, S.J. (1980), "Mathematical Models of Humoral Immune Response," *Modeling and Differential Equations in Biology (Lecture Notes in Pure & Applied Mathematics)*, Ed. T. A. Burton (New York: Marcel Dekker), 13–50.
14. Mohler, R.R. (1973), *Bilinear Control Processes* (New York: Academic Press).
15. Mohler, R.R. (1987), "Foundations of Immune Control and Cancer," *Recent Advances in System Science*, Ed. A. V. Balakrishnan (New York: Optimization Software, Inc.), (to appear).
16. Mohler, R.R., C. Bruni, and A. Gandolfi (1980), "A Systems Approach to Immunology," *Proc. IEEE* **68(8)**, 964–990.
17. Nicholls, D.F., and B. G. Quinn (1982), "Random Coefficient Autoregressive Models: An Introduction," *Lecture Notes in Statistics* (New York: Springer-Verlag), Vol. 11.

18. Roitt, I. (1974), *Essential Immunology* (Oxford and London: Blackwell Scientific), 162.
19. Rubinow, S.I. (1975), *Introduction to Mathematical Biology* (New York: John Wiley and Sons).
20. Smith, M.E., and W. L. Ford (1983), "The Recirculating Lymphocyte Pool of the Rat: A Systematic Description of the Migratory Behavior of Circulating Lymphocytes," *Immunol.* **49**, 83–94.
21. Subba Rao, T., and M. M. Gabr (1984), "An Introduction to Bispectral Analysis and Bilinear Time Series Models," *Lecture Notes in Statistics* (New York: Springer-Verlag), Vol. 24.

Maturation of the Immune Response

Theoretical Immunology, Part One, SFI Studies in the Sciences of Complexity
Ed. A. S. Perelson, Addison-Wesley Publishing Company, 1988

STUART A. KAUFFMAN,† EDWARD D. WEINBERGER† and ALAN S. PERELSON‡
†Department of Biochemistry and Biophysics, School of Medicine, University of Pennsylvania, Philadelphia, PA 19104, and ‡Theoretical Division, Los Alamos National Laboratory, Los Alamos, NM 87545

Maturation of the Immune Response Via Adaptive Walks On Affinity Landscapes

1. INTRODUCTION

Our aim in this article is to apply Darwin's idea of natural selection to the maturation of the immune response. In so doing, we shall introduce recent results concerning the character of adaptive walks via fitter mutants to local optima, where such walks occur in rugged fitness landscapes with many peaks, ridges, and valleys. In the present context, the *fitness landscape* in question is determined by the affinity with which each possible antibody molecule in an organism's repertoire binds to a given immunizing epitope. By an *adaptive walk*, we mean a random walk on the affinity landscape where a step can only be taken if it leads to a higher affinity. *Maturation of the immune response* refers to the observation that after immunization the average affinity of serum antibody for the immunizing epitope increases.

The article is organized in eight sections. In the second section, we review some of the major facts concerning clonal selection and the maturation of the immune response, and introduce a simple model of the maturation process. In the third section, we introduce the concept of an affinity landscape. In the fourth section, we develop a theoretical framework in which to study the character of adaptive walks that lead to B cells which secrete antibodies of high affinity for the immunizing epitope. In this framework, we are led to ask how many mutational steps occur on an adaptive walk from an initially selected "roughed in" antibody to an antibody of high affinity, how many alternative high-affinity antibodies are accessible via adaptive walks from the same initial antibody, how similar such optimal antibodies are to one another, what the waiting time to find successively higher affinity mutant variant V regions is, and so forth. In the fifth section, we introduce a new model, somewhat analogous to models of spin glasses, to study the statistical character of adaptive walks on rugged landscapes, where the ruggedness of the landscape is "tunable" via a parameter measuring how strongly amino acids in the primary sequence of a V region are coupled to one another in determining the overall affinity of the V region for the epitope in question. In the sixth section, we consider the implications of population genetic concepts for the theory of clonal selection, and note classical and new results suggesting a limitation in the power of selective forces to achieve or hold a population of cells under clonal selection at or near antibody sequences of very high affinity in the face of a given somatic mutation rate to the V region. In the seventh section, we return to a more careful correlation of theory with data, and show that our framework allows us to predict the number of somatic mutations that typically occur during maturation of the immune response, the time for maturation to occur, the number of divergent antibodies of high affinity for a given epitope, and the waiting times for new germ-line variants to be selected and mature to yet higher affinity antibody secreting B cells. In the final section we discuss injunctions to experimentalists and theoreticians for steps to improve this initial crude model. We also raise the issue of whether a somatic mutation theory of affinity maturation is compatible with the idea of an idiotypic network.

2. CLONAL SELECTION AND MATURATION OF THE IMMUNE RESPONSE

When an immunocompetent organism is exposed to an antigen and mounts an immune response, the complex sequence of events which ensues includes binding of the antigen to immunoglobulin molecules on the surface of B cells. Generally, those B cells whose immunoglobulin receptors match epitopes on the incoming antigen with sufficient affinity, proliferate and differentiate into antibody secreting cells. This process whereby antigen selects those B cells which will ultimately grow and secrete antibody constitutes clonal selection (Burnet, 1959). The process relies on having a diverse set of B cells, each cell expressing a single immunoglobulin gene product on its surface as a receptor for antigen, and then secreting this immunoglobulin as antibody upon antigen challenge.

The antigen specificity of an immunoglobulin is determined by the amino acid sequence of its heavy (H) and light (L) chain variable region domains. The diversity of the variable region domains is generated, in part, by the combinatorial assembly of five different V gene segments during each B cell's differentiation (reviewed in Honjo, 1983; Yancopoulos and Alt, 1986). A complete heavy chain V domain results from the joining of V_H, diversity (D) and H chain-joining (J_H) gene segments. Similarly, the light chain V domain is created by the joining of V_L and J_L gene segments. Variations occur in the precise sites of the joining events, and there can be insertion of random nucleotides at the junctional sites (Tonegawa, 1983). Each of these gene segments is chosen from a repertoire of several to several hundred alternatives, thus creating a large number of possible, alternative heavy and light variable regions. The association of independently made heavy and light chains adds further combinatorial diversity. Honjo (1983) estimates the minimal diversity in the mouse generated by these mechanisms alone to be 5.1×10^7, whereas Berek et al. (1985) estimate this diversity at 10^9.

In addition to *germ-line diversity* resulting from the multiple copies of V, D and J segments, *combinatorial diversity* resulting from the use of different combinations of gene segments, and *junctional diversity* resulting from variations in the joining of the gene segments, *somatic mutation*, which results in random nucleotide replacements, plays a major role in immunoglobulin diversification. In principle, somatic mutation which amplifies the potential V region diversity can lead to almost limitless variability. From the analysis of clonally related cells, it now appears that there exists a special hypermutation system which specifically alters bases in the V region at a rate of approximately 10^{-3} per base pair per generation, a rate approximately six orders of magnitude higher than the spontaneous mutation rate (cf. McKean et al., 1984; Clarke et al., 1985; Manser et al., 1985; Sablitzky et al., 1985a, b). Hypermutation of this type occurs during the T-cell-dependent differentiation of naive B cells into memory B cells. At other stages of differentiation somatic mutation may occur at lower rates, as has been shown for a pre-B cell line (Wabl et al., 1985). As explained below hypermutation is thought to play a major role in the maturation of the immune response (cf. Manser et al., 1985; Wysocki et al., 1986; Siekevitz et al., 1987).

ANTIGEN SELECTION THEORIES

The cellular and molecular mechanisms by which the maturation of the immune response occurs are still being uncovered. Classical theories suggest that competition for limited amounts of antigen may drive a selection process (cf. Siskind and Benacerraf, 1969). The argument goes as follows: the amount of antigen bound to cell surface immunoglobulin depends upon the product of the antigen concentration, C, and the affinity of the receptor for antigen, K. During the course of an immune response the antigen concentration C should decrease. If there is a critical amount of bound antigen required to stimulate a B cell into antibody production, then as C falls only those cells with increasing K values will remain stimulated and continue to produce antibody. Because the antibody secreted by a cell has the same affinity for the antigen as its receptor, one should see the average affinity of serum antibody increase during the course of an immune response. According to this theory, based on clonal selection, smaller and smaller subsets of preexisting B cells (i.e., those with higher and higher affinity) are selected by antigen during the course of the immune response.

Bell (1970, 1971) developed a mathematical model of clonal selection based on the above principles and was able to obtain considerable agreement between his theoretical predictions of affinity maturation and experiments of Eisen and Siskind (1964) and Siskind et al. (1968). Bell's model involved precursor B cells, proliferating B cells, plasma cells, B memory cells, antibody and antigen. Each of the first five populations were characterized by its affinity for the antigen and then broken down into J subpopulations. In his analysis of the model, Bell typically examined $J = 17$ different affinities and thus he needed to solve a system of $5J + 1 = 86$ ordinary differential equations numerically. Some further comments on Bell's model and other classical models can be found in Weinard and Conrad (this volume).

A SIMPLE ANALYTICAL MODEL

The following simple model, although not as detailed as Bell's, can be used to capture the essence of the antigen selection idea of affinity maturation (Perelson and Manser, manuscript in preparation). Assume the growth rate, r, of a B cell is a sigmoidal function of KC, i.e.,

$$r = r(KC) = \frac{r_{\max}(KC)^n}{\theta^n + (KC)^n}, \tag{1}$$

where $r_{\max}$ is the maximal rate of B cell growth, θ is a constant determining the value of KC at which growth is half maximal and n is the Hill coefficient which determines the steepness of the sigmoid. Further assume that B cells grow and die, i.e.,

$$\frac{dB}{dt} = r(KC)B - \mu B, \tag{2}$$

where B is the concentration of B cells with affinity K and μ, the death rate of B cells, is a constant. Let us assume that antigen is non-growing and decays exponentially, i.e.,

$$C(t) = C_0 e^{-\gamma t}, \tag{3}$$

where C_0 is the initial antigen concentration and γ is the antigen elimination rate. For purposes of illustration we assume γ is constant, although a more realistic model would make γ a function of the concentration of antibody and the antibody affinity. With r and C given above, Eq. (2) reduces to a linear time-varying equation whose solution is

$$B(t) = B_0 e^{-\mu t} \left[\frac{1 + \left(\frac{KC_0}{\theta}\right)^n}{1 + \left(\frac{KC_0}{\theta}\right)^n e^{-\gamma n t}} \right]^{\frac{r_{\max}}{\gamma n}}, \tag{4}$$

where B_0 is the initial concentration of B cells with affinity K.

A source of B cells can be added to Eq. (2) and various subpopulations of B cells all with differing affinities can be considered. Also antibody secretion by growing B cells can be included and the average serum affinity then predicted once an initial distribution of B cells of differing affinities is specified. Here we shall just illustrate the point that decaying antigen will select cells of high affinity. To do so, we will assume $\mu = 0$. Since all cells die at the same rate this will not influence the selection process, however it will allow us to make asymptotic estimates. From Eq. (4) we now see that at long times the B cell population will reach a level

$$B = B_0 \left[1 + \left(\frac{KC_0}{\theta}\right)^n \right]^{\frac{r_{\max}}{\gamma n}}. \tag{5}$$

Thus, if we compare two B cell populations, B_1 and B_2 , both of which initially had one cell, but with one having an affinity of 10 times the other (i.e., $K_1 = 10K_2$), with both cells being stimulated (say $K_2C_0 \gg \theta$), then at long times

$$\frac{B_1}{B_2} = \left(\frac{K_1}{K_2}\right)^{\frac{r_{\max}}{\gamma}}. \tag{6}$$

As typical values one might pick $r_{\max} = .08\ \mathrm{h}^{-1}$, corresponding to a 12 hour generation time, and $\gamma = .024\ \mathrm{h}^{-1}$, corresponding to a 40-hour characteristic antigen retention time. With these parameter values $B_1/B_2 = 10^3$; that is, there will ultimately be one thousand more high-affinity cells than low-affinity ones. Choosing $\mu \neq 0$ and looking at the dynamics of the response shows that the B cell populations only increase transiently, but during the transient, the high-affinity cells will reach a peak population level approximately three orders of magnitude larger than the low-affinity cells. From this one can conclude that during the course of a response the average antibody affinity will increase. Thus if one hypothesizes, as we did in Eq. (1), that the growth rate of a cell is an increasing function of KC, then decreasing antigen levels will select out high-affinity subpopulations. The crucial experimental point that still needs to be elucidated is whether or not this hypothesis is correct and whether a sigmoidal function of the type given by Eq. (1) is the correct descriptor of the B cell growth rate.

SOMATIC MUTATION THEORIES

Doria (1982) points out that certain patterns of affinity changes that are observed are not consistent with the classical theory. For example, the antigen selection theory would predict that low doses of antigen would lead to antibodies of higher affinity than would be found after giving high doses. However, Siskind et al. (1968) observe that that in response to a low dose of antigen, the affinity is lower than that observed in response to a high dose. Further, it has been found that the affinity does not rise monotonically during a response as is predicted by the classical theory, but rather it rises, reaches a maximum value and then falls (Urbain, et al., 1972; Werblin, et al., 1973). These discrepancies between theory and experiment plus recent evidence, obtained by the techniques of molecular biology, principally mRNA sequencing of hybridomas, indicate that a more complex process than simple antigen selection underlies affinity maturation. Below we summarize some of that evidence.

Early in the primary response antibodies seem to derive directly from germ-line genes with little or no somatic point mutation evident (Kaartinen et al., 1983; Manser et al, 1985; Wysocki et al., 1986). Junctional diversity can account for most of the variation seen. Late in the primary response or in secondary responses the majority of antibodies no longer directly correspond to germ-line varieties, but show extensive somatic point mutations. This lead Berek et al. (1985) to claim that the antibodies produced during a primary and secondary response to the same antigen are structurally very different. The patterns of diversification seem to differ in the response to different antigens. Milstein and coworkers have extensively studied the response to the hapten phenyl oxazolone (Kaartinen et al., 1983; Griffiths et al., 1984; Berek et al., 1985, 1987). Much of their work is summarized in Milstein (1986). They find that seven days after primary immunization, most of the antibody (73%) to oxazolone is derived from a single set of germ-line V genes ($V_H - Ox1$ and $V_\kappa - Ox1$). The few differences between the antibodies derive from junctional diversity. At day 14 after primary immunization, the same germ-line genes still dominate the response (54% of hybridomas express this gene combination); however, in sharp contrast to the findings at day 7, the antibodies express a small number of point mutations which appear to be responsible for a significant increase in affinity for the hapten. The avidity of day 7 antibodies were tightly clustered around the mean avidity. At day 14 there is much more scatter, i.e., some antibodies have a much higher avidity whereas others have much lower avidity (Kaartinen et al., 1983). During the secondary response to oxazolone, only 18% of hybridomas expressed the V_H-$Ox1$–V_κ-$Ox1$ gene combination. However among those that did, a small further increase in point mutations was evident. Surprisingly, these additional point mutations did not necessarily lead to an improvement in the affinity for the hapten (Berek, et al., 1985). The most important feature of the secondary response was a shift to other germ-line genes which were a source of variants with further improved affinity.

Based on these findings, Milstein suggests the maturation of the immune response to oxazolone proceeds in three stages. In the first, most of the antibody reflects a germ-line gene combination chosen for its capacity to bind the antigen with good affinity. In the second stage, cells expressing this gene combination proliferate and mutate. Cells expressing the unmutated form of the germ-line genes disappear and are largely replaced by cells expressing somatic variants with higher affinity for the antigen. Although mutant forms of the V_H-$Ox1$–V_κ-$Ox1$ combination constitute most of the primary response hybridomas, a shift to alternate germ-line combinations begins to become apparent. In the third stage, as the first type of germ-line gene combinations and their somatic mutants reach a limit in the affinity that they achieve, new germ-line gene combinations and their somatic mutants dominate the response. Thus, the V_H-$Ox1$–V_κ-$Ox1$ combination, although useful to start with, seems to become increasingly difficult to mutate towards very high-affinity antibodies. Other germ-line combinations, which initially may not generate as good antibodies, may be a better starting point from which high-affinity antibodies can ultimately be produced. Because low-affinity clones will not grow as well as high-affinity ones, one may have to wait a long time before these clones grow large enough to have a high probability of generating a high-affinity variant by point mutation. Another reason that new germ-line gene combinations ultimately dominate the response is that combinations other than V_H-$Ox1$–V_κ-$Ox1$ which bind to Ars with high affinity may be rare. One may need to wait until the bone marrow generates a cell with the appropriate gene combination. After this occurs, this newly arisen clone could come to dominate the response.

The pattern observed in the oxazolone system is not universal. Wysocki et al. (1986) studied the immune response to the hapten p-azophenylarsonate (Ars). Early in the response (days 3–9), anti-Ars antibodies were encoded for by a single $V_{H'}$ gene segment, termed $V_H Id^{CR}$, in combination with an assortment of D, J_H, V_L and J_L gene segments. These antibodies had low affinity for Ars (5×10^4 to $5 \times 10^5 M^{-1}$) and were not altered by somatic mutation. Later in the response (days 10-21; see Fish and Manser, 1987) and during the secondary response, the large combinatorial diversity seen early in the primary response disappeared. A majority (70 - 95%)of the hybridomas expressing $V_H Id^{CR}$ used a single choice of D, J_H, V_L and J_L. Thus in contrast to the oxazolone system in which the response diversifies with time, the response to Ars focuses until it is dominated by a single gene segment combination, called the canonical combination. Extensive somatic mutational changes occurred to the canonical combination (3% of $V_H Id^{CR}$ nucleotides were mutated) accompanied by an affinity increase of about 10-fold. Interestingly, B cells which had the highest affinity for Ars during the early primary response used the canonical V region. Thus one can view the process of affinity maturation as one in which clones with the highest affinity for antigen grow and dominate the primary response. After a sufficiently large pool of such cells has been generated somatic mutation operates on their V region genes to generate new V regions. Those with even higher affinity for Ars continue to grow and dominate the response.

The difference in the pattern of the response to oxazolone and to Ars may be due to differences in the number of germ-line genes that are available in the repertoire. In the Ars system where many gene combinations are available, a competition between them may occur to find the fittest. This combination is then subject to somatic diversification. In the oxazolone system, this initial competition is absent since most of the antibody produced uses the V_H-$Ox1$–V_κ-$Ox1$ combination. Generalizing from both systems we see that the simple antigen selection model was incomplete in that it assumed that all antibodies were initially present in the repertoire and subject to selection. Rather we now know that the germ-line provides "roughed in" V regions which somatic mutation then operates on. Further, because the germ line carries gene segments which are combinatorially assembled, the potential germ-line repertoire may be larger than the expressed repertoire at any time. For example, if a mouse with 10^8 lymphocytes can express 10^9 combinations, the estimate given by Berek et al. (1985), then not all combinations can be present at once. In such circumstances, new germ-line combinations may constantly be generated and subject to selection and somatic mutation. Further, during the course of a response a given germ-line combination may generate a large number of variants. However the affinity may ultimately reach a maximum, with further mutations not providing any improvement. The simple analytical model of the maturation process given by Eqs. (1)-(3) could be improved upon by allowing new variant B cells to enter the population at random times. A detailed simulation of this process is possible using the simulation techniques developed by Farmer, Packard and Perelson (1986) and discussed in Perelson (this volume, Part 2). Analytical results should also be possible.

3. THE AFFINITY LANDSCAPE

Implicit in the idea of maturation of the immune response, and indeed of clonal selection itself, is the idea of an *affinity landscape*. To make this idea precise, consider the potential antibody repertoire of an organism (i.e., all the different antibodies its genome is capable of assembling by combinatorial and mutational events) and the actual repertoire of an organism and the affinity of each antibody molecule for a given epitope. For each antibody molecule, we can write down a single number, its affinity, and may think of that affinity as a measure of the "fitness" of that antibody molecule with respect to that specific epitope. The next idea, implicit in maturation and clonal selection, is that the repertoire of antibody molecules of an organism can be represented in an ordered sequence in which each point represents one antibody molecule; and its neighbors are those antibody molecules that differ from it in a single amino acid. Thus, in general, for a protein with 100 amino acids and allowing all 20 kinds of amino acids, any position might be altered to 19 other amino acids; hence, such a protein has 1900 *1-mutant neighbors*. Therefore, we are led to conceive of the ordered space of all antibody molecules in the potential repertoire of an organism, in which the affinities of those antibodies for a specific epitope can be thought of as an affinity landscape, or of a fitness landscape using Sewell Wright's helpful concept (Wright, 1932). The great virtue of thinking in terms of this affinity landscape is that maturation of the immune response via somatic point mutation can be thought of as a walk (or more precisely as a set of walks) from one antibody molecule through a succession of neighboring variant antibody molecules of successively higher affinity. Since the affinity landscape will, in general, have many ridges, peaks, and valleys, it becomes obvious that the character of walks to peaks of high affinity will depend very strongly on the character of the landscape itself. Understanding the character of such landscapes and their implications for maturation of the immune response is the main purpose of this article.

PRELIMINARY EXAMPLES

Before considering affinity landscapes, it is worthwhile mentioning the generality of the ideas that we discuss here via several examples. Consider the space of all peptides of length 100. Peptides differ with respect to many functional properties, for example the capacity to catalyze a specific reaction. These differences can be used to ascribe a "fitness landscape" across the space of peptides. Evolution of peptides with respect to improved catalytic function can thus be thought of as an adaptive walk on this fitness landscape (Smith, 1970; Ninio, 1979; Eigen, 1987).

The idea of a fitness landscape is relevant to a wide variety of fields in which optimization problems occur, including physics and computer science. Our second example, the traveling salesman problem, is a famous problem from computer science. Here N cities exist with roads of specified length between pairs of cities. The problem consists in finding the shortest path starting with one city, passing through each other city exactly once, and returning to the initial city. Any possible solution is a circuit through the N cities; thus there are $N!$ solutions if we relax specification of the initial city. Each legitimate circuit can be deformed to $N(N-1)/2$ "neighboring" circuits by exchanging the positions of two cities in the circuit. The "fitness" of a circuit is just its length. An adaptive walk consists in beginning at a specific circuit, trying 1-mutant neighboring circuits, and moving to a new trial solution circuit if it is shorter.

As a third example, consider the space of all possible idiotypic networks, each with its unique set of B and T cells which mutually interact with one another. Each such network can be viewed as a 1-mutant neighbor of other potential networks in the repertoire of the organism generated by altering one B or T cell type. An adaptive walk across the space of idiotypic networks in the organism's repertoire would consist in assigning a fitness to each network with respect to a fixed spectrum of antigenic inputs and assessing the transitions between networks by this fitness landscape.

The natural questions that arise concerning any fitness landscape include these: (1) How many local optima are in the space with respect to 1-mutant moves, 2-mutant moves, etc? (2) If an adaptive process starts at one "entity" in the space and can move only to fitter neighbors, what is the expected number of improved variants passed on the way to a local optimum; that is, what is the expected length of an adaptive walk? (3) How many alternative fitter variants are 1-mutant neighbors to any entity and how does this number vary as fitness increases? Alternatively stated, if adaptation can only occur via fitter 1-mutant variants, how many ways can the process "branch" at each step uphill? (4) Correspondingly, how many alternative local optima are accessible to an arbitrary initial entity via adaptive walks? (5) Is the global optimum generally attainable? (6) If neutral mutations are allowed, or the adaptive process can pass via less fit variants, how does that affect the character of such adaptive walks?

Given any specific adaptive landscape, the tools of population genetics provide a conceptual armamentarium to ask many familiar questions (Ewens, 1979). Thus, given specified fitness differentials, can a haploid or diploid population of cells or organisms, dispersed initially across the space in some way and driven by mutations occurring at some frequency, be "pulled" by selective forces to local or global optima? Can such a population be maintained by such selective forces at or near such an optimum in the face of mutations? If near, how near? How long does it take the population to flow to such an optimum? Mutation and selection acting on a population has certain similarities to simulated annealing (See Kirkpatrick, 1983, for an introduction to optimization using simulated annealing and Weinberger, 1987, for a discussion of the relationship between simulated annealing and evolution). How can the tools of population genetics be used to understand the theory of simulated annealing and how can statistical mechanics be used to understand population genetics?

4. ADAPTIVE WALKS ON UNCORRELATED LANDSCAPES

Consider the affinity landscape of all antibody molecules in an organism's potential repertoire for a specific epitope. We expect that the affinities of a given antibody and those of its 1-mutant neighbors, obtained by replacing one specific amino acid in the variable domain, will be correlated. That is, if the first antibody has high affinity for an epitope, we expect a large fraction of its 1-mutant neighbors will have high affinity. The degree of correlation in affinities among 1-mutant V region neighbors is unknown and shall emerge as a critical experimental question. However, it is clear that the properties of this affinity landscape will have properties that are intermediate between a fully uncorrelated landscape and a fully correlated one. It is therefore useful to present an analysis of both of these extreme cases.

Consider first a fully uncorrelated affinity landscape. (The opposite extreme will be presented in the next section, as part of a more general model.) In such a hypothetical landscape, the affinity of any antibody for a given epitope is completely independent of the affinities of each of its 1-mutant neighbors. A convenient general picture of such a landscape is one in which the affinity of each antibody is drawn at random from some underlying distribution of affinities.

In following the mutational changes that occur to the V region of a stimulated B cell, we have chosen to keep track only of those mutational steps which increase the fitness of the antibody. In essence then, we will be following a cell lineage that leads to the fittest antibodies. Lineages in which antibody affinities decrease may also occur, but ultimately such clones will either be eliminated from the population or will be "diluted out" as more fit clones preferentially proliferate. A generalized version of the theory to be presented here would be needed to follow clones in which the affinity decreases.

In answering questions about adaptive walks, the actual affinity differences between antibodies is not as important as the relative rank orders of the affinities of neighboring antibody molecules. A considerable simplification is obtained by replacing the actual affinities of the ensemble of antibody molecules by their rank orders. Then in an uncorrelated landscape among, say, 10^8 different antibody molecules, each antibody is assigned at random a rank order from the worst (1) to the best (10^8).

THE NUMBER OF LOCAL OPTIMA

A benefit of considering the uncorrelated rank order affinity landscape is that there is a single general theory of adaptive walks on such a landscape. To make ideas concrete, consider V regions comprised of N amino acids, each amino acid being one of B (i.e., 20) types. In this simplified space of V regions, each N amino acids long, there are B^N possible V regions. The rank order affinity of each is assigned at random. Each V region has exactly $D = N(B-1)$ 1-mutant neighbors, obtained by changing a single amino acid to any of the other $B-1$ possible amino acids. The chance that a randomly chosen V region has a higher affinity than any of its D 1-mutant neighbors is

$$P_m = \frac{1}{D+1} = \frac{1}{N(B-1)+1} , \tag{7}$$

Since the total number of V regions is B^N, the expected number of local optima is just

$$M_1 = \frac{B^N}{D+1} = \frac{B^N}{N(B-1)+1} . \tag{8}$$

These expressions for the numbers of local optima with respect to 1-mutant variants generalize trivially to 2-mutant or j-mutant search ranges (Kauffman and Levin, 1987).

The implications of this analysis are that (1) the expected number of local optima in uncorrelated spaces is very large, and (2) local optima are nearly as abundant as the number of V regions. As an example, assume that there are a total of a 100 amino acids in both the heavy- and light-chain variable regions that contribute significantly to antibody diversity, and each amino acid can be any of 20 possible types, then $P_m = 1/1901 \approx 5.3 \times 10^{-4}$. The total number of V regions in the space is $20^{100} \approx 1.3 \times 10^{130}$ and the mean number of local optima $M_1 = 6.6 \times 10^{126}$.

The probability that a V region reached during an adaptive walk is a local optimum is related to its rank order, X, and T the highest, or top rank order. For a V region of rank X reached after the first step of an adaptive walk to be an optimum, it must be fitter than all of its D 1-mutant neighbors. The probability of this occurring is a special case of the hypergeometric distribution and is given by

$$P_m(X) = \frac{\binom{X-1}{D-1}}{\binom{T-1}{D-1}}, \tag{9}$$

The factor $D-1$ arises because after an adaptive walk has taken a step, the V region arrived at is fitter than at least one of its D neighbors, the one from which it has just arrived. The numerator is the number of ways of choosing $D-1$ random integers from the $X-1$ possible integers that are less than X. The denominator is the total number of ways of choosing the $X-1$ rank orders. For large X and large T, $P_m(X)$ is approximately $(X/T)^{D-1}$ (Kauffman and Levin, 1987). The probability P_m remains low for small values of X/T, then rises sharply as X/T approaches 1. Thus, another implication of our theory is that in uncorrelated landscapes most local optima are clustered among highly fit V regions.

THE LENGTHS OF ADAPTIVE WALKS

Perhaps the most surprising feature of adaptive walks in uncorrelated landscapes is how short such walks typically are. Consider the worst V region. All of its D neighbors are fitter. Recall that adaptive processes move to the first fitter variant encountered in the mutational process. The rank orders of the D fitter 1-mutant neighbors are distributed at random between rank order, 2, and the highest rank order, T. Thus, on average, the affinity of the first fitter variant found will be approximately half way between the lowest rank order V region and the top. Thus, on average, on the first improvement step, the process will step halfway to the top. Consider the next improvement step. On average, half the 1-mutant neighbors are fitter, and these are distributed at random and uniformly from the present rank order to the top. Thus, on the second improvement step, the process will again step half way from the present rank order to the top, T and, on average, will arrive at a relative rank order of 3/4. By the same argument, at each step the process on average moves halfway from its current location to the top. This leads to a simple approximate expression for the probability that an adaptive walk continues for k steps without encountering a local optimum. By the argument given above in the calculation of P_m we see that the probability that an adaptive walk has reached a local optima is the relative rank order reached raised to the $(D-1)$ power. Further, the probability an adaptive walk continues for at least one more step is one minus the probability of being at a local optima. If we let $P_j = Pr$ (number of steps in an adaptive walk $> j$) then $P_j = [1-(\frac{1}{2})^{D-1}][1-(\frac{3}{4})^{D-1}][1-(\frac{7}{8})^{D-1}]\ldots$, i.e.,

$$P_j = \prod_{R=1}^{j} \left[1 - \left(\frac{2^R - 1}{2^R} \right)^{D-1} \right]. \tag{10}$$

This formula is only an approximation because it assumes that the walk is at its mean position at each step. An exact formula has been derived and will be published elsewhere (Macken and Perelson, to appear).

The distribution of P_j values will have a long tail. We thus characterize the walk length by the median of this distribution, i.e., the value of j at which $P_j \approx 0.5$. Note that if $R = \log_2(D-1)$, the corresponding term in Eq. (10) is $1-(1-1/(D-1))^{(D-1)}$, which for large D is well approximated by $1 - 1/e = 0.63$. Similarly, the preceding terms are approximately $1 - (1/e)^2 = 0.86$ and $1 - (1/e)^4 = 0.98$. Thus, there is very little probability of the walk stopping more than one or two steps before the $R = \log_2(D-1)$ step and high probability that it will stop on that step or at most one or two more steps. The critical conclusion is that, in uncorrelated landscapes, adaptive walks are very short. The median number of improvement steps before reaching a local optimum is on the order of $\log_2 D$.

Consider the implications for the lengths of walks in peptide space, with peptides length 10, using all 20 amino acids. Then in an uncorrelated space, the number of 1-mutant neighbors of a peptide is 190, and the walk length is on the order of 7 or 8. If we consider peptides of length 100, then the number of neighbors to each peptide is 1900, and walk lengths increase, but only slightly, to about 11.

An adaptive walk in an uncorrelated landscape might branch in a number of alternative directions since, in general, more than one neighbor of a peptide has higher fitness. Because the predecessor on an adaptive walk is less fit than the current peptide, the expected number of fitter neighbors to a peptide of rank order X/T is simply $(T-X)(D-1)/T$. Thus, the number of alternative pathways toward fitter variants decreases linearly as rank order increases. From this it is easy to derive an upper bound on the number of local optima that are accessible from the lowest ranked entity. On average, after a single improvement step, $(D-1)/2$, or almost half of its neighbors, are fitter. After the next improvement steps, 1/4, 1/8,... of the neighbors are fitter. Thus, there can be $N_{op} = D \times D/2 \times D/4 \ldots \times D/D$ alternative steps. (Here for simplicity we have assumed that $\log_2 D$ is an integer.) Since many of these branching adaptive walks may rejoin one another, N_{op} is an overestimate of the number of local optima accessible from the lowest ranked V region. There are, on average, $\log_2 D$ terms in this series. Multiplying these terms leads, after minor rearrangement, to (Kauffman and Levin, 1987)

$$N_{op} = D^{(\log_2 D - 1)/2}. \tag{11}$$

While Eq. (11) is an upper bound, it demonstrates a further central feature of walks in affinity landscapes: only a tiny fraction of local optima are accessible from any initial point via adaptive walks. For peptide space with 20 amino acids and peptides length 100, the total number of local optima is 6.6×10^{126}, while the gross upper bound on the number of accessible optima from any single peptide is about 1.7×10^{16}. Thus at most 1 in 10^{110} of the local optima are accessible.

THE TIME TO REACH A LOCAL OPTIMA

The expected time to reach a local optimum is also calculable. Let the adaptive process "try" a 1-mutant variant each time unit, and begin with the lowest ranked peptide. Then the mean number of tries to find an improvement step is 1 the first time. Since on average half the neighbors of the first improved variant are fitter, the number of tries to find the second improved variant is 2. This then doubles each time to 4, 8,... for $\log_2 D$ steps. The sum of this series is $D-1$. Then, with one trial mutant per time unit, it takes on the order of D tries to attain a local optimum.

5. ADAPTIVE WALKS ON CORRELATED LANDSCAPES: THE N-k MODEL

Real antibody molecules with varying V regions undoubtedly do not constitute an uncorrelated affinity landscape with respect to binding of any specific epitope. Similar V regions certainly will tend to have similar affinities. As noted above, a major injunction to experimentalists is to discover just what the true correlation structure of affinity landscapes looks like by assessing the affinities of all 1-mutant variants of antibodies for a specific epitope, as a function of the affinity of the unmutated antibody molecule.

While a major task lies ahead in actual discovery of the statistical correlation structure of real affinity landscapes, we use here a general and flexible model, introduced by Kauffman (1988), that is meant to allow us to examine an entire family of rugged landscapes, and to "tune" the ruggedness of the landscape in a way that has direct molecular interpretation. We hope that true affinity landscapes can be located as typical members of this parameterized family of rugged landscapes. Hence, analysis of the family as a whole, we hope, will shed light on the still unknown properties of true affinity landscapes.

Consider a peptide of length N. Suppose that for each position within the peptide the contribution to the overall "fitness" or function of the peptide depends upon the amino acid in that position and the amino acids in k other positions, where k is a parameter of the model. Thus, a total of $k+1$ amino acids determine the fitness of any single position in the peptide. In assigning fitnesses we will impose a symmetry constraint: if the amino acid at position i bears directly on the fitness contribution of an amino acid at position j, then the relation is reciprocal; the amino acid chosen at position j bears on the contribution of an amino acid at position i.

Given N and k, we must decide how to select the k amino acid positions which bear on the overall fitness contribution of a particular amino acid position. For example, the k might be taken to be the $k/2$ nearest neighbors on each side of the given site, or they might be selected at random throughout the N potential sites. The selection of the k among neighboring or distant sites models how functional dependence may vary with distance along the peptide and its consequences for statistical features of adaptive landscapes.

We assign fitness contributions at random as follows: for each position i in the peptide, there are $k+1$ positions that bear on the fitness contribution of that site. There are a total of $B^{(k+1)}$ possible combinations of amino acids in the $k+1$ sites. For each of these possibilities, assign a "fitness" at random from the uniform distribution between 0 and 1. This completes the assignment of fitness contributions to the ith site. Repeat the process for each of the N sites in the peptide. For any specific peptide, the total fitness is obtained by summing the fitness contribution of each site and dividing by N. Thus, the fitness of any peptide is the sum of N-independent, identically distributed, random variables chosen from the uniform distribution and lies between 0 and 1. For large N, the central limit theorem implies that the distribution of fitnesses will approach a normal distribution with mean $1/2$ and variance $1/12N$. These facts that will be useful in the analysis that follows.

We now consider some special cases of the N-k model. For simplicity of analysis we focus our attention on peptides composed of two types of amino acids, i.e., $B = 2$.

$k = 0$: A HIGHLY CORRELATED ADDITIVE FITNESS MODEL

Changing k tunes the ruggedness of the fitness landscape. Consider first the $k = 0$ case. Here the contribution of each amino acid to the overall function of the peptide is independent of all other amino acids. By chance, either the 1 or the 0 amino acid in each position is fitter. Then the overall fittest peptide is the one in which all N positions have the fitter amino acid present. This case is identical to an N locus, two allele, haploid, additive fitness model in population genetics. As there, each locus, or position, makes a contribution to overall fitness that is simply added to the contributions of all other loci and is, on average, inversely proportional to the total number of loci.

The fittest peptide is, of course, a global optimum. Further, there are no other local optima. Instead, each suboptimal peptide lies on a connected pathway of improved variants to the global optimum. By successively changing any less fit amino acid to the more favored amino acid the fitness of the peptide improves. Furthermore, note that any peptide and all its 1-mutant variants have nearly the same overall fitness—the fitnesses differ by at most $1/N$. The landscape is thus highly correlated. In short, in the $k = 0$ case, the landscape has a single global optimum, all peptides which are suboptimal lie on connected pathways leading to that optimum in N or fewer steps, and neighbors in the space have very similar fitness values.

A number of additional features of $k = 0$ landscapes can be determined. First, the expected fitness of a peptide composed of two type of amino acids is $2/3$ at the single global optimum. This follows simply from the fact that for each position in the peptide, the two possible amino acids are assigned fitness values at random from the uniform distribution on the interval [0, 1]. Although the average fitness of each amino acid is 1/2, when one compares the fitnesses of the two possible amino acids at a given position, one needs to chose two random numbers. The average value of the minimum of two uniform random numbers between zero and one turns out to be 1/3, and the average value of their maximum turns out to be 2/3. Thus, at each position in the peptide, the average lower fitness value is 1/3, and the average upper fitness value is 2/3. (For peptides composed of B amino acids, this result generalizes, as can be seen by the use of order statistics (Feller, 1971), and the expected fitness of the global optimum can be shown to $B/(B+1)$).

A second feature of $k = 0$ landscapes is that the expected fitness of the global optimum is independent of N (i.e., 2/3 or $B/B+1$). Additionally, the variability in the value of the optimum in inversely proportional to N. The variance of the fitness of the global optimum is $\frac{1}{N}\left[\frac{B}{B+2} - (\frac{B}{B+1})^2\right]$, which for $B = 2$ is $\frac{1}{18N}$.

An adaptive walk which begins with a randomly chosen peptide will, on average, begin with a peptide having fitness 1/2. On average, $N/2$ of the amino acids will be the less favored ones at their specific loci. Thus, walks to the global optimum via fitter 1-mutant variants require on average $N/2$ changes and, hence, are on average of length $N/2$. Thus, a third feature of $k = 0$ landscapes is that *average walk lengths are linearly proportional to* N. Also, any single-step improvement on average increases the contribution of that locus from the less favored to the more favored value (i.e., from 1/3 to 2/3) and, hence, increases the overall fitness by $1/3N$. Finally, every step towards the optimum reduces the number of positions with the less fit amino acids by 1; hence, it reduces the number of fitter 1-mutant neighbors by 1. Thus, as a walk approaches the global optimum, the number of ways uphill decreases linearly.

We can also determine the statistics of the length of walks on $k = 0$ landscapes (including the time spent searching for a fitter variant). For simplicity, we present the analysis for $B = 2$. We start with a peptide that has c "correct" or optimal amino acids so that $N - c$ mutations must be made to reach the optimal peptide. We assume that mutations are made to randomly selected sites, including those sites that are already occupied by the correct amino acid. If a mutation occurs at one of the latter sites, however, the fitness of the peptide would decrease, and thus we do not take a step. The probability of a successful step is $(N - c)/N = p$, so that the probability of finding the first fitter variant on the nth try is $p(1 - p)^{n-1}$. This is a geometric distribution and has mean $1/p = N/(N - c)$ and variance $(1 - p)/p^2 = Nc/(N - c)^2$. A reasonable measure of the fluctuation about the mean is the coefficient of variation, which is the ratio of the standard deviation to the mean. In the present case, this ratio is $(c/N)^{1/2}$. Note that, for $c \approx N$, this ratio is approximately 1, so that the fluctuations are quite pronounced.

Having determined the statistics of the number of mutations required to take a step on the adaptive walk, we now determine the mean and variance of the number of mutations required to reach the optimum. After each step, c increases. Thus, the average number of mutations required to reach the optimum starting at $c = N/2$, is

$$\begin{aligned}\sum_{c=N/2}^{N-1} \frac{N}{N-c} &\approx N \int_{1/2}^{1-1/N} \frac{dx}{1-x} \\ &= N \ln \frac{N}{2}.\end{aligned} \tag{12}$$

The variance in the number of mutations required to reach the optimum starting at $c = N/2$ is

$$\begin{aligned}\sum_{c=N/2}^{N-1} \frac{Nc}{(N-c)^2} &\approx N \int_{1/2}^{1-1/N} \frac{x\,dx}{(1-x)^2} \\ &= N(N - \ln N - 2 + \ln 2).\end{aligned} \tag{13}$$

The coefficient of variation is, to leading order in N, $(\ln N/2)^{-1}$, so that the fluctuations in the number of mutations required to find the optimum are also relatively large unless N is large.

k = N − 1: AN UNCORRELATED LANDSCAPE MODEL

The $k = N - 1$ case, in which the fitness contribution of each amino acid depends on all of the amino acids in the peptide corresponds to an uncorrelated landscape. Now, alteration of a single amino acid at any position in the peptide changes the overall fitness contribution of each of the N amino acids to a different random value between 0 and 1. In short, changing one amino acid yields a peptide whose overall fitness is totally independent of the fitness of the original peptide. Such a landscape is fully uncorrelated.

We already have derived many properties of uncorrelated landscapes based on the fitness rank orders in section 4. In contrast to $k = 0$, walk lengths are short, of the order of $\log_2 N$, rather than $N/2$ (recall that for $B = 2$, $D = N$). The number of optima is very large. Each improvement step moves, on average, halfway to the top ranking peptide; hence the number of fitter 1-mutant neighbors decreases by half at each improvement step, rather than decreasing linearly.

A truly critical feature of $k = N - 1$ models is that the fitness of accessible optima are not independent of N. Instead, as N increases, the central limit theorem implies that fitnesses of local optima (as well as all other fitnesses) decrease toward the mean fitness, 1/2. This phenomenon constitutes a kind of "complexity catastrophe." In uncorrelated landscapes (with the further condition that fitness be normalized in some sense over the contributions of many subparts such that the central limit theorem applies), as N increases, the fitness of local optima recede toward the mean of the space. This important result implies that even if selection is strong enough to pull an adapting population to accessible local optima, those optima become mediocre as N increases. The probability distribution of fitnesses of local optima is approximately that of the maximum of $N+1$ independent Gaussian random variables, each with mean 1/2 and variance $1/(12N)$. This follows from the fact that the fitness of a peptide and its N mutant neighbors are all random variables chosen from an approximately normal distribution. The mean of the distribution of fitness optima for various N are shown in Table 1.

We will show in section 7 that affinity maturation is not occurring on an uncorrelated landscape. Thus, the immune system apparently operates in such a way as to avoid the "receding optima" complexity catastrophe.

$0 < k < N$: TUNABLY RUGGED AFFINITY LANDSCAPE MODELS

As k varies from 0 to $N-1$, the ruggedness of the landscape varies and fitnesses go from being highly correlated to uncorrelated. Before commenting in general on the alterations which can be expected to occur as k increases and as N increases, it is worth noting that this N-k model is also a model of *epistatic* interactions in population genetics. Recall that in population genetics, additive models assume that the fitness contribution of the allele at each locus is independent of, hence additive to, the fitness contributions at other loci. More generally, the fitness contribution of the allele at the ith locus depends upon the alleles present at some other loci. Those dependencies are termed epistatic interactions. In general, then, altering k in the N-k model studies how the "richness" of epistatic interactions can be expected to govern the ruggedness of fitness landscapes. We know of no general results in this area, except the following estimate of local fitness maxima: by definition, if a bit string is fitter than all of its neighbors, it is impossible to flip a bit on that string and obtain a fitter variant string. For the nearest neighbor model, each bit not only *affects* the fitness contribution of $k+1$ bits (including itself), but the contribution of each bit is also *affected by* $k+1$ bits. In the random neighbor model, the number of bit contributions that are affected by each bit is a random variable with mean $k+1$. Accordingly, we can think of an improvement step in both models as the substitution of $k+1$ old fitness contributions by $k+1$ new fitness contributions that have a larger sum. If k is sufficiently large relative to N, the sum of the replaced fitness contributions is close to $(k+1)F$, where F is the current fitness of the string (When $k = N-1$, the sum of the replaced fitness contributions is NF.). The $k+1$

TABLE 1 Average Fitness Values of Local Optima for k=N-1 (standard deviations shown in parentheses)

N:	8	16	24	48	96
	0.65 (.06)	0.63 (.04)	0.62 (.03)	0.60 (.02)	0.57 (.01)

TABLE 2 Estimated Fitnesses of Local Optima for Various N and k Values

k\N	16	24	48	96
4	.70	.72	.76	.80
8	.65	.67	.70	.72
16	–	.62	.64	.66
24	–	–	.62	.63
48	–	–	–	.60

new fitness contributions are $(k+1)$-independent random samples of the interval $[0,1]$. The central limit theorem tells us how to estimate the distribution of the sum of the new fitness contributions: the sum is roughly normally distributed with mean $(k+1)/2$ and variance $(k+1)/12$. A neighbor is fitter if the sum of the new contributions is greater than $(k+1)F$, which occurs with probability

$$\left[\frac{6}{\pi(k+1)}\right]^{1/2} \int_{(k+1)F}^{\infty} e^{\frac{6}{k+1}(x-\frac{k+1}{2})^2}\, dx. \tag{14}$$

Because there are N bits that can be flipped in the search for fitter neighbors , the expected number of fitter neighbors is N times this probability. We can estimate the mean fitness of a local optimum by observing that a local optimum is characterized by being fitter than all its neighbors. The mean fitness of local optima is, therefore, F such that the expected number of fitter neighbors is less than one, or F such that

$$\frac{1}{N} > \left[\frac{6}{\pi(k+1)}\right]^{1/2} \int_{(k+1)F}^{\infty} e^{\frac{6}{k+1}(x-\frac{k+1}{2})^2}\, dx. \tag{15}$$

One can solve this equation for F numerically. The results for various N and k values (including k small enough to invalidate the assumptions of the calculation) are shown in Table 2.

We have also studied the N-k model by performing a series of Monte Carlo simulations of uphill walks from an initial, random configuration for a variety of values of N and k. Each simulation proceeded by identifying all of the fitter neighbors of the current configuration and then randomly selecting a new configuration from these fitter neighbors. The simulations stop when no fitter neighbors are found (i.e., when a local optimum had been attained). We used two different kinds of neighborhoods. The first, restricted to even values of k, took the neighborhood to be the $k/2$ nearest loci to the left and to the right of the locus in question. To ensure that all loci, including the first and last, have k neighbors, we assumed periodic boundary conditions; that is, we assumed that the loci are arranged in a circle. The second rule took the neighborhood for each locus to be itself and k other loci selected randomly over the entire string. Because it is not possible to assign neighborhoods randomly and still enforce the symmetry condition stated above, we dropped this condition in simulations using random neighborhoods.

The means and standard deviations of 100 simulations for each of the listed N and k values are given in Tables 3–6. For ease of comparison, we have included the values for the $k = N - 1$ cases in square brackets at the table entries where $N = k$. The standard deviation values for all entries are shown in parentheses. Note that Tables 3 and 4 have three values listed for certain values of N and k. Below the values for the uniform distribution, we report the results of simulations in which we chose the fitnesses per locus from two distributions other than the uniform distribution. The first of these alternatives is the "tent" distribution, $p_{te}(x)$, defined for $0 \leq x \leq 1$ by

$$p_{te}(x) = \begin{cases} 4x, & \text{if } 0 \leq x \leq 1/2 \\ 4(1-x), & \text{if } 1/2 < x \leq 1 \end{cases} \tag{16}$$

TABLE 3 Mean Fitnesses of Local Optima for Various N and k Values (Monte Carlo Results Assuming Nearest Neighbor Interactions)

k\N	8	16	24	48	96
0	.65 (.08) .62 (.06) .70 (.11)	.65 (.06)	.66 (.04)	.66 (.03)	.67 (.02) .62 (.02) .70 (.03)
2	.70 (.07) .65 (.05) .75 (.08)	.70 (.04)	.70 (.03)	.70 (.02)	.71 (.02) .65 (.01) .75 (.02)
4	.70 (.06)	.71 (.04)	.70 (.04)	.70 (.03)	.70 (.02)
8	[.66 (.06)] [.62 (.04)] [.72 (.06)]	.68 (.04)	.68 (.03)	.69 (.02)	.68 (.02)
16	–	[.66 (.04)]	.66 (.03)	.66 (.02)	.66 (.02)
24	–	–	[.63 (.03)]	.64 (.02)	.64 (.02)
48	–	–	–	[.60 (.02)]	.61 (.02)
96	–	–	–	–	[.58 (.01)] [.55 (.09)] [.59 (.01)]

TABLE 4 Mean Walk Lengths to Local Optima for Various N and k Values (Monte Carlo Results Assuming Nearest Neighbor Interactions)

k\N	8	16	24	48	96
0	5.5 (1.2) 5.0 (1.5) 4.8 (1.6)	10.0 (1.9)	13.6 (2.2)	25.3 (3.4)	50.0 (4.6) 49.1 (5.0) 49.5 (5.2)
2	5.1 (1.9) 4.8 (2.0) 4.5 (1.8)	9.1 (3.2)	12.2 (3.1)	23.5 (4.6)	46.2 (6.6) 44.0 (6.4) 44.7 (6.6)
4	4.6 (1.8)	7.6 (2.5)	10.4 (2.9)	20.3 (3.9)	38.3 (6.1)
8	[3.7 (1.5)] [3.0 (1.5)] [3.0 (1.5)]	5.7 (2.3)	8.7 (3.0)	16.3 (4.3)	28.7 (5.3)
16	–	[4.3 (1.7)]	5.6 (2.0)	11.2 (3.1)	20.1 (3.9)
24	–	–	[4.5 (1.4)]	8.7 (2.9)	15.8 (3.8)
48	–	–	–	[5.0 (2.0)]	10.0 (3.1)
96	–	–	–	–	[6.1 (2.4)] [4.9 (2.0)] [5.5 (2.1)]

with mean 1/2 and variance 1/24. The second is the "trough" distribution, $p_{tr}(x)$, defined for $0 \leq x \leq 1$ by

$$p_{tr}(x) = \begin{cases} 2(1-2x) & \text{if } 0 \leq x \leq 1/2 \\ 2(2x-1), & \text{if } 1/2 < x \leq 1 \end{cases} \tag{17}$$

with mean 1/2 and variance 3/24.

We note that for all three distributions the simulation results show the same qualitative behavior as a function of N and k. Naturally, the choice of "tent," "uniform," or "trough" distributions alters the actual fitness values of the accessible optima seen in Tables 3 and 5, since the "tent" distribution tends to contract the deviations toward the mean, 0.5, whereas the "trough" distribution tends to expand the deviations away from 0.5 compared to the uniform distribution. However, results which depend only on the rank orders, such as the mean walk length to a local optima, are similar for all three distributions (see Tables 4 and 6). For all three distributions, for $k = 0$, walks to optima are about $N/2$ steps, and for fixed N, as k increases from 0 to $N-1$, the lengths of walks decrease from approximately $N/2$ to roughly $\log_2 N$ (Tables 4 and 6).

TABLE 5 Mean Fitnesses of Local Optima for Various N and k Values (Monte Carlo Results Assuming Random Interactions)

k\N	8	16	24	48	96
0	.65 (.08)	.65 (.06)	.66 (.04)	.66 (.03)	.66 (.02)
2	.68 (.08)	.68 (.05)	.69 (.04)	.69 (.03)	.69 (.02)
4	.69 (.07)	.70 (.05)	.71 (.03)	.72 (.03)	.72 (.02)
8	[.68 (.06)]	.69 (.04)	.70 (.04)	.71 (.03)	.72 (.02)
16	–	[.64 (.04)]	.65 (.03)	.67 (.02)	.69 (.02)
24	–	–	[.63 (.03)]	.65 (.02)	.66 (.02)
48	–	–	–	–	–
96	–	–	–	–	–

The Monte Carlo results are consistent with the analytical predictions given above for $k = 0$. Recall in that case there is a single global optimum of fitness 2/3, independent of N. The results in the first row of Table 3 show optima with fitness values between .65 – .67 for all choices of N. Our analytical results for $k = N - 1$ are also in quite good agreement with the simulations (compare Table 1 with Tables 3 and 5). For $0 < k < N - 1$, we note that the characteristics of the landscapes, in terms of the height of optima and the number of steps to optima, seem to vary smoothly between the bounding properties sketched for the uncorrelated and completely correlated cases. Lastly, it is worth recalling that the fitness predictions

TABLE 6 Mean Walk Lengths to Local Optima for Various N and k Values (Monte Carlo Results Assuming Random Interactions)

k\N	8	16	24	48	96
0	5.5 (1.2)	10.0 (1.9)	13.6 (2.2)	25.3 (3.4)	50.0 (4.6)
2	5.5 (1.8)	9.1 (2.6)	13.7 (3.1)	25.6 (4.5)	49.1 (6.1)
4	4.8 (2.0)	8.6 (2.8)	13.2 (3.2)	24.2 (5.4)	48.0 (6.8)
8	[3.6 (1.5)]	6.6 (2.3)	9.9 (3.0)	19.6 (5.2)	37.6 (6.6)
16	–	[4.2 (1.7)]	5.9 (2.3)	11.7 (4.0)	24.8 (5.6)
24	–	–	[4.5 (1.8)]	9.1 (2.6)	17.6 (4.5)
48	–	–	–	–	–
96	–	–	–	–	–

for $k > 0$ (Table 2) are for averages over all local optima, but the distribution of fitness optima obtained by Monte Carlo is weighted by the probability that the observed optima will be reached by an adaptive walk. That the two averages agree for high k suggests that the local features of the landscape—at least for high k—are such that local optima are sampled uniformly by adaptive walks starting from random initial positions.

For all three distributions, the predicted receding optima complexity catastrophe occurs for $k = N - 1$ since the fitness values of attainable optima do fall toward the mean fitness, 0.5. But further, Table 3 suggests that this complexity catastrophe occurs not only for $k = N - 1$, but also for $k = N/2$, $k = N/4$, etc. That is, the numerical results suggest that if k increases linearly with N, then for large N, the attainable optima recede toward 1/2. If complex systems were to exhibit this property, then as they became more complex, the attainable optima would recede toward the mean of the space from which the systems are drawn. Is it possible to prevent this complexity catastrophe?

Table 3 shows sufficient conditions to avert the receding optima complexity catastrophe. If k is small and fixed, then as N increases, the fitness values of the attainable optima do not decrease, but rather stay approximately constant. We conjecture that evolution may have "fixed" k to attain proteins and which evolve on "good" landscapes. Table 3 shows another feature. For fixed N, optima for small k (e.g., $k = 2, 4$) are actually higher than for $k = 0$! Thus, a small degree of coupling between sites yields better landscapes than no coupling at all. However, if k is increased even further the fitness values of optima decrease, with the uncorrelated case, $k = N - 1$ having the lowest fitness optima. Table 5 shows the results for the uniform distribution, but with the constraint to neighbors relaxed. Here the k sites which affect each site are drawn at random from among the N. Nevertheless, the major features noted above in Table 3 are retained in Table 5. The major difference is that, for fixed small k, as N increases, the attainable optima appear to increase slowly in fitness value. In Table 3, the optima may increase, but no statistically significant trend can be established. The basic similarities in Tables 3 and 5 strongly suggest that the properties found depend on very robust ordering features of the general N-k model. Thus, given N and k alone, most of the features of the landscape are already nearly determined. If so, this implies that the number of "parts," N, and their richness of coupling, k, are among the most important determinants of landscape ruggedness.

CONDITIONS ON N AND k FOR RECOMBINATION TO BE HELPFUL

Although we have no direct evidence that recombination in the V region of stimulated B cells plays a role in somatic diversification, we do know that recombination or its equivalent (combinatorial association of gene segements) occurs during B cell development. Thus, recombination among germ-line gene segments provides new antibodies which then undergo somatic point mutation. Because of interest in the immune system and, more generally, in peptide and protein evolution, and also because of the possibility that recombination of two partially useful domains may lead to more rapid improvement, we are led to ask under what conditions recombination between sequences in a landscape within the N-k model will be helpful. Consider the $k = 0$ case. Here, if the left half of one peptide is of high fitness while the right half of a second is of high fitness, breaking and joining of the two highly fit halves leads immediately to a more highly fit peptide—and does so in fewer steps than via accumulation of point mutations. In the $k = N - 1$ case, joining the two "good" halves via recombination will yield a peptide with a random fitness; hence, there is no obvious reason to expect recombination to be helpful. Of course, recombination does constitute a mutational process which jumps long distances across protein space and, thus, may help adaptation by jumping beyond local optima. But that is a different issue.

Consider N large and k small, with the k chosen to be nearest neighbor positions. We expect that locally coadapted clusters of amino acids will arise along the peptides (cf. Stadnyk, 1987). Hence, any local optimum should have a number of such coadapted domains along its length. If the domain sizes are small with respect to N, then breaking and joining the left and right halves of two peptides should leave most domains unaltered in each half and should disrupt only one or two domains at the join. That disruption should be ameliorated by subsequent walks to new local optima. Hence, recombination should be useful if the disruption at the recombinational join is less than the gain from the two good halves—and can in any event be fixed by subsequent coadaptation of the amino acids spanning the join. Conversely, if the k sites are scattered at random among the N, it is not obvious whether recombination is helpful other than by exploring the space. Only further work will tell.

6. POPULATION GENETICS: THE LIMITATIONS ON SELECTION

In the present section we explore some ideas derived from population genetics. In previous sections we examined the evolution of a single clone of cells under the assumption that we always moved to a fitter variant. Here we relax these assumptions and consider the evolution of a population of clones, some of which can increase in fitness and others which decrease in fitness due to mutation.

The behavior of such a population will be governed by the statistical features of the fitness landscape upon which adaptation occurs, by the rate of finding fitter variants, by the fitness differentials between mutants and wild type, by the size of the adapting population, by the initial dispersion of the population across the landscape, and, most critically, by the mutation rate. A major point we wish to make is that, in the face of a fixed mutation rate, an adapting population may not be able to "climb to" or remain clustered about local optima.

Consider the $k = 0$ case with $B = 2$. As noted above, this corresponds to an N-locus, two-allele, haploid additive genetic model. Let the assumed constant rate of mutation from the more favored amino acid to the less favored amino acid equal the reverse mutation rate, and let the mutation rates be equal for all positions. As the number of amino acids, N, increases, the fitness contribution of each to the overall peptide decreases linearly. Thus, the selection pressure to keep the favored amino acid at a given position decreases as N increases. Since the mutational force tending to change from the more favored to the less favored state is constant but the fitness contribution of each locus decreases as N increases, it is to be expected that there is a critical value of N above which selection is no longer strong enough to hold a population at the optimum. This is in fact the case. A classical genetic result shows that the expected number of amino acids in the peptide that are in the less favored state is proportional to N^2 (Ewens, 1979). For a given mutation rate and N smaller than a critical length, selection will pull a population to the optimum and hold it in a narrow mutant distribution around that optimum. For N larger than a critical value, selection is too weak a force, and the population diverges from the optimum and accumulates some fraction of less fit amino acids. Furthermore, since the balance of mutation and selection leaves the population at some mean "distance" from the global optimum, the population is constrained by that balance to wander near a surface of constant distance from the global optimum. The variants on this surface are near neutral mutants of one another since all are the same distance from the global optimum. Finally, due to founder effects assuring that future generations derive from only a small cohort or a single member of the present generation, the population that wanders across the surface tends to be a coherent cloud, rather than diffused neutrally all over the surface.

These classical results, later rediscovered by Eigen and Schuster (1979) and named a complexity catastrophe by them, imply a limit in the length of a sequence that selection can pull to a local optimum in the face of constant mutation, at least within the confines of an additive fitness model. Thus, if the rate of fitness decrease due to somatic mutation during maturation of the immune response is higher than the rate of increase in fitness due to selective differentials between mutant variants, this additive model strongly suggests that clonal selection will not be able to hold an adapting population of B cells exactly at a local optimum. Rather, the population will accumulate a mutant spectrum and mill around in a neighborhood of the optimum. As Eigen and Schuster have pointed out, however, the optimal sequence may typically be inferred from the sequences in the mutant spectrum from the most common amino acid at each position.

Thus, we have seen two distict types of complexity catastrophes. The first discussed earlier refers to the fact that in uncorrelated landscapes ($k = N - 1$), the accessible optima recede toward the mean fitness $1/2$, even under conditions where selection is strong enough to always pull a population to a fitter variant. The second, just discussed, is a "mutation–selection complexity catastrophe," which at least applies to an additive fitness model ($k = 0$), and shows that as the length of a string N increases selection ultimately becomes too weak to hold the population at the global optimum.

DOES THE MUTATION–SELECTION COMPLEXITY CATASTROPHE EXTEND TO RUGGED LANDSCAPES

The mutation–selection complexity catastrophe arises in fully correlated landscapes ($k = 0$) because the fitness contribution per locus decreases inversely with N, while the mutation rate per locus is constant and the number of sites that can be mutated away from the global optimum increases with N. Eventually, the gradient of selective forces per locus becomes too flat to pull a population uphill against mutational forces. Does this classical result for $k = 0$ landscapes extend to rugged landscapes? The general answer is unknown.

Consider the uncorrelated case of a $k = N - 1$ landscape. For large N, almost any single mutant of a highly fit peptide will lower the fitness to nearly the average in the space, 1/2. However, as N increases, due to the "receding optima" complexity catastrophe, all accessible optima fall slowly toward the mean fitness, 1/2. As the difference between the height of a local optima and 1/2 decreases, the fitness decrease from the optimum per mutation will decrease. But our simulation results indicate that the rate at which the height of these optima fall with increasing N is sufficiently slow that the net fall in fitness per mutation is substantially larger than in the $k = 0$ case. This implies that as N increases that the selective force tending to pull the population uphill in the face of mutation remains rather high in the $k = N - 1$ case when compared with the $k = 0$ case. Hence, the mutation–selection complexity catastrophe may be delayed to higher N or may be avoided.

Even more importantly, for fairly small values of k, and large values of N, the landscape is rugged but correlated. We believe, but do not yet know, that this results in a landscape with a large number of peaks whose "sides" in most directions are very steep, hence selection forces along those directions remain strong. In more detail, since for small k, as N increases, the local optima remain high and avoid the receding optima complexity catastrophe, we suspect that those landscapes retain accessible optima where the fitness contribution per mutant bit does not decrease as N increases, or at most decreases only slightly. Thus, in such landscapes, the persistence of high optima with steep sides as N increases should help selection hold adapting populations at those high optima in the face of mutation. This would suggest that evolution "tunes" k with respect to N to attain good landscapes. In turn, this suggestion invites analysis of actual V regions to determine how many amino acids any given amino acid actually interacts with directly by criteria of adjacency.

These questions suffice to make it clear how little we yet know about the relation between the structure of fitness landscapes and the consequent behavior of adapting populations upon them. But the questions, at the theoretical level, are surely answerable and must underpin our understanding of any adaptive process, including that of maturation of the immune response, to which we now turn in more detail.

7. CORRELATION WITH IMMUNOLOGICAL DATA

Immediate application of the body of theory discussed above to maturation of the immune system lies in six areas: (1) The number of mutations accumulated on an adaptive walk from an initial "roughed in" antibody, derived directly from a germ-line variant, to a mature high-affinity antibody. (2) The time required to reach such an optimum. (3) The divergence to alternative optimal antibodies by maturation from the same germ-line variant in response to the same antigenic stimulus. (4) The similarity between these divergent optimal sequences. (5) The phenomenon of "milling around" or further accumulation of mutations in the V region that do not further increase the affinity of mature immune response antibodies. (6) The waiting time distribution for emergence of newly matured antibodies derived from an independent germ-line variant late in the immune response. None of these applications is extensively developed yet. Our intent is to show that application of ideas derived from consideration of the structure of affinity landscapes and population genetics gives insight into the maturation of the immune response.

THE NUMBER OF MUTATIONAL STEPS TO OPTIMAL ANTIBODIES

The V region, in the heavy or light chain, consists of 110-120 amino acids embedded in the primary structure of the entire antibody heavy or light chain. Thus, the DNA coding sequence of the V region for either chain is on the order of 350 bases. Selection acts on the intact antibody molecule. Thus the antibody V region (heavy and light chain) has $2 \times 3 \times 350$, or about 2000 1-mutant neighbors, given the four nucleotide code. Taking account of the fact that only about .75 of all substitutions lead to amino acid changes (Lewin, 1987), we assume the V region has about 1500 1-mutant neighbors. In the idealization that the landscape is entirely uncorrelated, this implies that the median length of an adaptive walk is on the order of $\log_2 1500$ or about 10. Thus, about 10 point mutations should accumulate in the antibody V region. There also will be considerable variation from antibody to antibody about this value. This prediction must be modified by the fact that 10 is the estimated walk length from the "worst" antibody to a local optimum, whereas the "roughed in" germ-line variant from which the walk begins is clearly already of some modestly high rank order. In fact, as we discuss next the observed mean number is about 6–8.

How many base changes do we actually see during an immune response? From the mutation rate, 10^{-3} base pairs per generation, we can estimate a mean. Assuming an 18 h generation time and that the V region of the heavy and light chain corresponds to 2000 base pairs, we expect 2 mutations every 18 h, or about 9 mutations at day 7 of a primary response and 14 mutations at day 14. Antibodies seen during a secondary response should show the total accumulation of mutations that occurred during the primary response. Because very few mutations are observed early in the primary response (i.e., in IgM as compared with IgG), it has been suggested that the mutation mechanism may not operate uniformly throughout the primary response, but may be turned on only after the first week or so (T. Manser, private communication). If this were the case, then of order 7 mutations might be expected by the end of the primary response.

The available data suggest that the typical number of nucleotide changes during maturation of the immune response is on the order of 10 (Bothwell et al., 1982; Tonegawa, 1983; Heinrich et al., 1984; Berek et al., 1985; Clark et al., 1985), although there is great variability from antibody to antibody. For example, Crews et al. (1981) studying the V_H gene encoding the response to phosphorylcholine find between 1 and 8 residues have changed when hybridomas are compared with the germ-line sequence. Bothwell et al. (1982) found 3 mutations in a $\lambda 1$ light chain and 6 mutations in a $\lambda 2$ light chain V region. McKean et al. (1984) studying the V_κ region of antibodies raised against the major antigenic determinant on influenza virus found 7 or 8 replacements compared with the germ-line sequence. Clarke et al. (1985) also studying secondary response antibodies raised against influenza found three hybridomas with 20, 12 and 19 V_H-coding region mutations and 9, 8, and 15 V_κ-coding region mutations. The authors note that precursors to these clones may have been stimulated by unknown antigens, and therefore accumulated mutations before exposure to influenza. Wysocki et al. (1986) studying the response to Ars find between 1 and 7 nucleotide replacements in the V_H region in secondary response antibodies. Berek et al. (1987) find that in a tertiary response to phenyloxazole, V_H-Ox1 had about 8 mutations, while in V_κ-Ox1 it seemed as if an optimum of about 10 mutations had been reached.

AFFINITY LANDSCAPES ARE CORRELATED

Despite the apparent capacity of the hypothesis of a fully uncorrelated landscape to predict the number of mutations which accumulate during maturation of the immune response, i.e., about 10, the success is actually spurious in a way which invites a more useful and detailed use of the N-k model. *A priori*, we would expect the fitnesses of real V regions to be correlated with respect to single amino acid substitutions. But, further, our results lead to a strong indirect proof of this fact. Typically only 1 out of every 10^5 B cells responds to a given epitope (cf. Press and Klinman, 1974; Cancro et al., 1978). Those that respond are B cells whose antibodies already have modestly high affinity for the antigen, compared with randomly chosen antibody molecules. In terms of rank orders they are, in fact, in the top 0.001% of all antibodies. Recall each antibody V region only has on the order of 1500 1-mutant variants. Were it the case that V regions were truly uncorrelated, then any beginning antibody, already the best per 100,000, would be extremely unlikely to have any 1-mutant neighbors among 1500 randomly chosen ones which were of higher affinity. That is, the beginning "roughed in" antibodies of modest affinity would already be local optima, or at most one step from such optima. Therefore, given the fact that on the order of 10 mutations accumulate during maturation of the immune response, we conclude that the landscape cannot be uncorrelated. Examination of the N-k model reveals that for $N = 96$, as k decreases from $N-1$ toward 0, walk lengths increase from $\log_2 N$ to $N/2$. Thus, correlated landscapes have longer walks than uncorrelated landscapes, and even walks starting at relatively high rank order could continue for the observed 10 steps.

Our task, then, is to try to use the N-k model to discover and characterize the kind of correlated landscape which occurs for real V regions. In future work we expect to proceed in three steps, which lead to directly testable consequences. The first step is to use the N-k model to estimate k. We know the length of the V region is roughly 200 amino acids and all 20 types of amino acids are utilized. Thus, we can use the N-k model for the corresponding values of N (200) and B (20), to find that value of k such that adaptive walks which begin with strings which are already in the top 0.001% take on average 10 adaptive steps before attaining a local optima. This is, of course, mere curve fitting. However, once k is determined the model will predict the fraction of the 1500 1-mutant neighbors of the beginning antibody which have higher affinity than that starting antibody. Further, the model will predict how the fraction of higher affinity 1-mutant variant antibodies dwindles as each adaptive step accumulating another point mutation occurs.

The second step in using the N-k model, having fixed N, B and determined a value for k, is to analyze the expected number of alternative local optima which are accessible via adaptive walks through fitter variants from an initial roughed in antibody which is in the top 0.001%. This is open to comparison with data from inbred strains of mice exposed to the same epitope, where typically more than a single V region variant of the same "canonical" V region is found after maturation Thus, fixing N and k, should allow predictions about the number of accessible optima.

The third step is based on analyzing the similarity, or overlap, in amino acid sequences between the different local optima accessible from one beginning V region. If sequences are found in which adjoining stretches of amino acids all have changed, one can presume that the entire stretch was subject to selection, and thus one could argue that the k sites that influence a given site are taken from neighboring sites. Alternatively, if the amino acid changes are distributed seemingly at random throughout the V region then the k sites chosen at random may provide a good model. More likely, however, other patterns will arise that reflect the statistical features of the three-dimensional folding pattern of the V region; both nearby and distant amino acids may lie close to one another and, hence, influence the antibody's affinity.

In short, our more mature hope for the N-k model is that it will allow us to build a realistic model for the statistical structure of affinity landscapes, based on N, k, and the distribution of k among the N. We also hope that the N-k model will inspire experiments to discover the true statistical features of affinity landscapes.

WAITING TIME FOR THE EMERGENCE OF NEWLY MATURED GERM-LINE VARIANTS

Late in the maturation of the immune response, one may find new variants that have matured by accumulation of somatic mutations from a different germ-line variant than that which predominated early in the immune response. This is the case during the immune response to oxazolone (Berek et al., 1985). The theory we have developed gives a simple way to think about this phenomenon.

Consider a set of independent adaptive walks on an uncorrelated rugged landscape. Let the first walk initiate at a point and walk to a local optimum. Now initiate a second walk from a different random initial point, and walk to a second optimum. What is the chance that the second optimum is better than the first? More generally, if a succession of walks are initiated, what is the expected waiting time for such walks to find successively higher local optima? The answer, is derivable from our theory above or the theory of record values (Feller, 1971), using the distribution of the local optima. On average, the first walk ends on an optimum in the middle of that distribution. Thus, the chance that the next walk ends on a higher optimum is 1/2. If it succeeds, which on average takes two tries, it will, on average, lie halfway between that optimum and the global optimum. Thus, by a now-familiar argument, each higher local optimum will double the expected waiting time to find a yet higher local optimum. Thus, whatever the time taken to select one germ-line variant and complete its maturation process, one would expect $2^{10} \approx 1000$ fold that time to find the tenth improved germ-line form.

8. DISCUSSION

The theory of adaptive walks presented in this paper allows one to follow the mutations that occur in a single clone which constantly improves in its fitness. Thus, the theory can make predictions about the number of mutations that should be seen in highest affinity clones generated during an immune response, the time it takes to replace one antibody by a higher affinity variant, and the degree of similarity between antibodies that have reached a locally optimum affinity. The theory does not address questions at the population level and, thus, does not predict the number of clones of a given affinity. In the theory, no cognizance is taken of the fact that clones which grow large become a larger target for mutational change and also are more likely to be sampled in hybridoma experiments. The theory also ignores interactions among lymphyocyte populations, such as idiotypic interactions, which might cause one cell line to enhance or inhibit the proliferation of another B cell line. Also ignored is the phenomenon that clones which have attained a higher affinity may preferentially continue to expand leading to the classical phenomenon of affinity maturation via antigen selection. Including such interactions and treating the population dynamics of the various clones is a complex task which still remains.

Maturation of the immune response is easy to understand within the context of clonal selection. The classical antigen selection argument given earlier in this paper, coupled with somatic mutational refinements goes a long way towards explaining this process. Even so, the theory of clonal selection has been questioned. Jerne (1974) has proposed that the immune response should not be viewed as a set of isolated clones each responding independently to antigen, but rather as a set of interacting clones communicating with each other via idiotypic–anti-idiotypic interactions. At mutation rates of 10^{-3} per base pair per generation, at least one of the two daughter cells generated at every division should show a mutation in the antibody V region. Shouldn't such large scale changes destroy any existing idiotypic network or at least prevent regulation of newly arising clones? If the repertoire is complete, i.e., if every possible shape can be recognized by other clones in the system, then newly arising clones will still be regulated. Perelson and Oster (1979) estimate that a population of 10^6 or more different clones, made at random, will be complete, so this may not be a problem. Further, it is possible that somatic mutations that affect the affinity of an antibody leave the idiotype of the antibody unchanged and thus do not affect the network structure. In fact, an idiotypic family of antibodies is generally encoded by a single V_H-V_L gene segment combination, or a small group of highly related V_H and V_L gene segments, combined with D and J_H gene segments that can be quite diverse (Manser et al., 1985). Thus mutations that effect D or J_H can change the affinity and leave the idiotype unaltered. More quantitative estimates are required to understand the interplay of somatic mutation and the structure of idiotypic networks. Simulation models, such as the one of Farmer, Packard and Perelson (1986) described by Perelson in this volume, may play an important role in addressing such questions.

ACKNOWLEDGEMENTS

We thank Catherine Macken for critically reading and commenting on the manuscript. This work was performed under the auspices of the U.S. Department of Energy. S. A. K. and E. D. W. were supported by grant N00014-85-K-0258 from the Office of Naval Research.

REFERENCES

1. Bell, G. (1970), "Mathematical Model of Clonal Selection and Antibody Production," *J. Theoret. Biol.* **29**, 191–232.
2. Bell, G. (1971), "Mathematical Model of Clonal Selection and Antibody Production. II.," *J. Theoret. Biol.* **33**, 339–378.
3. Berek, C., G. M. Griffiths and C. Milstein (1985), "Molecular Events during Maturation of the Immune Response to Oxazolone," *Nature* **316**, 412–418.
4. Berek, C., J. M. Jarvis, and C. Milstein (1987), "Activation of Memory and Virgin B Cell Clones in Hyperimmune Animals," *Eur. J. Immunol.* **17**, 1121–1129.
5. Bothwell, A. L. M., M. Paskind, M. Reth, T. Imanishi-Kari, K. Rajewsky and D. Baltimore (1982), "Somatic Mutation of Murine Immunoglobulin Lambda Light Chains," *Nature* **298**, 380–382.
6. Burnet, M. (1959), *The Clonal Selection Theory of Acquired Immunity* (Cambridge: Cambridge University Press).
7. Cancro, M. P., W. Gerhard and N. H. Klinman (1978), "The Diversity of the Influenza-Specific Primary B-Cell Repertoire in Balb/c Mice," *J. Exp. Med.* **147**, 776–787.
8. Clark, S. H., K. Huppi, D. Ruezinsky, L. Staudt, W. Gerhard and M. Weigert (1985), "Inter- and Intraclonal Diversity in the Antibody Response to Influenza Hemagglutinin," *J. Exp. Med.* **161**, 687–704.
9. Crews, S., J. Griffin, H. Huang, K. Calame and L. Hood (1981), "A Single V Gene Segment Encodes the Immune Response to Phosphorylcholine: Somatic Mutation is Correlated with the Class of the Antibody," *Cell* **25**, 59–66.
10. Doria, G. (1982), "Immunoregulatory Implications of Changes in Antibody Affinity," *Regulation of Immune Response Dynamics*, Eds. C. DeLisi and J. R. J. Hiernaux, (Boca Raton, FL: CRC Press), Vol. II.
11. Eigen, M. (1987), "Macromolecular Evolution: Dynamical Ordering in Sequence Space," *Emerging Syntheses in Science*, Santa Fe Institute Studies in the Sciences of Complexity series, Ed. David Pines (Reading, MA: Addison-Wesley), vol. 1.
12. Eigen, M. and P. Schuster (1979), *The Hypercycle* (New York: Springer-Verlag).
13. Eisen, H. N. and G. W. Siskind (1964), "Variations in Affinities of Antibodies during the Immune Response," *Biochemistry* **3**, 996–1008.
14. Ewens, W. J. (1979), *Mathematical Population Genetics* (New York: Springer-Verlag).
15. Farmer, J. D., N. H. Packard and A. S. Perelson (1986), "The Immune System, Adaptation, and Machine Learning," *Physica* **22D**, 187–204.
16. Feller, W. (1971), *An Introduction to Probability Theory and its Applications* (New York: Wiley), vol. 2, 2nd ed.

17. Fish, S. and T. Manser (1987), "Influence of the Macromolecular Form of a B Cell Epitope on the Expression of Antibody Variable and Constant Region Structure," *J. Exp. Med.* **166**, 711–724.
18. Heinrich, G., A. Traunecker and S. Tonegawa (1984), "Somatic Mutation Creates Diversity in the Major Group of Mouse Immunoglobulin κ Light Chains," *J. Exp. Med.* **159**, 417–435.
19. Honjo, T. (1983), "Immunoglobulin Genes," *Ann. Rev. Immunol.* **1**, 499–528.
20. Jerne, N. K. (1974), "Towards a Network View of the Immune System," *Ann. Immunol. (Inst. Pasteur)* **125C**, 373–389.
21. Kaartinen, M., G. M. Griffiths, A. F. Markham and C. Milstein (1983), "mRNA Sequences Define an Unusually Restricted IgG Response to 2-Phenyl-oxazolone and its Early Diversification," *Nature* **304**, 320–324.
22. Kauffman, S. A. (1988), *Origins of Order: Self Organization and Selection in Evolution*, to appear.
23. Kauffman, S. A. and S. Levin (1987), "Towards a General Theory of Adaptive Walks on Rugged Landscapes," *J. Theoret. Biol* **128**, 11–45.
24. Kauffman, S. A. and R. G. Smith (1986), "Adaptive Automata based on Darwinian Selection," *Physica D* **22D**, 68–82.
25. Kirkpatrick, S., C. D. Gelatt, Jr. and M. P. Vecchi (1983), "Optimization by Simulated Annealing," *Science* **220**, 671–680.
26. Lewin, B. (1987), *Genes* (New York: Wiley), 3rd ed.
27. Manser, T., L. J. Wysocki, T. Gridley, R. I. Near and M. L. Gefter (1985), "The Molecular Evolution of the Immune Response," *Immunol. Today* **6**, 94–101.
28. McKean, D., K. Huppi, M. Bell, L. Staudt, W. Gerhard and M. Weigert (1984), "Generation of Antibody Diversity in the Immune Response of BALB/c Mice to Influenza Virus Hemagglutinin," *Proc. Natl. Acad. Sci. USA* **81**, 3180–3184.
29. Milstein, C. (1986), "From Antibody Structure to Immunological Diversification of Immune Response," *Science* **231**, 1261–1268.
30. Ninio, J. (1979), *Approaches Moleculaires de l'Evolution* (Paris: Masson).
31. Perelson, A. S. and G. F. Oster (1979), "Theoretical Studies of Clonal Selection: Minimal Antibody Repertoire Size and Reliability of Self–Non-Self Discrimination," *J. Theoret. Biol.* **81**, 645–670.
32. Press, J. L. and N. R. Klinman (1974), "Frequency of Hapten-Specific B Cells in Neonatal and Adult Mouse Spleens," *Eur. J. Immunol.* **4**, 155–159.
33. Siskind, G. W., P. Dunn and J. G. Walker (1968), "Studies on the Control of Antibody Synthesis II. The Effect of Antigen Dose and of Suppression by Passive Sntibody on the Affinity of Antibody Synthesized," *J. Exp. Med.* **127**, 55–66.
34. Siskind, G. W. and B. Benacerraf (1969), "Cell Selection by Antigen in the Immune Response," *Immunol. Rev.* **10**, 1–50.
35. Smith, J. Maynard (1970), "Natural Selection and the Concept of Protein Space," *Nature* **225**, 563–564.

36. Sablitzky, F., G. Wildner and K. Rajewsky (1985a), "Somatic Mutation and Clonal Expansion of B Cells in an Antigen-Driven Immune Response," *EMBO J.* **4**, 345–350.
37. Sablitzky, F., D. Weisbaum and K. Rajewsky (1985b), "Sequence Analysis of Non-Expressed Immunoglobulin Heavy Chain Loci in Clonally Related, Somatically Mutated Hybridoma Cells," *EMBO J.* **4**, 3435–3437.
38. Stadnyk, I. (1987), "Schema Recombination in Pattern Recognition Problems," *Proc. 2nd International Conference on Genetic Algorithms and their Applications, MIT, Cambridge, MA.*
39. Tonegawa S. (1983), "Somatic Generation of Antibody Diversity," *Nature* **302**, 575–581.
40. Urbain, J., A. Van Acker, C. De Vos-Cloetens and G. Urbain-Vasanten (1972), "Increase and Decrease in Binding Affinity of Antibodies during the Immune Response," *Immunology* **9**, 121–136.
41. Wabl, M., P. D. Burrows, A. von Gabain, C. Steinberg (1985), "Hypermutation at the Immunoglobulin Heavy Chain Locus in a Pre-B-Cell Line," *Proc. Natl. Acad. Sci. USA* **82**, 479–482.
42. Weinberger, E. D. (1987), *A Stochastic Generalization of Eigen's Model of Natural Selection, Ph.D. Thesis, Courant Institute of Mathematical Sciences, New York, NY*, available from University Microfilms, Ann Arbor, Michigan.
43. Werblin, T. P., Y. T. Kim, F. Quagliata and G. W. Siskind (1973), "Studies on the Control of Antibody Synthesis. III. Changes in Heterogeneity of Antibody Affinity during the Course of an Immune Response," *Immunochem.* **24**, 477–492.
44. Wright, S. (1932), "The Roles of Mutation, Inbreeding, Crossbreeding and Selection in Evolution," *Proc. 6th Int. Congress on Genetics* **1**, 356–366.
45. Wysocki, L., T. Manser and M. L. Gefter (1986), "Somatic Evolution of Variable Region Structures during an Immune Response," *Proc. Natl. Acad. Sci. USA* **83**, 1847–1851.
46. Yancouplos, G. D. and F. W. Alt (1986), "Regulation of the Assembly and Expression of Variable-Region Genes," *Ann. Rev. Immunol.* **4**, 339–368.

RICHARD G. WEINAND and MICHAEL CONRAD
Department of Computer Science, Wayne State University, Detroit, Michigan 48202

Affinity Maturation and Ecological Succession: An Analogy Suggested by Computer Modeling

ABSTRACT

Affinity maturation, the progressive increase of antibody affinity for antigen, is not a logical consequence of the clonal-selection model, and is, in fact, inconsistent with clonal selection unless additional mechanisms are assumed. We show, through computer modeling, that all the essential phenomena of affinity maturation, including its dose dependency, can be duplicated in detail in the framework of clonal selection by assuming two classes of lymphocytes, generalists and specialists. The generalists are analogous to the pioneer species that dominate in the early stages of ecological succession, while the specialists are analogous to the highly adapted, conservative species that dominate in the later stages of evolution.

1. INTRODUCTION

The immune system, like natural evolutionary systems, undergoes maturation in time. In the case of natural evolutionary systems — collection of plants and animals —this development is known as ecological succession. On a time scale that is long enough for new species to emerge, ecological succession becomes identical with evolutionary succession. In the case of the immune system, ecological succession has its equivalent in the phenomenon of affinity maturation — the successional development of antibody species.

The purpose of this paper is to explore this analogy between ecological succession and affinity maturation. The main element of the analogy is that the plants and animals in an ecological community often become more specifically adapted to their environment and to each other in the course of succession, while antibody species become more specific to the antigenic challenge and may even develop specificities towards one another. Our hypothesis is that the detailed course of affinity maturation can be understood by postulating two classes of lymphocytes, generalists and specialists. The generalists are low-threshold cells capable of binding a wide range of antigens, analogous to pioneer species in an ecological community. The specialists are high-threshold cells with a narrow affinity range, analogous to the conservative species that dominate in the later stages of succession.

Our model is clearly an expansion of the well-known analogy between the clonal-selection theory of Burnet and the Darwinian mechanism of natural selection (Jerne, 1955). The Darwinian model implicates within itself a collection of self-reproducing organisms, and in this respect, organic evolution is essentially an ecosystem process (Conrad, 1983). The clonal-selection model utilizes a collection of self-reproducing cells, and to some extent, it is useful to view this community of cells as an ecological community subject to an antigenic environment. We have developed detailed computer models of cell interactions in the immune system with this analogy in mind, and have found that these models can account for the time course and dose dependency of affinity maturation in a manner that is both robust and rather simple in terms of the assumptions made.

2. AFFINITY MATURATION: THE CONCEPTUAL DIFFICULTY

How antigen dose and time after immunization influences the affinity of antibody to an antigen has been systematically studied by a number of authors, beginning with Eisen and Siskind (1964; see also, Siskind et al., 1968; Steiner and Eisen, 1967; Werblin et al., 1973). In these experiments, rabbits were given injections of antigen in a variety of does, and the quantity and affinity of antibodies to this antigen were measured at various intervals. The common finding was that the immune response over a range of antigen doses had three important characteristics:

1. The average affinity increased progressively with time.
2. This increase was greater for low doses of antigen (dose dependency).
3. The average affinity early in the response (1 to 2 weeks) was approximately the same low value for all antigen doses.

Any acceptable model of immunity should be able to account for these facts. This, however, has proved to be surprisingly difficult. A number of excellent computer models of the immune system have been developed that incorporate the basic mechanisms known to operate in the immune system and which are polyclonal in character (Bell, 1970, 1971a,b; Bruni et al., 1975). In general, these models demonstrate the competence of the clonal-selection mechanism to defend the organism from antigenic insults. However, when the parameters of the model are tuned to fit empirical rate constants, it becomes impossible to account for the facts of affinity maturation (1–3 above) without introducing highly *ad hoc* assumptions.

Let us consider why this is the case. According to the antigen selection hypothesis (Siskind and Benacerraf, 1969), the quantity of antigen available early in the response is sufficient to stimulate a wide range of B cells, including those of relatively low affinity for the antigen. As the quantity of antigen becomes progressively reduced over time, those B cells with the highest affinity receptors for the antigen are preferentially selected to proliferate, differentiate, and secrete antibody, thereby increasing the average antibody affinity as the response progresses.

Computer models incorporating the antigen selection hypothesis give responses of the form illustrated in Figure 1 under typical assumptions regarding binding, triggering, homogeneity, and choice of realistic rate and time constants. The diagram illustrates the response to a range of doses, and it is clear that the response fails to meet the basic facts (primarily 2 and 3). All doses produce some degree of maturation, as predicted by the hypothesis (fact 1). But, for low doses of antigen, the antibody affinity early in the response is already at or near its maximum, and very little maturation can result (in contradiction to facts 2 and 3). *This occurs because low doses of antigen are capable of stimulating only the highest affinity cells.* This basic conceptual difficulty arises even when more sophisticated assumptions regarding binding and triggering are used in computer models (Bell, 1971a; DeLisi and Perelson, 1976).

The basic question is thus: for low doses of antigen, where do the low-affinity antibodies that characterize the early phase of the response come from? Siskind and Benacerraf (1969) recognized that this was a problem, and suggested that statistical factors based on the number of cells involved could play a role. It is reasonable to assume, as we do in our model, that a typical distribution of B cells contains far more cells of low affinity for an antigen than cells of high affinity. All things equal, this should favor the low-affinity cells early in the response, simply due to their greater quantity. At first sight, this suggests that it might be possible to account for the dose-dependency relationship in terms of a balance between the opposing pressures of equilibrium (based on affinity) and statistical factors (based on cell quantity). However, if statistical factors become increasingly important at low antigen doses, then they would become even more important as the availability of antigen decreases over the time course of the response. Affinity would decrease rather than increase (apart from the possible offsetting effect of a change in the relative distribution of high- and low-affinity cells).

It is possible to repair some of these discrepancies by introducing *ad hoc* alterations on the parameters of the model. For example, Bell (1970) discovered that

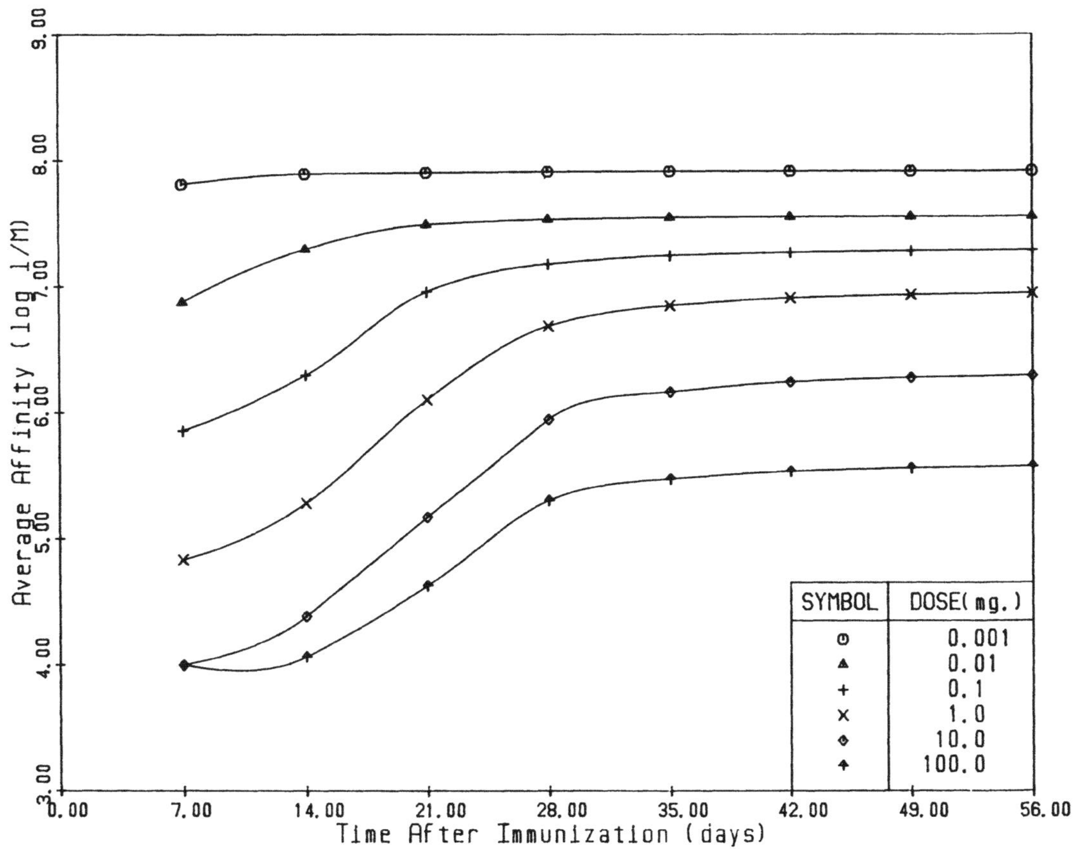

FIGURE 1 Dose response curves implied by the antigen selection hypothesis. The curves were generated by the base model described in this paper, but are typical of those in Bell (1970, 1971a).

it was possible to reproduce a more realistic time development of maturation by assuming that the time release of the antigen is protracted over a significant period of time (1000 hrs.) by an adjuvant. Low doses of antigen, however, still produced only high-affinity antibodies early in the response. He also showed (1971a) that the addition of controls on mechanisms of proliferation could improve results, though without correcting the inverted dose-dependency relation. In a model that used a continuous distribution of B cell affinities, Bruni (1975) was able to reproduce the dose-dependency relation, but only by introducing the assumption of a low basal rate of antibody production from all immunocompetent cells, *regardless of their affinity for the antigen*.

We should also note that the problems highlighted above are bound up with the tenability of basic assumptions about the specific role of antigen in immunological control. It seems reasonable to assume that the time-varying concentration of antigen should serve as a major parameter controlling affinity maturation. But this runs into a problem with experimental facts regarding the timing of antigen decline and affinity maturation (see Davis et al., 1967). This problem may be solved in computational models by introducing an external time-release mechanism to delay the presentation of antigen over an appropriate time period (i.e., several weeks), as in the adjuvant assumption discussed above. While this hypothesis should not be excluded at this stage, it is not clear why the presence of a prolonging agent can always be assumed; and it is not clear why slow presentation of the antigen would not, in some cases, lead to tolerance rather than a heightened response. We point these issues out not to suggest that the time-varying concentration of antigen is unimportant in controlling the immune response, but rather to indicate that it is worth considering whether more general mechanisms could be present in the framework of clonal selection that both yield the basic facts of affinity maturation and fit more naturally to the basic elements of timing in the immune system.

We cannot close this section without a comment on the important role that computer models have played. Without them, it would have been impossible to clearly discern the incompatibility of an undressed clonal-selection mechanism with the basic facts of affinity maturation. This is due to the fact that the difficulties are not purely in the logic of the situation; they can be repaired by unrealistic choices of rate constants. What computer modeling has shown is that the clonal-selection model in its simplest form is incompatible with affinity maturation under realistic assumptions about physiological rate constants.

3. GENERALIST-SPECIALIST HYPOTHESIS

We now consider how the clonal-selection mechanism might be dressed up to account for the facts of affinity maturation. Our hypothesis, to be called the generalist-specialist mechanism, is based on two primary assumptions:

A1. B cells are divided into two classes, generalists and specialists. Generalist cells have surface receptors with a low affinity for a very wide range of antigens. Specialist cells, on the other hand, have receptors with a very high affinity for a specific antigen, and a very low affinity for all others.

A2. Generalist cells are stimulated at a lower threshold of antigen concentration than are specialist cells, and differentiate to become antibody secreting plasma cells with little or no proliferation.

The generalists, as indicated in our introductory remarks, are similar to a pioneer species in ecological succession in that they respond easily and rapidly, but produce antibodies that are not highly adapted to the specific antigen. The specialists, on the other hand, are more like conservative species that dominate in a more mature ecosystem in that they respond slowly, but produce highly adapted antibodies.

We also make the important assumption that B cell proliferation and differentiation are controlled primarily by the extent of initial contact between the antigen and immunocompetent B cells, as expressed by the crosslinking of B cell surface receptors. Each triggered cell is assumed to undergo a number of proliferative cycles that is determined by the amount of crosslinking that has occurred. In general, high-affinity cells bind and crosslink more antigen than low-affinity cells. As a consequence, they proliferate for a longer period of time. Proliferation is assumed to proceed at a constant rate until complete, at which point the entire clone differentiates into terminal, antibody-secreting, plasma cells. This mechanism of proliferative control is similar to a theoretical optimal strategy proposed by Perelson et al. (1976).

What advantage might the generalist-specialist strategy offer, aside from the fact that it provides a simple and natural account of the facts of affinity maturation in the context of clonal selection? The answer, we believe, is that it is better to respond to an antigen rapidly with a generalist antibody that works in a somewhat sloppy way than with a barrage of specific antibodies that hardly work at all. Maintaining a basal rate of specific, but unsuitable, antibodies allows for a proper dose-dependency relation, but it does not contribute to the defense of the organism. Responding initially to antigens with generalist antibodies allows the organism to survive the initial attack long enough to martial highly specific antibodies that are effective for the particular antigen in question. This is the whole point of affinity maturation.

The question might be raised, why not recruit specialist antibody from the start? This is unfeasible, since the specialist clones in a previously uncontaminated organism would have to undergo a series of proliferative cycles before they could produce sufficient antibodies to play a significant defensive role. Generalist clones can be initially large enough to produce a significant quantity of beneficial antibody right from the start, without requiring proliferation. Given this whole set-up, it becomes quite workable for the organism to depend on the time-varying concentration of the antigen as a significant control parameter for regulating the detailed course of the response.

4. THE COMPUTER MODEL

Now, let us briefly describe a polyclonal computer model of the humoral immune response to T-dependent antigens that incorporates the above assumptions, and serves to test whether they, indeed, account for the phenomena of affinity maturation. The model was actually developed to study immunological control in general, and as such, originally included the components represented in Figure 2. But, like previous investigators, we soon found that the use of experimental rate constants led to inconsistencies with affinity maturation, and that these were not dependent on the details with which control loops are modeled. Thus, in the version of the model to be described, we do not include the interactions among T cells, macrophages, and antigen separately, but assume that these processes produce a supply of B cell growth factor (GF) that is a saturation-limited function of the antigen dose. Also, we do not include memory cells at this time since we are concerned with the response to a new antigen. Mechanisms for the production of memory cells can easily be incorporated without significantly affecting the dynamics of the primary response. We also do not include the more recently observed phenomenon of somatic mutation, noting that it does not occur early in the response and can not, therefore, have an effect on early antibody affinity (Wysocki et al., 1986).

The specific assumptions are as follows. Immunocompetent B cells are considered to be arrested in some phase of the cell cycle before DNA synthesis. In the absence of an antigenic stimulus, the total B cell population is maintained at a constant level by balanced rates of cellular formation, maturation and catabolism. The B cells are distributed into groups of cells that cover a wide spectrum of affinities. A few groups have a very high affinity for a particular antigen, and many more groups have progressively lower affinities for that antigen. In response to an antigenic challenge administered as a single dose in a homogeneous compartment, the antigen binds with B cell surface receptors and crosslinking occurs. B cells that undergo a sufficient degree of crosslinking are triggered, and express surface receptors for growth factor. This growth factor is assumed to be a clonally non-specific soluble factor secreted by T cells through interactions among T cells, macrophages, and antigens (treated as a lumped system). Triggered B cells bind B cell growth factor and are induced to undergo a number of proliferative cycles. When proliferation is completed, the cells differentiate to become terminal, antibody-secreting plasma cells. Secreted antibodies bind with available antigen, forming immune complexes that are cleared from the system by phagocytic cells. These assumptions are formalized in the set of equations listed in the Appendix. Model parameters (Appendix, Table 1) were chosen to correspond as closely as possible to the experimental studies of Eisen and Siskind (1964), Steiner and Eisen (1967), Werblin and Siskind (1972), and Werblin et al. (1973). The experimental model used in all these investigations was the rabbit. Dinitrophenyl bovine gamma-globulin (DNP-BGG) was used as the antigen, and the quantity and affinity of anti-DNP antibodies were measured at periodic intervals.

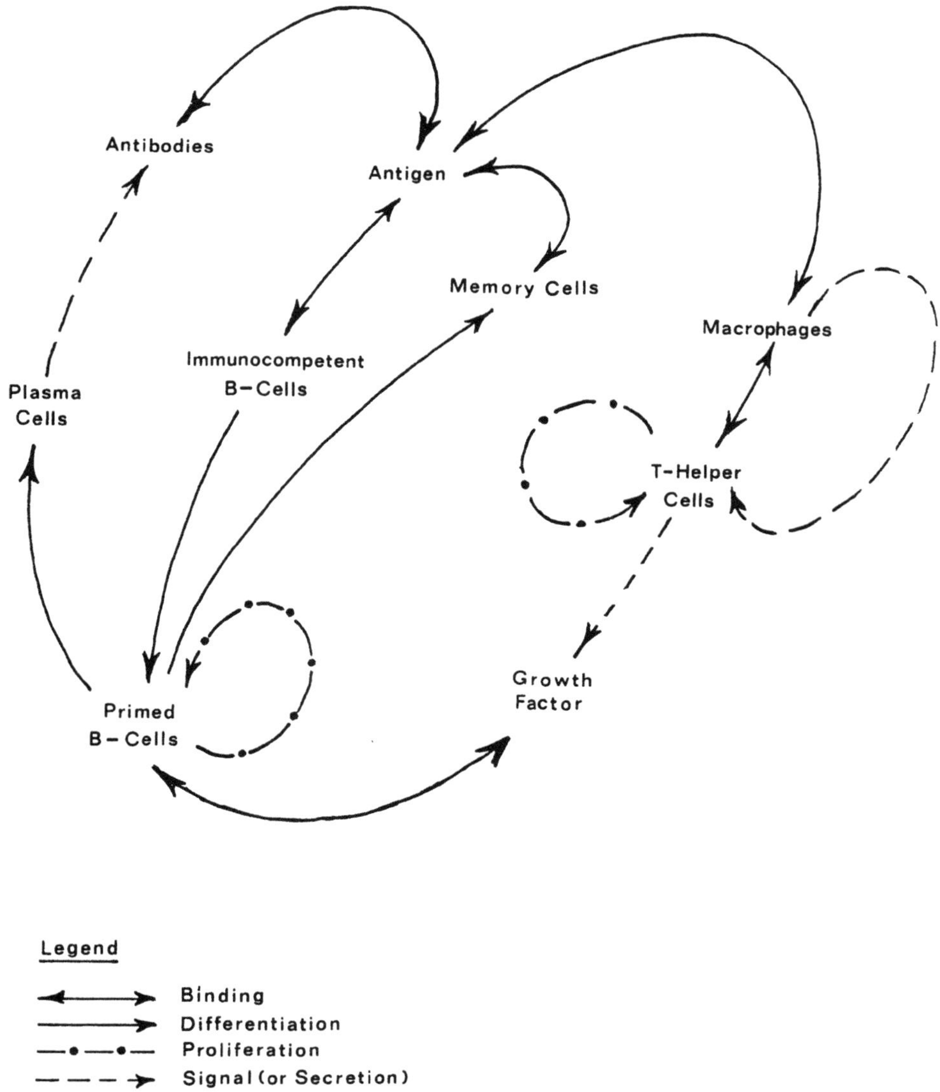

FIGURE 2 Major components of the model.

The generalist-specialist model is the same as the base model described except for the addition of a second generalist class of B lymphocytes. These cells form a single clone with a low affinity for a wide variety of antigens and with a low activation threshold. On activation, generalist lymphocytes differentiate without proliferation to become antibody-secreting plasma cells.

5. EXPERIMENTAL RESULTS

We have performed extensive experiments with a number of versions of the model. Here, we concentrate specifically on the behavior exhibited when the presence of a generalist clone is assumed. Results of experiments over a range of antigen doses are shown in Figure 3.

At all immunizing doses of antigen, the antibodies secreted early in the response are only from the generalist cells. As a consequence, the early antibody affinity is always low. At the lowest antigen dose, the response is somewhat abortive, and only low-affinity antibodies are produced throughout the response. This occurs because the antigen dose is just high enough to reach threshold binding and crosslinking of the highest affinity cells, and little or no proliferation occurs. The continued application of additional small quantities of antigen could, in this case, lead to the induction of low zone tolerance. At slightly higher doses of antigen, only those specialist cells with the highest affinity for the antigen will be stimulated. After several rounds of proliferation, these cells differentiate to become antibody secreting plasma cells and begin to secrete high-affinity antibodies. The low-affinity antibodies secreted early in the response by the generalist cells, followed by greater quantities of high-affinity antibodies later in the response from specialist cells, result in the maximally maturing low-dose response observed in (natural) experimental systems. For a moderate to high dose, on the other hand, specialist cells encompassing a wide range of affinities for the antigen are stimulated. This leads to the production of the greater quantity of antibodies needed to deal with the larger dose of antigen, and it also results in a lower average affinity of antibodies secreted by the specialist cells. The final average affinity is thus lower, and the degree of maturation correspondingly lower. At very high doses of antigen, inhibition of crosslinking occurs on the highest affinity cells, and maturation of the response is, therefore, limited. These results are in excellent agreement with the three defining characteristics of affinity maturation stated in section 2.

We also used the model to investigate two other mechanisms that could conceivably play an important role in affinity maturation. The first is lymphocyte circulation between the blood and the lymph. When an antigen is injected subcutaneously, the response is observed to be highly localized in the draining lymph nodes of the injection site. As a consequence, only a small fraction of the potentially reactive cells are initially exposed to the antigen. As B and T cells circulate between the blood and the lymph in the nodes, all the reactive cells would have the opportunity to encounter the antigen. Circulation thus provides a natural mechanism for delaying the contact between the reactive cells and the antigen, rather than introducing an artificial mechanism for delaying the presentation of antigen to the reactive cells as in Bell (1970, 1971a) and Bruni (1975). We modeled this effect by simply maintaining two compartments: an infected compartment and an uninfected compartment. Antigen is injected into the infected compartment, which contains only a small fraction of the B cells. The remainder of the B cells are presented to this compartment at an exponentially declining rate, and within 4 to 6

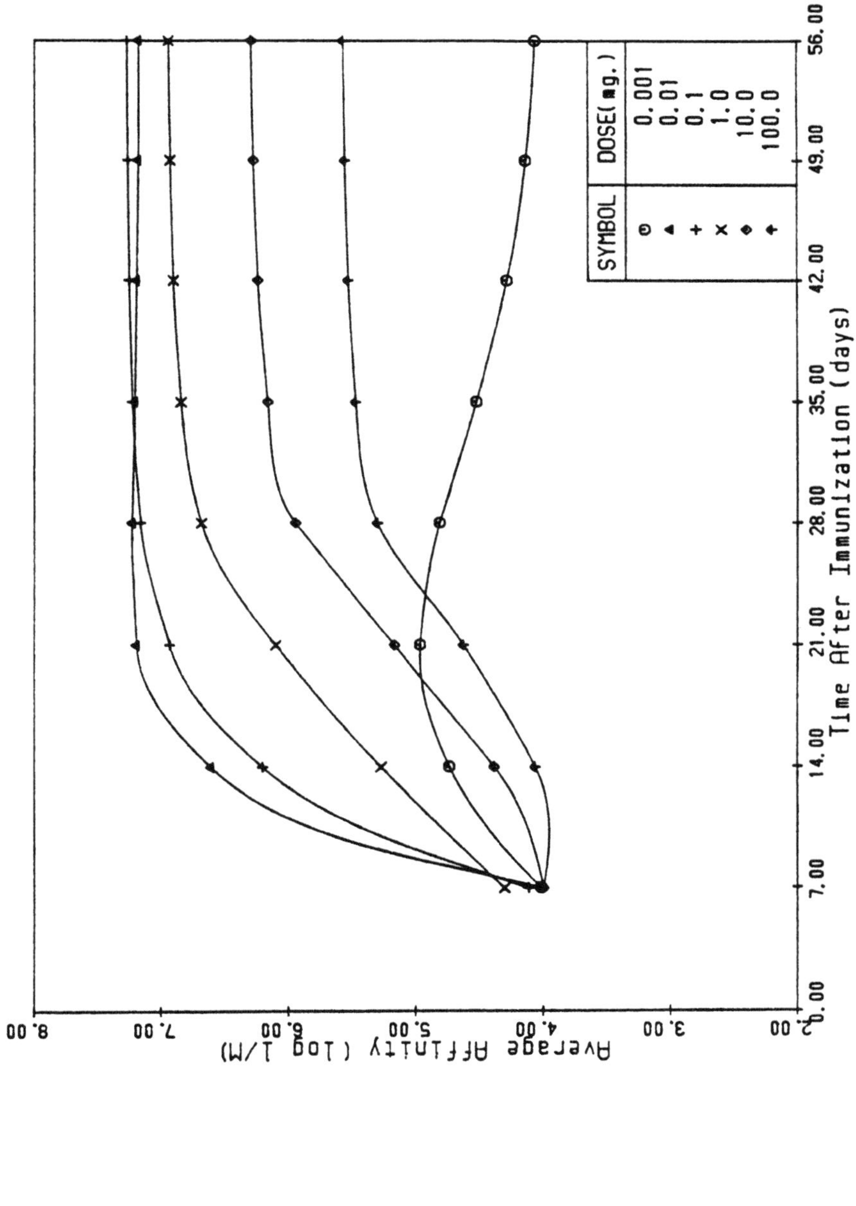

FIGURE 3 Dose response curves for the generalist/specialist hypothesis.

weeks virtually all of the cells have been exposed. The compartmental model was tested with the same six antigen-dose levels as the generalist-specialist model. The only noticeable effect (as compared to the base model, without generalist cells) was a slight dampening of the response. In essence, the same quantity of antigen presented to an initially smaller quantity of B cells resulted in the binding of a smaller fraction of the total antigen, and in a greater fraction of bound antigen per cell. These two factors had somewhat offsetting effects on the final outcome, and as a consequence, the overall effect was minimal. A more complete representation of the circulatory system might, however, produce different results.

The second mechanism concerned another possible solution to the more basic problem of producing a base of low-affinity antibodies early in the response, especially in response to a low dose of antigen. If the B cell triggering threshold was a stochastic parameter distributed about some mean value, then a low dose of antigen could stimulate at least some low-affinity cells as well as those of high affinity. When this idea was tested in the model, however, we found that the threshold variability had to be unrealistically large to produce a significant effect.

6. SUMMARY AND DISCUSSION

Of the models we have tested only the generalist-specialist model reproduces the full set of phenomena associated with the affinity maturation. We recall the main assumptions of the model. The first is that lymphocytes fall into two classes, specialists and generalists. Specialists are the lymphocytes previously assumed in models of the immune system. Generalists respond quickly and provide a base of comparatively low-affinity antibody early in the response. The second assumption is that B cell proliferation and differentiation are controlled principally by the initial quantity of antigen crosslinked with B cell receptors, giving a proliferative advantage to the highest affinity cells and delaying the secretion of high-affinity antibodies.

Experimental support for two classes of lymphocytes can be found in Werblin et al. (1973). This study followed the distribution and heterogeneity of antibody affinity as a function of time after immunization. These investigators found that

> "...initially a symmetrical distribution of heterogeneous low-affinity antibody molecules is present [our generalist antibodies]. With time the population becomes skewed toward the high-affinity end of the distribution, the average affinity increases, and the bulk of the antibody comes to be present in a subpopulation of high-affinity antibody of relatively restricted heterogeneity [our specialist antibodies]."

Importantly, they also reported that "low affinity subpopulations were found to persist throughout the course of the immune response, and they were not present in detectable amounts in normal serum." These findings point to the existence of the two classes of lymphocytes we have assumed in our model.

Let us finally return to our initial analogy between the immune system and an evolutionary ecosystem. The generalists, we can recall, are like pioneer species with a flexible genetic system (cf. Mayr, 1963). These species are capable of adapting rapidly and usually dominate in the early stages of ecological succession. The specialists are like the genetically conservative species that are specifically adapted to the environment and that dominate in the later stages of succession. The pioneer species often serve to convert a disturbed or damaged environment into one more hospitable to life processes. Similarly, the generalist lymphocytes serve as a first line of defense against antigenic disturbances to the organism.

We can put this analogy in a more general context. In the case of an ecological system evolution by variation of genes and selection of phenotypes provides the mechanism through which organisms adapt to their environment. The environment may include the physical environment or other organisms, and in the course of succession species may either arise *de novo* through genetic mechanisms (on a long time scale) or spring into prominence from a dormant state (on a short time scale). In the case of the immune system, the problem is to adapt antibodies to antigens. The antigens may be physical or biotic elements of the environment within the organism, or may even include other antibodies, as in the case of the idiotype–anti-idiotype networks of Jerne (1974). The mechanism of adaptation is some combination of germ-line and somatic variation of antibody genes and clonal selection on the basis of the fitting of antigen to the tertiary structure of the antibody (its phenotype). The need for such an internalized evolutionary system is, at least in part, connected to the fact that the microbial world evolves at a much faster rate than can complex organisms such as vertebrates. The immune system provides an internalized evolutionary mechanism that compensates for this fact (Conrad, 1985).

The immune system, as part of an organism, is clearly subject to the formative power of natural selection in a way that evolutionary ecosystems taken as a whole cannot be. This allows for the development of many control loops which are specifically designed to meet certain needs of the organism, such as those of defense, immune surveillance of body components, and separation of self and not-self generally. Ecosystems as a whole, even in the final stages of succession, are much more loosely organized. The only requirement they need to fulfill in order to persist is that of stability. Nevertheless, in the light of the functional similarity between the two systems, it is not surprising that strategies and patterns of development found in ecology would also occur in immunity.

APPENDIX

I. LIST OF SYMBOLS

For each clone $i = 0, n$ (0 = generalist clone; $1, n$ = specialists)

N_{di} = No. of developing B cells
N_{mi} = No. of mature (immunocompetent) B cells
N_{ti} = No. of triggered B cells
N_{pi} = No. of immature plasma cells
N_{si} = No. of antibody-secreting plasma cells
R_{ui} = No. of unbound receptors on N_{mi} cells
R_{bi} = No. of bound receptors on N_{mi} cells
R_{gi} = No. of growth factor receptors on N_{ti} cells
N_{Abi} = No. of free antibody molecules
N_{Cxi} = No. of antigen-bound antibody molecules
r_{fi} = Forward rate constant for antigen binding
r_{ri} = Reverse rate constant for antigen binding
S_i = Synthesis rate of new B cells
N_{Ag} = Quantity of free antigen
c_j = Set of constants; $j = 1, 14$

II. MODEL EQUATIONS

Mature B cells (N_{mi}) have a fixed number of receptors (c_3) for antigen on their cell surfaces. In the presence of antigen, these receptors bind antigen reversibly and reach equilibrium according to

$$N_{Ag} R_{ui} r_{fi} = R_{bi} r_{ri}; \quad i = 0, n \tag{1}$$

Mature B cells are triggered when a threshold fraction(c_4) of their receptors are bound with the antigen, as expressed in

$$\frac{R_{bi}}{c_3 N_{mi}} > c_4 \tag{2}$$

Thus, for cells that meet the triggering threshold,

$$\frac{dN_{ti}}{dt} = \frac{N_{mi}}{c_5}, \tag{3}$$

$$\frac{dN_{mi}}{dt} = \frac{-N_{mi}}{c_5}, \tag{4}$$

$$\frac{dR_{bi}}{dt} = \frac{-R_{bi}}{c_5} \tag{5}$$

and

$$\frac{dR_{ui}}{dt} = \frac{-R_{ui}}{c_5}. \tag{6}$$

Triggered cells express receptors for GF in proportion to their quantity of bound antigen, as given by

$$\frac{dR_{gi}}{dt} = \frac{R_{bi}}{c_5}. \tag{7}$$

Each triggered cell proliferates, expanding to a clone size that is proportional to the quantity of its expressed receptors. The rate of proliferation is determined by a fixed mitotic cycle time plus a variable delay that is a saturation-limited function of GF receptor concentration. This is given by

$$gf_i = \frac{R_{gi}}{c_3 N_{ti}}, \tag{8}$$

$$dN_{ti} = \frac{N_{ti}}{c_6 + c_7 \left(\frac{c_8}{c_8 + gf_i} \right)}. \tag{9}$$

When fully proliferated, triggered cells differentiate to become immature plasma cells which, in turn, mature after a time delay to become antibody-secreting plasma cells. Plasma cells secrete antibody and are catabolized after a time delay. These relations are expressed by

$$\frac{dN_{ti}}{dt} = \frac{-N_{ti}}{c_9}, \tag{10}$$

$$\frac{dN_{pi}}{dt} = \frac{N_{ti}}{c_9} - \frac{N_{pi}}{c_{10}}, \tag{11}$$

$$\frac{dN_{si}}{dt} = \frac{N_{pi}}{c_{10}} - \frac{N_{si}}{c_{11}}, \tag{12}$$

and

$$\frac{dN_{Abi}}{dt} = c_{12}N_{si} - \frac{N_{Abi}}{c_{13}}. \tag{13}$$

Secreted antibodies bind antigen reversibly and reach equilibrium, modifying Eq. (1) to become

$$N_{Ag}(R_{ui} + N_{Abi})r_{fi} = (R_{bi} + N_{Cxi})r_{ri}; \quad i = 0, n. \tag{1'}$$

Finally, antigen-antibody complexes are removed from the system after a time delay as expressed by

$$\frac{dN_{Cxi}}{dt} = \frac{-N_{Cxi}}{c_{14}}. \tag{14}$$

III. INITIAL CONDITIONS

We assume that the system is initially in a steady state before the introduction of antigen. The clones of developing B cells (N_{di}) and mature B cells (N_{mi}) are maintained at a constant quantity by balanced rates of synthesis, maturation and catabolism, and all other cell populations are zero. These conditions are expressed by

$$\frac{dN_{di}}{dt} = S_i - \frac{N_{di}(0)}{c_1} = 0, \tag{16}$$

and

$$\frac{dN_{mi}}{dt} = \frac{N_{di}(0)}{c_1} - \frac{N_{mi}(0)}{c_2}. \tag{17}$$

We select a set of initial values for $N_{mi}(0)$ and calculate $N_{di}(0)$ and S_i from

$$N_{di}(0) = \frac{c_1 N_{mi}(0)}{c_2} \tag{18}$$

and

$$S_i = \frac{N_{di}(0)}{c_1}. \tag{19}$$

In selecting $N_{mi}(0)$, we assume that the highest affinity clone contains the smallest number of cells, and each clone of successively lower affinity contains a progressively larger quantity of cells by a constant factor (c_0), as expressed by

$$N_{m(i+1)}(0) = c_0 N_{mi}(0). \tag{20}$$

IV. MODEL PARAMETERS

We assume that the rabbit contains 10^8 specialist B cells that are capable of reacting with the antigen. These B cells are grouped into 40 separate clones, and all cells in a clone are identical. The affinity constants for these clones are spaced logarithmically in the interval 10^4 to 10^9 M^{-1}. The number of generalist cells is assumed to be 10^6, with an affinity constant of 10^5 M^{-1}. The remaining model parameters are given in Table 1.

TABLE 1 Model Parameters

$c_0 = 1.20$	Clonal quantity factor
$c_1 = 100$ hrs.	Mean maturation time of N_{di} cells
$c_2 = 336$ hrs.	Mean lifetime of N_{mi} cells
$c_3 = 10^5$	No. of antigen receptors/B_m cell
$c_{4a} = 0.001$	Triggering threshold—specialists
$c_{4b} = 0.0001$	Triggering threshold—generalists
$c_5 = 24$ hrs.	Mean time for B cell triggering
$c_6 = 24$ hrs.	Minimum N_{ti} cell division time
$c_7 = 48$ hrs.	Proliferation delay factor
$c_8 = 0.02$	Delay saturation factor
$c_9 = 48$ hrs.	Mean N_{ti} differentiation time
$c_{10} = 100$ hrs.	Mean N_{pi} maturation time
$c_{11} = 96$ hrs.	Mean N_{si} lifetime
$c_{12} = 7.2E6$	Antibody secretion rate/N_{si} cell/hr.
$c_{13} = 200$ hrs.	Mean free antibody lifespan
$c_{14} = 8$ hrs.	Mean antigen-antibody complex removal time

REFERENCES

1. Bell, G. I. (1970), "Mathematical Model of Clonal Selection and Antibody Production," *J. Theor. Biol.* **29**, 191–232.
2. Bell, G. I. (1971a), "Mathematical Model of Clonal Selection and Antibody Production II," *J. Theor. Biol.* **33**, 339–378.
3. Bell, G. I. (1971b), "Mathematical Model of Clonal Selection and Antibody Production III: The Cellular Basis of Immunological Paralysis," *J. Theor. Biol.* **33**, 379–398.
4. Bruni, C., M. A. Giovenco, G. Koch, and R. Strom (1975), "A Dynamical Model of the Humoral Immune Response," Math. Biosciences **27**, 191–211.
5. Conrad, M. (1983), *Adaptability* (New York: Plenum Press).
6. Conrad, M. (1985), "Evolutionary Adaptability of Biochemical Systems (Concept of the Superinflammatory Response)," *Proc. 16th FEBS Congress*, Part A, 495–509.
7. Davis, B. D., R. Dulbecco, H. N. Eisen, H. S. Ginsberg, and W. B. Wood, Jr. (1967), *Microbiology* (New York: Harper & Row).
8. DeLisi, C., and A. S. Perelson (1976), "The Kinetics of Aggregation Phenomena I. Minimal Models for Patch Formation on Lymphocyte Membranes," *J. Theor. Biol.* **62**, 159–210.
9. Eisen, H. N., and G. W. Siskind (1964), "Variations in Affinities of Antibodies During the Immune Response," *Biochemistry* **3**, 996–1008.
10. Jerne, N. K. (1955), "The Natural Selection Theory of Antibody Formation," *Proc. Natl. Acad. Sci. USA* **41**, 849–857.
11. Jerne, N. K. (1974), "Towards a Network Theory of the Immune System," *Ann. Immunol. Inst. Pasteur* **125C**, 373–385.
12. Mayr, E. (1963), *Animal Species and Evolution* (Cambridge, Massachusetts: Harvard University Press).
13. Perelson, A. S., M. Mirmirani, and G. F. Oster (1976), "Optimal Strategies in Immunology I. B-Cell Differentiation and Proliferation," *J. Math. Biol.* **3**, 325–367.
14. Siskind, G. W., and B. Benacerraf (1969), "Cell Selection by Antigen in the Immune Response," *Adv. Immunol.* **10**, 1–50.
15. Steiner, L. A., and H. N. Eisen (1967), "Sequential Changes in the Relative Affinity of Antibodies Synthesized During the Immune Response," *J. Exp. Med.* **126**, 1161–1183.
16. Werblin, T. P., and G. W. Siskind (1972), "Distribution of Antibody Affinities: Technique of Measurement," *Immunochemistry* **9**, 987–1011.
17. Werblin, T. P., Y. T. Kim, F. Quagliata, and G. W. Siskind (1973), "Studies on the Control of Antibody Synthesis II. Changes in Heterogeneity of Antibody Affinity During the Course of the Immune Response," *Immunology* **24**, 477–492.

18. Wysocki, L., T. Manser, and M. L. Gefter (1986), "Somatic Evolution of Variable Region Structures During an Immune Response," *Proc. Natl. Acad. Sci. USA* **83**, 1847–1851.

Index

www.ingramcontent.com/pod-product-compliance
Ingram Content Group UK Ltd.
Pitfield, Milton Keynes, MK11 3LW, UK
UKHW022207190926
475702UK00003B/249
* 9 7 8 0 3 6 7 0 9 1 1 1 8 *